AF180178

GaN Technology

Maurizio Di Paolo Emilio
Editor

GaN Technology

Materials, Manufacturing, Devices and Design for Power Conversion

Springer

Editor
Maurizio Di Paolo Emilio
Power Electronics News
Pescara, Italy

ISBN 978-3-031-63237-2   ISBN 978-3-031-63238-9 (eBook)
https://doi.org/10.1007/978-3-031-63238-9

© The Editor(s) (if applicable) and The Author(s), under exclusive license to Springer Nature Switzerland AG 2024
This work is subject to copyright. All rights are solely and exclusively licensed by the Publisher, whether the whole or part of the material is concerned, specifically the rights of translation, reprinting, reuse of illustrations, recitation, broadcasting, reproduction on microfilms or in any other physical way, and transmission or information storage and retrieval, electronic adaptation, computer software, or by similar or dissimilar methodology now known or hereafter developed.
The use of general descriptive names, registered names, trademarks, service marks, etc. in this publication does not imply, even in the absence of a specific statement, that such names are exempt from the relevant protective laws and regulations and therefore free for general use.
The publisher, the authors and the editors are safe to assume that the advice and information in this book are believed to be true and accurate at the date of publication. Neither the publisher nor the authors or the editors give a warranty, expressed or implied, with respect to the material contained herein or for any errors or omissions that may have been made. The publisher remains neutral with regard to jurisdictional claims in published maps and institutional affiliations.

This Springer imprint is published by the registered company Springer Nature Switzerland AG
The registered company address is: Gewerbestrasse 11, 6330 Cham, Switzerland

If disposing of this product, please recycle the paper.

Preface

In the ever-evolving realm of electronic devices, the pursuit of more efficient, compact, and high-performance components remains relentless. Gallium Nitride (GaN) technology emerges as a formidable contender, poised to redefine the landscape of power electronics. Gallium nitride (GaN) stands as a beacon of innovation in the quest for sustainable energy solutions.

This volume stands as a comprehensive exploration of GaN technology, encompassing its foundational principles, manufacturing processes, diverse applications spanning various industries, and the challenges and future trends shaping its trajectory.

Authored by Dilder Chowdhury, Chap. 1 serves as an introductory gateway into the realm of GaN technology, offering a historical overview and underscoring its significance in contemporary electronics. Chapter 2, penned by Chien-Chung Lin and Shyh-Chiang Shen, delves into the intricate material properties of GaN and its related compounds, laying the groundwork for understanding its supremacy in power electronic systems.

Sonu Daryanani, Aurore Constant, Cristina Tringali, and Ferdinando Iucolano dissect the manufacturing processes of GaN technology in Chap. 3, shedding light on the challenges and methodologies driving its advancement.

Chapter 4, a collaborative effort by Florin Udrea, Kengo Ohmori, Ming Su, Michael de Rooij, Sam Abdel-Rahman, Walter Balzarotti, Doug Bailey, Andrew Smith, Oliver Häberlen, Tim McDonald, Mark Davidson, Troy Baker, Ryo Takeda, Takamasa Arai, and Stephen Oliver, offers a comprehensive exploration of GaN technology's pivotal role in revolutionizing power electronics, covering material properties, device configurations, layout considerations, packaging, and thermal solutions, as well as reliability testing and characterization.

Authored by Michael de Rooij, Alejandro Pozo, Stephen Colino, John Glaser, Marco Palma, Martin Wattenberg, Max Zafrani, Alex Lidow, Tony Marini, Robert Strittmatter, Scott Durkin, Jim Honea, Alex Q. Huang, Emad Nazerian, and Peng Han, Chap. 5 delves into the diverse applications of GaN technology across various sectors, from DC-DC conversion to space applications.

Chapter 6, featuring contributions from Nihit Bajaj, Daniel Makus, Rafael Garcia, Igor Spinella, Paul Wiener, Junyang Luo, Matthias Kasper, Guilherme Bueno Mariani, Hubie Noto, Thierry Bouchet, Peter Vaughan, Rémi Comyn, Dieter Liesabeths, Dima Novo, Denis Marcon, Francesco Di Domenico, Alex Zahabizadeh, and Stephen Oliver, offers a deep dive into GaN technology's transformative potential across different industries and applications, through case studies and expert insights. Last section of Chaps. 5 and 6 will analyze general aspects of foundry services.

Authored by Patrick Le Fèvre, Geoff Haynes, Kennith Kin Leong, Vladimir Odnoblyudov, Cem Basceri, and Han Wui Then, Chap. 7 presents a comprehensive overview of the challenges and future trends in GaN technology, exploring its transformative potential and ongoing efforts to overcome hurdles and drive widespread adoption.

Chapter 8, penned by Gerald Deboy and Matthias Kasper, examines the competitive landscape and technological advancements in GaN-based power devices, providing insights into their classification, advantages, and applications.

In Chap. 9, authored by Prabhu Karunakaran and Navdeep Saboo, readers are presented with an in-depth analysis of the burgeoning market of GaN semiconductor technology, analyzing market trends, challenges, opportunities, and future projections.

Together, these chapters offer a holistic view of GaN technology, from its foundational principles to its diverse applications and market prospects, serving as an invaluable resource for researchers, engineers, and enthusiasts alike.

With its remarkable efficiency and reliability, GaN technology is revolutionizing industries such as chargers, adapters, and data centers. GaN's quick response to changes suits motor drives, renewables, and electric vehicles, while its ability to operate at elevated temperatures ensures reliability.

Gallium nitride power devices have the potential to replace legacy silicon power devices in many power-conversion applications. The much-improved conduction and switching figures of merit (FOMs) with GaN lead to improved efficiencies, higher power densities, and the possibility to operate at much higher switching frequencies, leading to system-level improvements in cost and performance.

Pescara, Italy

Maurizio Di Paolo Emilio

Acknowledgments

This book is dedicated to the trailblazers in the tech, semiconductor, and power systems design sectors, who tirelessly pursue the vision of energy efficiency amidst an ever-growing global demand for power. As the thirst for energy escalates, exerting immense pressure on Earth's finite resources, the quest for efficiency becomes paramount to fueling the continuous evolution of our technology-driven world.

Gallium nitride-based devices are leading the charge in this effort, and they hold great promise for overcoming a key obstacle that has so far prevented renewable energy sources from being widely used. Data centers, electric vehicles, and renewable energy systems are just a few of the vital power-dependent businesses where energy efficiency has become an absolute necessity.

The aim of this book is to provide a comprehensive overview of GaN technology, delving into its origins, device manufacturing, and electrical characteristics that render it ideal for power electronics systems across various vertical markets, and exploring its practical applications. Furthermore, we endeavor to analyze its current market penetration and anticipate its potential to emerge as a mainstream technology in the future.

The realization of this project owes much to the invaluable contributions of technology experts, and stakeholders within the power electronics industries. I extend my heartfelt gratitude to the engineers, companies, academics, and industry analysts who graciously shared their insights and expertise on all facets of GaN technology. Thanks to their unwavering support and willingness to impart knowledge that this endeavor has come to fruition.

Contributors

- Dilder Chowdhury, Nexperia
- Chien-Chung Lin, National Taiwan University
- Shyh-Chiang Shen, Georgia Institute of Technology
- Sonu Daryanani, Technical Writer

- Aurore Constant, STMicroelectronics
- Cristina Tringali, STMicroelectronics
- Ferdinando Iucolano, STMicroelectronics
- Florin Udrea, Cambridge GaN Devices
- Kengo Ohmori, Rohm Semiconductor
- Ming Su, Rohm Semiconductor
- Michael de Rooij, Efficient Power Conversion
- Sam Abdel-Rahman, Infineon Technologies
- Walter Balzarotti, Infineon Technologies
- Doug Bailey, Power Integrations
- Andrew Smith, Power Integrations
- Oliver Häberlen, Infineon Technologies Austria AG
- Tim McDonald, Infineon Technologies
- Mark Davidson, semiconductor executive
- Troy Baker, Bolt Semiconductor
- Ryo Takeda, Keysight Technologies
- Takamasa Arai, Keysight Technologies
- Stephen Oliver, Navitas Semiconductor
- Alejandro Pozo, Efficient Power Conversion
- Stephen Colino, Efficient Power Conversion
- John Glaser, Efficient Power Conversion
- Marco Palma, Efficient Power Conversion
- Martin Wattenberg, Infineon Technologies
- Max Zafrani, Efficient Power Conversion
- Alex Lidow, Efficient Power Conversion
- Robert Strittmatter, Efficient Power Conversion
- Tony Marini, dtm Associates, Engineering Consultants
- Scott Durkin, Nexperia
- Jim Honea, Nexperia
- Alex Q. Huang, The University of Texas at Austin
- Emad Nazerian, The University of Texas at Austin
- Peng Han, The University of Texas at Austin
- Nihit Bajaj, Infineon Technologies
- Daniel Makus, Infineon Technologies
- Rafael Garcia, Infineon Technologies
- Igor Spinella, Eggtronic
- Paul Wiener, Infineon Technologies
- Junyang Luo, Infineon Technologies
- Matthias Kasper, Infineon Technologies
- Guilherme Bueno Mariani, Infineon Technologies
- Hubie Noto, Silanna Semiconductor
- Thierry Bouchet, Wise Integration
- Peter Vaughan, Power Integrations
- Rémi Comyn, KnowMade
- Dieter Liesabeths, VisIC Technologies

- Dima Novo, VisIC Technologies
- Denis Marcon, Innoscience Europe
- Francesco Di Domenico, Infineon Technologies
- Alex Zahabizadeh, Texas Instrument's GaN business
- Patrick Le Fèvre, PRBX
- Geoff Haynes, Consultant
- Kennith Kin Leong, Infineon Technologies Austria AG
- Vladimir Odnoblyudov, Qromis
- Cem Basceri, Qromis
- Han Wui Then, Intel Corporation
- Gerald Deboy, Infineon Technologies
- Prabhu Karunakaran, Frost & Sullivan
- Navdeep Saboo, Frost & Sullivan

Many thanks to all of you!

Additionally, special thanks to the Springer team for their assistance and guidance throughout this journey.

As we embark on this journey through the field of GaN technology, we invite our readers to join us in exploring its transformative potential. Together, let us unravel the intricacies of this innovative technology, understand its implications for power electronics, and envision its role in shaping a more sustainable and efficient future.

In the pursuit of progress, collaboration is paramount. May this book serve as a beacon of knowledge, igniting conversations and sparking innovations that propel us toward a brighter, more energy-efficient tomorrow.

Contents

About the Book

The book is a comprehensive exploration of Gallium Nitride (GaN) technology, highlighting its foundational principles, manufacturing processes, diverse applications across industries, challenges, and future trends. It features contributions from a multitude of experts in the field, each focusing on specific aspects of GaN technology:

Dilder Chowdhury introduces the historical context and significance of GaN in contemporary electronics.

Chien-Chung Lin and Shyh-Chiang Shen delve into the material properties of GaN and related compounds.

Sonu Daryanani, Aurore Constant, Cristina Tringali, and Ferdinando Iucolano explore GaN manufacturing processes.

Florin Udrea, Kengo Ohmori, Ming Su, Michael de Rooij, Sam Abdel-Rahman, Walter Balzarotti, Doug Bailey, Andrew Smith, Oliver Häberlen, Tim McDonald, Mark Davidson, Troy Baker, Ryo Takeda, Takamasa Arai, Stephen Oliver, and others examine GaN's pivotal role in power electronics.

Michael de Rooij, Alejandro Pozo, Stephen Colino, John Glaser, Marco Palma, Martin Wattenberg, Max Zafrani, Alex Lidow, Tony Marini, Robert Strittmatter, Scott Durkin, Jim Honea, Alex Q. Huang, Emad Nazerian, Peng Han, and others discuss diverse applications of GaN technology.

Nihit Bajaj, Daniel Makus, Rafael Garcia, Igor Spinella, Paul Wiener, Junyang Luo, Matthias Kasper, Guilherme Bueno Mariani, Hubie Noto, Thierry Bouchet, Peter Vaughan, Rémi Comyn, Dieter Liesabeths, Dima Novo, Denis Marcon, Francesco Di Domenico, Alex Zahabizadeh, Stephen Oliver, and others explore GaN's transformative potential in various industries.

Patrick Le Fèvre, Geoff Haynes, Kennith Kin Leong, Vladimir Odnoblyudov, Cem Basceri, Han Wui Then, and others present an overview of challenges and future trends in GaN technology.

Gerald Deboy and Matthias Kasper analyze the competitive landscape and technological advancements in GaN-based power devices.

Prabhu Karunakaran and Navdeep Saboo provide an in-depth analysis of the GaN semiconductor market.

Contributors

Sam Abdel-Rahman Infineon Technologies Americas Corp., Morrisville, NC, USA

Takamasa Arai Keysight Technologies, Tokyo, Japan

Doug Bailey Power Integrations, San Jose, CA, USA

Nihit Bajaj Director, GaN Product Marketing, Infineon Technologies, San Jose, CA, USA

Troy Baker CTO, Bolt Semiconductor, Hummenlstown, PA, USA

Walter Balzarotti Infineon Technologies AG, Neubiberg, Germany

Cem Basceri QROMIS, Inc., Santa Clara, CA, USA

Thierry Bouchet Wise Integration, Hyères, France

Guilherme Bueno Mariani Infineon Technologies, Villach, Austria

Dilder Chowdhury Power GaN Technology at Nexperia, Manchester, UK

Stephen Colino EPC, El Segundo, CA, USA

Rémi Comyn Technology & Patent Analyst, KnowMade, Valbonne, France

Aurore Constant Research and Development Engineer, STMicroelectronics, Tours, France

Sonu Daryanani Technical Writer, Las Palmas de Gran Canaria, Spain

Mark Davidson Semiconductor Executive, Downington, PA, USA

Gerald Deboy Infineon Technologies Austria, Villach, Austria

Michael de Rooij EPC, El Segundo, CA, USA

Michael A. de Rooij EPC, El Segundo, CA, USA

Maurizio Di Paolo Emilio Power Electronics News, Pescara, Italy

Francesco Di Domenico Senior Principal System Architect, Infineon Technologies, Villach, Austria

Scott Durkin Nexperia, Goleta, CA, USA

Rafael Garcia System Architect for Onboard Chargers and DCDC Applications for Electric Vehicles, Infineon Technologies, Villach, Austria

John Glaser EPC, El Segundo, CA, USA

Oliver Häberlen Infineon Technologies Austria AG, Villach, Austria

Peng Han University of Texas, Austin, TX, USA

Geoff Haynes Consultant, Poole, UK

Jim Honea Nexperia, Goleta, CA, USA

Alex Q. Huang University of Texas, Austin, TX, USA

Ferdinando Iucolano Senior Design Architecture ENG II, STMicroelectronics, Catania, Italy

Prabhu Karunakaran Industry Analyst at Frost & Sullivan, Chennai, Tamil Nadu, India

Matthias Kasper Innovation Group Leader, Infineon Technologies, Villach, Austria
Infineon Technologies Austria, Villach, Austria

Patrick Le Fèvre POWERBOX (PRBX), Stockholm, Sweden

Kennith Kin Leong Infineon Technologies,, Villach, Austria

Alex Lidow EPC, El Segundo, CA, USA

Dieter Liesabeths VisIC Technologies, Munich, Germany

Chien-Chung Lin Department of Electrical Engineering, Graduate Institute of Photonics and Optoelectronics, National Taiwan University, Taipei, Taiwan

Junyang Luo System Architect for Charger/Adapter, Infineon Technologies, Singapore, Singapore

Daniel Makus Application and Business Development Director for xEV power conversion, Infineon Technologies, Neubiberg, Germany

Denis Marcon Innoscience Europe, Leuven, Belgium

Tony Marini EPC, El Segundo, CA, USA

Tim McDonald Infineon Technologies, Americas Corp, El Segundo, CA, USA

Emad Nazerian University of Texas, Austin, TX, USA

Hubie Noto Silanna Semiconductor, Union City, CA, USA

Dima Novo VisIC Technologies, Nes Ziona, Israel

Vladimir Odnoblyudov QROMIS, Inc., Santa Clara, CA, USA

Kengo Ohmori Rohm Semiconductor, Santa Clara, CA, USA

Stephen Oliver Navitas Semiconductor, Torrance, CA, USA

VP Corporate Marketing & Investor Relations, Navitas Semiconductor, Torrance, CA, USA

Marco Palma EPC, El Segundo, CA, USA

Alejandro Pozo EPC, El Segundo, CA, USA

Navdeep Saboo Industry Analyst at Frost & Sullivan, Chennai, Tamil Nadu, India

Shyh-Chiang Shen School of Electrical and Computer Engineering, Georgia Institute of Technology, Atlanta, GA, USA

Andrew Smith Power Integrations, San Jose, CA, USA

Igor Spinella Eggtronic, Modena, MO, Italy

Robert Strittmatter EPC, El Segundo, CA, USA

Ming Su Rohm Semiconductor, Santa Clara, CA, USA

Ryo Takeda Keysight Technologies, Tokyo, Japan

Han Wui Then Intel Foundry Technology Research, Intel Corporation, Hillsboro, OR, USA

Cristina Tringali Responsible for GaN Power Low V and RF Technology, STMicroelectronics, Catania, Italy

Florin Udrea Cambridge GaN Devices, Cambridge, UK

Peter Vaughan Director of Automotive Business Development, Power Integrations, San Jose, CA, USA

Martin Wattenberg Infineon Technologies Austria, Villach, Austria

Paul Wiener Strategic Marketing, Infineon Technologies, San Jose, CA, USA

Max Zafrani EPC, El Segundo, CA, USA

Alex Zahabizadeh Texas Instrument's GaN Business, Dallas, TX, USA

Chapter 1
Introduction to GaN Technology

Dilder Chowdhury

Gallium Nitride High Electron Mobility Transistors (GaN HEMTs) offer higher electron mobility and high critical electric field that facilitates high-speed faster response i.e., higher switching speed even at high temperatures. Robust higher band gap and higher critical electric field helps allow a lower electron traveling path and a lower on-resistance compared to Si- and SiC-based devices. Even though GaN thermal conductivity at room temperature is low compared to SiC, at high temperatures, GaN thermal conductivity is comparable to doped SiC. As a result, smaller devices are possible and further possibility of higher frequency operation are being demonstrated by most of the GaN-based commercially available power devices. Widespread adoption of Wide Bandgap (WBG) power electronic devices continues and GaN Power HEMTs are helping increased efficiencies, increased frequencies, reduced solution sizes, and eventually reduced system costs of a wide variety of power conversions. Enhancing the capability beyond what was possible with Si-based solutions.

1.1 Overview of Wide Bandgap

Wide Bandgap (WBG) materials have electronic energy band gaps significantly larger than one electron Volt (eV), setting them apart from traditional semiconductor materials like silicon (Si) with narrower bandgaps. The adoption of WBG materials in power electronics has been driven by their unique attributes that promise to revolutionize the field. In general, WBG materials have much greater energy gaps, breakdown electric fields, thermal conductivities, melting points, electron mobility,

D. Chowdhury (✉)
Power GaN Technology at Nexperia, Manchester, UK
e-mail: dilder.chowdhury@nexperia.com

© The Author(s), under exclusive license to Springer Nature Switzerland AG 2024
M. Di Paolo Emilio (ed.), *GaN Technology*,
https://doi.org/10.1007/978-3-031-63238-9_1

and velocities. Thanks to these properties, WBG semiconductor-based power devices can operate at substantially higher voltages, switching frequencies, and temperatures than Si.

For instance, a thinner drift layer with a greater doping concentration can be employed for WBG power devices at the same blocking voltage as the breakdown field is significantly larger than that of Si. When compared to Si majority carrier devices, unipolar devices with thinner blocking layers and higher doping concentration field-effect transistors (MOSFETs), have lower specific on-resistances. WBG devices will have a fast-switching speed due to the higher breakdown field, shorter drift region or channels, higher electron mobility, and velocity. First, in WBG unipolar devices like MOSFETs, a smaller chip size is made possible by lower on-resistance at the same breakdown voltage. Because of the thinner drift size and high electron mobility, the switching speed is high. Second, during switching transients, the charges are swept in and out of the channels at the high saturated drift velocity. Since WBG has a greater electron saturation drift velocity than Si, WBG devices switch at a faster rate.

Many III-V and II-VI compound semiconductors (i.e., semiconductors made up of elements from groups III, V, and II, VI of the periodic table) have large bandgaps. SiC and GaN are two potential candidates for WBG semiconductor materials, and they exhibit the best theoretical characteristics—such as high blocking voltage capability, high-temperature operation, and high switching speed—against other commercially available materials. As a result, SiC and GaN are the primary materials used in practically all WBG semiconductor power devices currently available on the market or in research and development.

1.2 Overview of GaN

As the journey for net-zero emission is getting momentum, the quest for better performance, higher efficiencies, lower energy losses, lower heat dissipation and wastages, smaller and more compact solution and form factors have never been so important. Progressive innovation brought Si-based technologies to its current state of sophistication, offering amazing performances breaking the 1-D material limits. Improvements in Si-based technology is still underway but at a much slower pace, contributing smaller evolution and application specific benefits! However, even with tremendous progress it made during the last 70 years it is now reaching its limit and to maintain the progress, the need for new materials has evolved. In this journey, Wide Band Gap (WBG) materials like Silicon Carbide (SiC) and Gallium Nitride (GaN) have shown their huge potential and are beginning to enter the power conversion markets. Currently, GaN power devices are in the forefront of this journey, demonstrating significantly superior performances compared to all other commercially available Si- and SiC-based products. Even though the material properties are well known as in Figs. 1.1 and 1.2, commercially available devices and products are

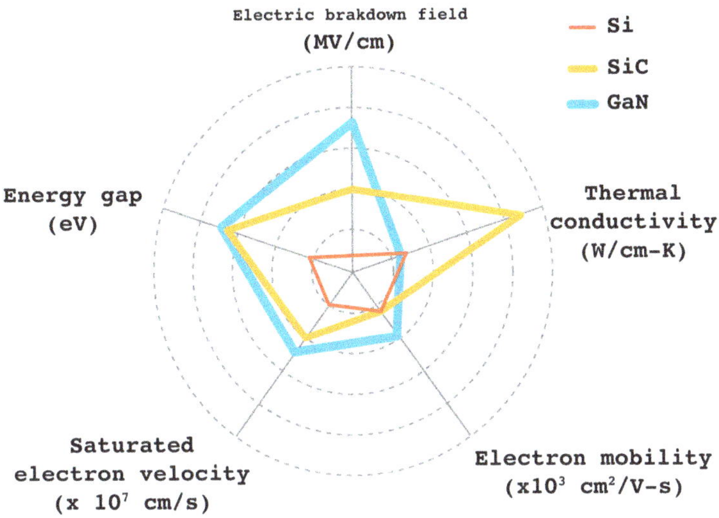

Fig. 1.1 Summary of Si, SiC, and GaN relevant material properties

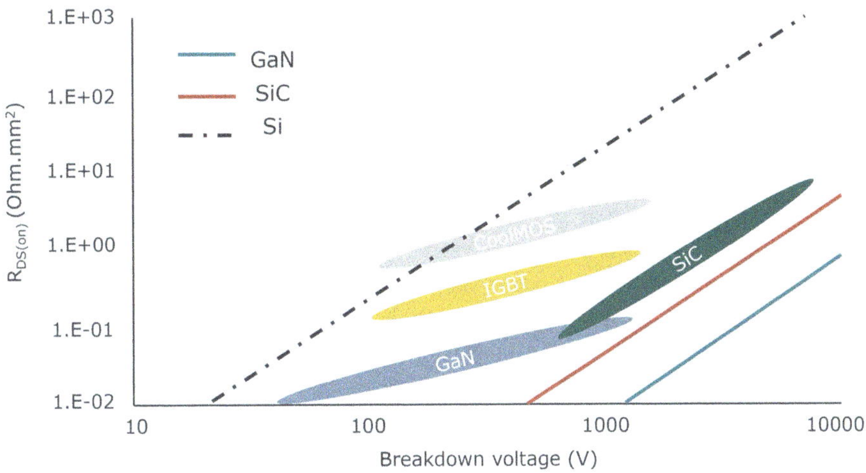

Fig. 1.2 Theoretical limits and currently demonstrated performance

limited, with many challenges yet to be overcome. Even at this early stage of development, GaN High Electron Mobility Transistors (HEMT) power devices already surpassed the performance of the other technologies, with continuously new and better generation products are coming. Hence, enabling improved efficiencies, reducing solution sizes, and improved power density at reduced cost.

1.3 General Features

GaN and SiC (silicon carbide) are both wide bandgap materials. While these materials offer outstanding performance, their characteristics, applications, and gate drive requirements are different. SiC can compete with IGBT transistors in high-power and extra high-voltage (above 650 V) applications. Likewise, GaN can compete with current MOSFETs and super junction (SJ) MOSFETs in power applications with voltages up to 650 V. Even though GaN potentially offers superior high-frequency and high-voltage performance than SiC due to its larger breakdown electric field and electron velocity, GaN wafer manufacturing technique remains a significant obstacle for vertical devices. Currently, lateral devices known as heterojunction field-effect transistors (HFETs), also known as high electron mobility transistors (HEMTs), offer a viable alternative that capitalizes on the high-mobility layer created at the heterojunction between GaN and AlGaN. HEMTs have been produced on Si, SiC, and sapphire substrates. Si is far less expensive than competing materials, and its conductive qualities confer distinct commercial and manufacturing advantages as a substrate.

Two variants of the GaN technology are GaN-on-silicon (GaN-on-Si) and GaN-on-silicon-carbide (GaN-on-SiC). GaN-on-SiC has contributed a lot to space and military radar applications. Today, RF engineers are looking for new applications and solutions to take advantage of GaN-on-SiC.

Due to the two-dimensional electron gas (2DEG) produced by the spontaneous and piezoelectric polarization at the heterojunction between undoped GaN and AlGaN, an HEMT does not need intentional doping and a p–n junction, unlike a MOSFET. First, a multilayer buffer or seed layer is formed on the substrate wafer to provide crystal lattice strain relief for the subsequent epitaxial deposition of a thick pure GaN layer. The AlGaN cap layer is next deposited, followed by the formation of the 2DEG layer between the two. The 2DEG is a thin layer of native electrons that generates a low-resistance path, normally on current route between the drain and the source. A gate is employed on the AlGaN cap layer to switch off the device by depleting the 2DEG when a negative voltage is supplied. Consequently, the HEMT is a depletion-mode (normally-on) device. Widespread ways for achieving normally off HEMTs include the non-insulated gate enhancement mode, the insulated gate enhancement mode, and the cascode.

Since many decades ago, most power electronic devices have been based on silicon, a semiconductor that can be made cheaply and is nearly free of defects. Silicon's theoretical performance has been reached nearly totally, revealing the material's limitations, such as its limited voltage blocking and heat transfer capabilities, efficiency, and non-negligible conduction losses. Wide bandgap (WBG) semiconductors, such as gallium nitride (GaN), outperform silicon in terms of efficiency, switching frequency, temperature, and voltage.

The bandgap of gallium nitride is 3.4 electron volts (eV), which is roughly three times that of the silicon, which is 1.1 eV. This indicates that more energy is required to move an electron from the valence band to the conduction band. Hence, the suitability of GaN in the higher breakdown voltages and enhanced stability at higher temperatures. GaN HEMTs are the desirable substitute for silicon power MOSFETs and Schottky diodes in the fields of high-efficiency power converters, since it significantly boost the efficiency of power conversion stages. Gallium nitride provides for significant advantages over silicon, including improved energy efficiency, smaller dimensions, lighter weight at lower total cost.

GaN HEMTs have outstanding $R_{DS(on)}$ and figure of merit values (FOM). The figure of merit may be four to ten times lower than that of a super junction (SJ) FET, depending on the voltage and current ratings. GaN is hence appropriate for high-frequency applications. Utilizing a GaN HEMT with a low RDS (on) value decreases conduction losses along with the reduced switching losses and boosts efficiency.

GaN HEMTs lack an inherent body diode; hence, there is no reverse recovery charge. These devices are inherently capable of reverse conduction and their properties vary based on the gate voltage. Reverse conduction capability may be advantageous over conventional IGBTs at the system level since antiparallel diodes are not needed.

GaN enables efficient operation at high switching frequencies by eliminating reverse recovery losses. For instance, the use of super junction MOSFETs in bridgeless totem-pole power factor correction (PFC) in continuous conduction mode (CCM) is not feasible due to their significant reverse recovery losses. By using GaN transistors as high-frequency switches, the reverse recovery loss is minimized and the associated switching loss is drastically decreased.

GaN has a higher breakdown electric field and, hence, a higher breakdown voltage than Silicon. GaN has a breakdown field of 3.5 MV/cm, whereas silicon's is just 0.23 MV/cm. This makes gallium nitride 10 times more high voltage withstand capability. GaN may be used by manufacturers and designers in power applications with a smaller footprint.

Gallium nitride's electron mobility is 2000 cm^2/Vs, much greater than Silicon's mobility of 1500 cm^2/Vs. Therefore, electrons in gallium nitride crystals may travel almost 30 percent quicker than in Silicon. High critical electric field and the high electron mobility at higher temperature allow gallium nitride with a significant advantage for usage in radiofrequency (RF) applications since it can withstand greater switching frequencies than Silicon.

The thermal conductivity of gallium nitride is similar to silicon (1.3 W/cm K against 1.5 W/cm K). GaN enables quicker switching rates than silicon slew rate (dV/dt of 100 V/ns or more). This significantly reduces switching losses. However, the great efficiency of GaN reduces the thermal loads of the circuit, allowing it to run at a lower temperature than silicon. This improves thermal management and eliminates the need for external heatsinks.

Enhanced GaN HEMTs do not always need a negative voltage to shut off. While a negative voltage shields the gate from voltage spikes, it increases the reverse conduction loss. Negative turn-off voltage is a method used sometimes to prevent spurious turn-on during power-on or transients. In reality, several commercially marketed eGaN gate drivers can operates without negative turn-off voltage since it increases reverse conduction losses and adds expense and complexity.

GaN HEMTs produced using GaN-on-Si technology are the optimal choice for medium power applications, allowing for high power conversion efficiency, compact form factor, and reasonable cost. Additionally, the tiny size and ability to operate at high frequencies allow for the reduction or deletion of costly and bulky passive components and heat sinks.

The benefit of GaN-on-Si technology is that silicon is easily available and inexpensive material, and silicon wafers are available on a huge scale. When one semiconductor is grown on top of another material, a significant number of crystalline defects are introduced into the developed crystalline layer. In the case of GaN-on-Si, the defect density is roughly 10^8–10^{10}/cm^2. These faults compromise the operating dependability of the gadget at high voltage. Consequently, GaN HEMTs with voltage ratings beyond 900 V have not been marketed commercially, and the majority of them have a breakdown voltage of about 650 V.

The recent availability of high-quality bulk GaN substrates has allowed for the creation of vertically structured GaN-on-GaN power devices with greater breakdown voltage, higher current handling capacity, and exceptionally small sizes. Vertical GaN offers compact-form factor devices with breakdown voltages between 100 V and 4 kV and switching frequencies of a few megahertz.

Vertically conducting devices need the substrate and drift zone to have the same material. This needs a bulk GaN substrate for GaN. While costly and not accessible in diameters bigger than 100 mm, bulk GaN substrates enable the manufacture of vertically conducting GaN transistors in materials with defect concentrations between 10^3 and 10^5/cm^2. This low defect density enables GaN transistors to have a high degree of dependability at voltages exceeding 1000 V.

Consequently, vertical GaN devices can handle the most demanding power applications, including data center power supply, servers, and power conversion applications in electric cars, solar inverters, motors, and high-speed trains. The high switching frequencies also permit the downsizing of large magnetic components and capacitors, while retaining a very high level of efficiency.

1.4 Historical Development

Gallium (Ga) was initially postulated by Dmitri Mendeleev in 1871, and Paul-Émile Lecoq de Boisbaudran made its discovery in Paris a few years later, in 1875. Subsequently, 57 years passed before the initial polycrystalline GaN material was produced through the process of ammonia (NH3) diffusion over liquid gallium. GaN material was synthesized in 1932 at the George Herbert Jones Laboratory through the reaction of gallium metal and ammonia at temperatures ranging from 900 to 1000 degrees Celsius. Furthermore, H.P. Maruska and J.J. Tietjen documented the expansion of single-crystal film of GaN in 1969, which was 37 years later. First-time use was made of hydride vapor phase epitaxy (HVPE) to deposit GaN onto a sapphire substrate. In 1972, a blue light detector based on GaN was devised by J.I. Pankove, E.A. Miller, and J.E. Berkeyheiser, a few years later.

Nevertheless, despite emitting blue luminescence at ambient temperature, its power efficiency was subpar.

The pursuit of commercially viable GaN applications was reignited in the 1980s with the advent of growth technologies such as metal-organic chemical vapor deposition (MOCVD) and molecular beam epitaxy (MBE). Presently, MOCVD and MBE are utilized in tandem to manufacture GaN devices, the specific method dependent on the intended use. Since previously stated, the initial development of GaN-based LEDs occurred in 1972 when GaN doped with magnesium produced the compound that emanates blue light in the form of an LED. However, despite being the initial LEDs to emit a blue-violet hue, their brightness was insufficient for practical applications.

Isamu Akasaki and Hiroshi Amano, two colleagues from Nagoya University, and Shuji Nakamura, who was a researcher at Nichia Chemical in Japan at the time, patented a GaN-based method for producing high-brightness blue LEDs in 1991. Subsequently, in 1993, they showcased these ultra-bright blue LEDs that were created by doping GaN with surplus magnesium.

The three principal colors required for LED-based illumination applications were red, green, and blue (RGB); however, the production of blue LED was technologically unfeasible at the time. By serving as the conclusive component in the development of RGB LEDs, the blue LED ultimately laid the foundation for white LEDs and color-changing LEDs.

The introduction of commercially available high-brightness blue LEDs marked a turning point in the electronics industry. It initiated the replacement of inefficient incandescent lighting and cathode ray tube (CRT) televisions and monitors with solid-state lighting, which was initiated by a renaissance of lighting applications in which blue LEDs served as the precursor. Following this, the advancement of Blu-ray optical discs was substantially influenced by blue LED technology based on GaN, which subsequently supplanted DVDs as the predominant method of data storage for visual media. So, in essence, Blu-ray represents a particular concept. The 2014 Nobel Prize in Physics was awarded to blue LED pioneers Akasaki, Amano, and Nakamura in recognition of the immense impact their contributions had on solid-state illumination, blue LED technology, and data storage.

GaN-based high electron mobility transistors were introduced by Eudyna Device Inc. in Japan in 2004 as depletion-mode transistors for RF applications. The HEMT architecture was designed in response to a 1975 phenomenon discovered by Takashi Mimura and his group. Following this, in 1994, M. A. Khan and his colleagues observed an exceptional degree of electron mobility in the vicinity of the AlGaN and GaN heterostructure interface, which they characterized as a two-dimensional electron gas (2DEG).

Additionally, the first GaN/AlGaN heterostructure produced via MOCVD was revealed by Khan's group. At ambient temperature, the electron mobility of this heterostructure was twelve times greater than that of bulk GaN with an equivalent thickness. These findings highlighted the potential of GaN in the field of power electronics. Two-Dimensional Electron Gas (Two-DEG), characterized by its high power densities, low on-resistance, and high current capabilities, significantly contributed to the enhancement of GaN-based power devices.

In order to generate a power gain benchmark in the multi-gigahertz frequency range, Eudyna monetized this endeavor. In 2005, Nitronex Corporation unveiled the initial depletion-mode RF HEMT transistor constructed from GaN grown on silicon substrates. At that juncture, GaN RF transistors started to gain traction in RF designs, particularly in infrastructure applications requiring their exceptional voltage and efficiency. Nevertheless, the implementation of GaN RF transistors was hindered by the expense of the devices and the inconvenient depletion-mode operation.

The development of enhancement-mode GaN transistors, alternatively referred to as GaN FETs, commenced with the synthesis of a thin GaN layer on the aluminum nitride (AIN) layer of a standard silicon wafer via the MOCVD method. This process entails the precise manipulation of gaseous intermediates within a chamber to generate the material layers. A buffer is provided by the AIN layer between the substrate and GaN.

Efficient Power Conversion (EPC) introduced the initial silicon wafers fabricated with enhancement-mode GaN, which were intended to serve as power MOSFET substitutes, in 2009. The advent of field effect transistors (FETs) facilitated the mass production of GaN at a reduced expense utilizing conventional silicon manufacturing facilities and technology. Fujitsu, MicroGaN, Texas Instruments, and Panasonic all quickly entered the league and began developing their own GaN devices.

The mode of enhancing GaN transistors were specifically engineered to substitute power MOSFETs in scenarios where the ability to transition quickly or convert power efficiently is of the utmost importance. Immediately following their conception, research commenced on GaN power integrated circuits (GaN ICs), which would combine a GaN FET, drive circuitry based on GaN, and circuit protection into a single surface-mount device.

The integration resulted in a gate-drive loop characterized by nearly zero impedance, which enhanced efficiency by eradicating the turn-off losses of the FET. The Hong Kong University of Science and Technology (HKUST) initiated the development of low-voltage GaN power integrated circuits (ICs), and the inaugural devices were showcased in 2015.

Then, in 2016, HRL Laboratories in Malibu, California showcased a GaN power integrated circuit (IC) that cost-effectively realized the complete potential of GaN electronics. Notably, prominent GaN player Navitas Semiconductor was established in 2013 in the parking lot of HRL atop a trailer. Gene Sheridan and Dan Kinzer, former HRL executives and co-founders, obtained a license from HRL for their GaN power electronics technology the following year.

Some of the players in the GaN market include EPC, IR, Transphorm, Fujitsu, Panasonic, Infineon, GaN Systems, NXP, ExaGaN, Navitas, Visic, TI, Nexperia, Microsemi, Innoscience, Wolfspeed, Power Integration, X-Fab, TSMC, ST micro, Toshiba, and onsemi.

There are mainly three different types of GaN HEMTs for power applications: enhancement-mode (E-mode), cascode, and direct drive cascode or cascode lite. E-mode GaN HEMTs are normally off devices, meaning that they require a positive gate voltage to turn on, which simplifies the circuit design and improves the reliability and safety. However, E-mode GaN HEMTs are more challenging to fabricate, as they require complex engineering of the GaN surface and interface. Cascode GaN HEMTs are normally on devices, meaning that they require a negative gate voltage to turn off, which increases the complexity and cost of the driver circuit. However, cascode GaN HEMTs are easier to drive, as they use a GaN HEMT in series with a low-voltage silicon MOSFET. Simple LV MOSFET drivers with high CMTI (>100 V/ns). Direct drive cascode or cascode lite GaN HEMTs are a hybrid between E-mode and cascode, as they use a high-voltage E-mode GaN HEMT in series with a low-voltage silicon MOSFET. This configuration need more complex driver circuit.

1.5 Significance in Modern Electronics

GaN HEMT with its significant GaN material property advantages and furthermore enhanced by the 2DEG in the quantum well formed at the GaN/AlGaN interface, revolutionized modern electronics. As the energy demand is increasing astronomically, the core expectations of the modern electronics and power conversion has grown. With higher efficiencies, smaller form factors, better precision and accuracy, better control, and stability. In this journey, Power GaN HEMT technology can perform most efficiently and bring the best for the purpose. Modern electronics and power conversion relies on lowering the conversion losses, more precision and accuracy, smart smooth function with reduced sizes and weights. Fundamental to the electronics is the switching in addition to the conduction losses. Within the switching perspective, turn on and turn off losses come into effect. Lower the losses, higher the efficiency and higher the possibility to increase the frequency. As the frequency increases, the size and cost of the passives like the inductors decreases and the capacitors can be replaced by low-cost smaller ceramic capacitors. Invariably, larger passives due to lower operating frequencies with other semiconductor technologies cause increased system costs, bulky big designs, and poor performance with significant material wastages.

Power GaN HEMT (Gallium Nitride High Electron Mobility Transistor) technology represents a significant advancement in power electronics, offering a plethora of benefits across various applications. Its emergence has revolutionized the field, primarily due to its exceptional performance, reduced losses, and simplified drive circuits.

One of the most notable advantages of GaN HEMT technology is its high performance. GaN devices exhibit superior characteristics compared to traditional silicon-based transistors, such as faster switching speeds and lower on-state resistance (RDS(on)). This enables GaN HEMTs to handle higher frequencies and power densities, making them ideal for high-performance applications where efficiency and speed are crucial.

Furthermore, GaN HEMTs are characterized by their minimal reverse recovery charge (Q_{rr}), resulting in reduced overshoot and undershoot during switching. This feature is particularly beneficial in applications where precise control of voltage and current is essential, leading to improved system stability and reliability.

Another key advantage of GaN HEMT technology is its ability to switch at high temperatures without a significant increase in losses. Traditional silicon-based devices often experience performance degradation at elevated temperatures, leading to increased power losses. In contrast, GaN HEMTs maintain their efficiency even in harsh operating conditions, making them suitable for a wide range of environments, including automotive, aerospace, and industrial applications.

Moreover, GaN HEMTs exhibit much lower conduction losses at less than full load, resulting in improved overall efficiency and reduced power consumption. This characteristic is particularly advantageous in applications with varying load conditions, where maintaining efficiency across different operating scenarios is critical.

The inherent properties of GaN HEMTs, such as lower gate charge and low Q_{GD} (gate-drain charge), contribute to reduced driver losses and simplified drive circuits. This simplification leads to cost savings and improved reliability in power electronic systems.

Additionally, the positive temperature coefficient of $R_{DS(on)}$ in GaN HEMTs makes them easy to parallel, allowing for scalability and flexibility in designing high-power systems. This feature simplifies the implementation of parallel configurations, leading to increased power handling capability without the need for complex circuitry.

Furthermore, the design and layout considerations for GaN HEMT-based systems play a crucial role in optimizing performance and system stability. Tight, optimized low inductance layouts for both the drive loop and power loop are essential for minimizing ringing and improving overall system efficiency.

Contribution of power GaN technology is enormous as it helps to reduce the solution sizes significantly through reduced wasted energy, increased efficiencies, and reduced cooling. It opens many new possibilities that were not possible before. New topologies, new application areas, and new horizons are opened, giving engineers the freedom for new ideas and possibilities like new motor technology, improved battery system, precision industrial automation, better functional accuracy, and improved efficiency in motor control.

As the power GaN FET technology has enabled significantly faster response means capability of higher dv/dt, di/dt and frequency, to take the full advantage, other associated technologies need to improve. Hence, opening more and more innovation in other areas such as motor technologies with better windings capable of withstand higher dv/dt, operating at higher frequencies to get the desired high efficiencies. Other innovation areas are passives like inductors, inductor core materials, and capacitors capable of higher frequencies with lower losses. Packaging of these devices and the power modules with robust low inductance, low parasitic and low thermal resistance is in itself a huge challenge, that will improve over time with significant innovation and investments for demanding applications. Specific application requirements, like the power to be delivered, the mission profile, the voltage ratings, the load profile, and environmental factors, like temperature, pressure, and humidity shape the materials used as well as optimization. Efficient heat extraction from the devices is another avenue that is receiving lots of attention due to the high-power concentration with smaller devices and smaller solutions. Top side cooling, both side cooling, using more thermally efficient materials with better CTE match are among a few for improving the system solution.

As the electrification with AI is significantly enhanced in every sector, the need for highly efficient appropriate power conversion solution with appropriate form factor is essential. More powerful solutions with less cooling helps bringing the system complexity down significantly. Competent design engineers can play a significant part in redefining the system solutions. Most of the limitations of semiconductor switches are taken to a new level with power GaN HEMT technologies and limitations imposed by the other components now need to work to enable significantly improved performance and solutions.

Power GaN HEMT technology is at its early stage of technological and commercial maturity compared to Si-based technologies. There is and there will be significant innovation and investment going forward to make this technology as the norm of the future adoption. Even at this early stage, the benefit is very clearly demonstrated. Further improvements are on the way. Since the introduction of Crypto currency, data mining, AI, all of these increasing the huge data handling requirements while there is strong need for net zero emission, forcing more and more energy demand and efficient power conversion. Application areas where the benefit is obvious is huge and wide. Automotive electrification (Hybrid and Battery EV), EV charging, Lidar, Automotive and domestic LED lighting, Industrial automation and equipment, Telecom infrastructure including Cellular base station, High speed computing, Server and storage, Consumer electronics, Data and communication networks, Uninterruptible power supplies (UPS), Energy conversion, Power utilities and power distribution, military, aerospace and aircraft, UAVs, motor drive and control, servo motors, Photovoltaic inverters, Industrial drilling and wielding, precision robotics, better and faster response detection among the few that can significantly benefit from the adoption of GaN technology. Many more applications are adding to the list as the new possibilities are opening. A new era is evolving and will revolutionize the power electronics solutions as the system solutions are becoming more and more optimized.

Dilder Chowdhury is based in Nexperia's MOS & GaN HQ in Manchester, UK. Dilder received his PhD in III-V Microelectronic Materials and Semiconductor Device Technology at The University of Manchester (UMIST). He has worked in semiconductor industries for over 27 years from Innovation Manager to International Marketing Managers role. First joining in Philips Semiconductors, then to NXP and now in Nexperia's Innovation team, then contributing to the release of Nexperia's first Power GaN products as Architect for Power GaN technology, before his current role as Strategic Marketing Director for the Nexperia Power GaN Business Group. He served as Assistant Professor in the EEE at BUET.

Chapter 2
GaN Material Properties

Chien-Chung Lin and Shyh-Chiang Shen

Semiconductors in general are solid state materials in amorphous, poly-crystalline, or single crystalline forms. The class of semiconductors distinguishes from electric conductors such as metallic materials in that semiconductors formed two groups of energy states, referred to as the conduction band and the valence band, respectively, in the energy space. These two bands are separated by certain level of energy, the so-called energy gap, in which gap no allowed energy levels exist in the intrinsic material. The presence of the energy gap allows electrons and holes to be controlled in a predictable way by a use of external perturbations such as the electric potential, heat, or photons. The manipulation of electrons within semiconductor chips constitutes the basis of today's microelectronic world. From computer processors to mobile phones, these microelectronic systems have become essential parts of our daily life. Today's business transactions and human communications cannot do without the aid of semiconductor-based integrated circuits and associated electronic systems.

The widespread use of microelectronic systems is facilitated by the abundance of electricity generated by a wide range of energy sources. The natural resources are limited, however, and there have been serious concerns in the ever-expanding use of electricity in recent years. Ways to enable efficient energy conversion in microelectronic systems have become a key to bring a greener future with less waste of electricity. In recent years, electrifying automobiles has attracted major interests in microelectronic worlds. Typical microelectronic systems are used in ordinary

C.-C. Lin (✉)
Department of Electrical Engineering, Graduate Institute of Photonics and Optoelectronics, National Taiwan University, Taipei, Taiwan
e-mail: chienchunglin@ntu.edu.tw

S.-C. Shen
School of Electrical and Computer Engineering, Georgia Institute of Technology, Atlanta, GA, USA

© The Author(s), under exclusive license to Springer Nature Switzerland AG 2024

M. Di Paolo Emilio (ed.), *GaN Technology*,
https://doi.org/10.1007/978-3-031-63238-9_2

environments where the temperature is in a normal operating range (−40 °C to 125 °C) Silicon-based or other low-bandgap materials such as GaAs and InP have become the workhorse for today's microelectronics systems. These materials have a bandgap energy of around 0.7 eV (for silicon) and less than 2.0 eV (for conventional compound semiconductors). For electronic systems used in harsh environments such as outdoor locations or within electric vehicles, high-temperature and high-power operation capabilities become highly desired. Electronic or optoelectronic devices made of low-energy-gap semiconductors in these harsh environments would require extra care. For example, silicon-based microelectronic systems in vehicles usually need special care such as additional cooling and ambient temperature control either by hardware or firmware implementations. The resulting systems become bulky and inefficient.

When energy utilization efficiency and harsh environment operation are put in context, wide-bandgap (WBG) semiconductors emerged as the materials of choice for next-generation electronic system implementation. As the name suggests, wide-bandgap semiconductors have a "wider" bandgap compared to conventional semiconductors such as silicon or GaAs. Devices made of WBG materials are less susceptible to the temperature fluctuation and can operate at much higher ambient temperature (>200 °C) without causing electronic circuit's malfunction. The added benefit of the WBG semiconductor is high dielectric breakdown strength. The critical electric field in today's wide-bandgap semiconductors are at least 10 times higher than that in silicon. Provided that WBG semiconductors can possess similar electronic conductivity, WBG's high breakdown strength allows for a reduction of physical dimension of the device for the same voltage rating. In turn, the devices could operate with less resistive loss and smaller reactance. Conversely, the WBG of the same device dimension in the could operate at higher voltage because of the high critical electric field. As a result, the high-temperature operating range, power handling capability, and the energy utilization efficiency of WBG-semiconductor-based power electronic systems are much better than the silicon-based counterparts. Semiconductor devices made of these WBG materials are highly suitable for high-voltage and high-temperature operations, making them highly sought material platforms for a wide variety of semiconductor device implementations in academia and in semiconductor manufacturing industry.

The most common WBG semiconductor materials systems are silicon carbide (SiC) and III-nitride (III-N). Silicon carbide comes with poly forms. Each form has its unique atomic structural arrangement with distinct mechanical, chemical, and thermal properties, as well as material stability. Commonly seen SiC semiconductor materials are in the 4H or 6H forms. III-N materials systems are another class of compound semiconductors that constitute of one or more Group-III elements in the Periodic Table combined with nitrogen atoms to form a semiconductor. The representative WBG III-N materials are gallium nitride (GaN), aluminum nitride (AlN), aluminum gallium nitride (AlGaN), and indium aluminum nitride (InAlN).

Since the last four decades, expansive academic research publications in both SiC- and GaN-related publications have been reported, as shown in Fig. 2.1. This

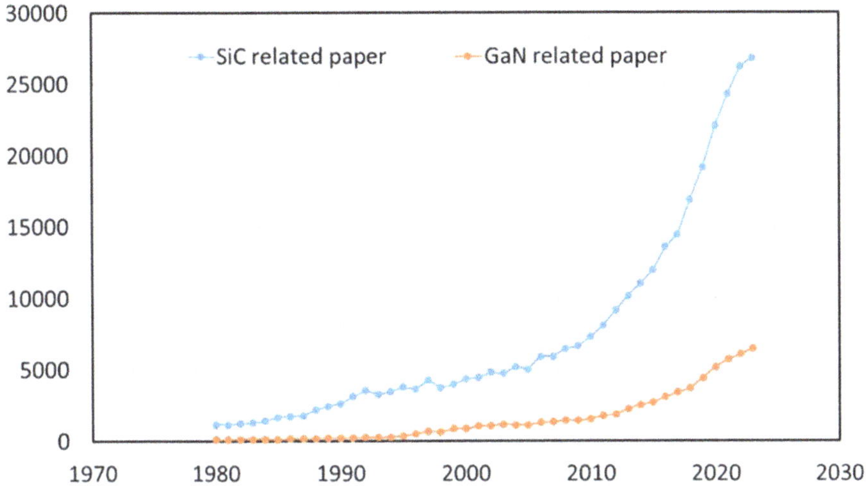

Fig. 2.1 A chart showing the cumulative number of technical journal and conference reports on SiC- and GaN-related publications, respectively, over the years. (Data source: https://app.dimensions.ai/)

trend was accelerated in the past decade due to the potential market penetration of electric vehicles (EVs) and concerns of global warming brought by the usage of fossil fuel. In the following sections, we will explore the benefits of GaN for power electronic systems from the perspective of material properties. A brief discussion on the epitaxial growth technologies, as well as the available substrate choices for III-N materials systems will be presented.

2.1 Crystalline Structure of GaN and Related Materials

There are three major classes of crystalline structures in today's semiconductor materials systems: Diamond, Zincblende, and Wurtzite structures. The representative materials for diamond structures are elemental semiconductors in the Group IV of semiconductors, i.e., silicon and germanium. Each atom in a diamond crystalline use its four valence electrons that form covalent bonds with the four nearest neighbors. Since each electron coordination has the same electrostatic strength, the most stable form of the diamond structure exhibit a tetrahedron in the unit cell. The resulting conventional unit cell of a diamond semiconductor has an atomic arrangement that is identical two face-centered-cubic (FCC) structures interlaced with each other, with one spaced apart by one-quarter of the lattice constant in all directions in a three-dimensional space, as shown in Fig. 2.2a. The semiconducting properties of these materials are formed either by external excitation such as thermal or photon events. The manipulation of the electric conductivity is also achieved by substituting some atoms in the lattice sites with elements other than Group IV elements,

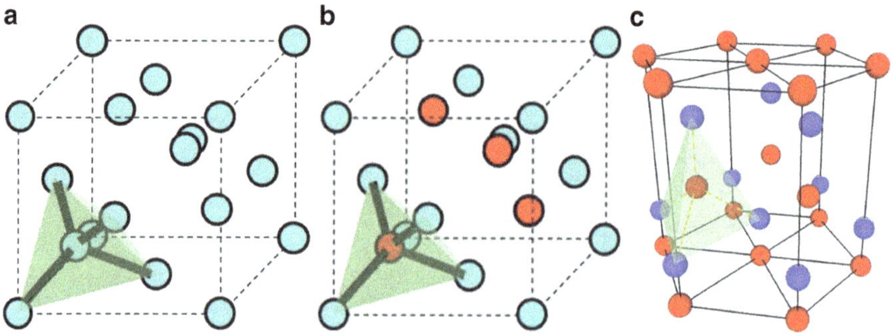

Fig. 2.2 The crystalline structure for (**a**) diamond lattice, (**b**) the zincblende lattice, and (**c**) the Wurtzite lattice

referred to as dopants. The p-type dopants can be found from the right-hand side of the periodic table of the semiconductor, or the acceptor dopants; the n-type dopants can be found on the right hand side of the period table, or the donor dopants.

The zincblende crystalline, a name given after the zinc sulphide mineral is a variant of the diamond structure. It is formed by interlacing two FCC of different elements, one by Group II (or Group III) elements and the other by Group VI (or Group V) elements to form a densely packed crystalline. The two FCCs inter-penetrate each other in the atomic arrangement that is similar to the diamond crystalline, except that each atom is electrostatically bonded to the nearest elements of different type in the crystalline. The unit cell of the zincblende forms a tetrahedron, as shown in Fig. 2.2b. Similar to the diamond crystalline, substitutional dopants facilitate the change of electronic conductivity and help create p-type or n-type materials in these semiconductors. This structure is commonly seen in conventional compound semiconductors such as GaAs and InP.

There is a distinct difference between diamond lattice and zincblende lattice in terms of electron property, though. Because the zincblende crystalline is formed by the bonding of dissimilar atoms, the inter-atomic bonding exhibits certain level of ionicity. In other words, the covalently bonded electrons are not equally shared between Group III and Group V atoms. A dipole moment is created as a result and a polarization field exists. This polarization is not negligible along the <111> direction in a zincblende structure: each plane is formed by the same groups of atoms (A-plane and B-plane) and alternates along this crystalline direction. Such charge polarization is intrinsic to the material, usually referred to as the *spontaneous polarization*. In a case when the lattice is under external pressure or intentional mechanical stress, a *piezoelectric* field can be created. The polarization effect is a unique feature of compound semiconductors. It must be considered in the design of heterostructure compound semiconductor devices, as it significantly changes material properties such as energy band structures and the resulting electron transport characteristics. The spontaneous polarization and piezoelectric polarization become significant in III-N.

The diamond and zincblende structures are considered as cubic crystalline structures. Although cubic GaN structure exists, it is still in an early-stage development as of today. Today's III-N materials are commonly seen in the form of Wurtzite crystalline structure, as shown in Fig. 2.2c. Similar to the zincblende structure, Group III atom in a Wurtzite structure is tetrahedrally bonded with the neighboring nitrogen atoms. Atoms of the same group (Group III or N) are aligned on the same basal planes that are perpendicular to the c-axis in a close-packed hexagon form. Compared to conventional cubic semiconductors, such crystalline creates not only structural difference but electronic transport property change, often times, new manufacturing challenges.

In Fig. 2.2c, the Wurtzite structure is defined by two unit vectors that is 60-degree apart on the basal plane for identifying specific in-plane directions, along with an orthogonal vector (the *c*-vector). Typical crystalline direction uses a 4-number Miller's indices representation: the first three indices define the information pertaining the basal plane and the fourth contains the c-axis information. For example, (0001) represents the *c* polar plane; (11(bar)00) represents the *m*-plane, etc. Each basal plane consists of the same group of atoms, i.e., Group III elements or nitrogen and alternate along the c-axis. Similar to the origin of the polarization charge along the <111> in the zincblende structure, a polarization dipole moment is seen along the c-axis. In III-N materials systems, inter-atomic bonding possesses more ionicity than that in As-based or P-based compounds. Such polarization is huge enough to alter the electron distribution in a semiconductor structure.

Augmented by the large piezeoelectric coefficient in III-N materials, the presence of the polarization charge induced by the spontaneous polarization and the piezoelectric polarization altogether plays a dominant role in the bandgap engineering of III-N heterostructures. It forms the theoretical basis of modern GaN high-electron-mobility transistor (HEMT) designs. For example, GaN HEMTs may consist of a lower bandgap material such as GaN for the electron channel, and a higher bandgap material (AlGaN) as the electron barrier in a transistor. The spontaneous polarization charge in (Al)GaN can be expressed as (reference: Wood and Jena [1])

$$P^{\mathrm{sp}}_{\mathrm{Al}_x\mathrm{Ga}_{(1-x)}N} = -0.09x - 0.034x + 0.019x(1-x), x \in [0,1]\mathrm{C}/\mathrm{m}^2$$

The piezoelectric polarization charge is a function of the basal strain and that exerted on the constituent materials in a heterostructure. The strain is created by the lattice constant mismatch of the specific material to the substrate. Let the strain of an epitaxial layer be defined as the parameter ϵ, where $\epsilon = (a_{\mathrm{substrate}} - a_{\mathrm{epi\ material}})/a_{\mathrm{epi\ material}}$, and a is the unstrained lattice constant. Using Vegard's law, the lattice constant for GaN is 3.189 Angstroms and that for $Al_xGa_{1-x}N$ is 3.189–0.077x Angstroms. Assume that the material is not mechanically relaxed and the crystalline is elastically strained, the theoretical piezoelectric polarization charge for AlN and GaN, respectively, can be expressed as:

$$P_{\mathrm{GaN}}^{pz} = -0.918\epsilon + 9.514\epsilon^2\ \mathrm{C/m^2}$$

$$P_{\mathrm{AlN}}^{pz} = -1.808\epsilon - 7.888\epsilon^2\ \mathrm{C/m^2}$$

To the first-order approximation, the piezoelectric polarization charge of $Al_xGa_{1-x}N$ is

$$P_{\mathrm{Al}_x\mathrm{Ga}_{(1-x)}N}^{pz} = x\,P_{\mathrm{AlN}}^{pz} + (1-x)\,P_{\mathrm{GaN}}^{pz}$$

It was found that such polarization charge density easily outnumbers the achievable doping level in conventional HEMT structure using modulation-doped schemes, GaN HEMT designs can leverage the huge amount of polarization-induced electron concentration, or the 2-dimensional electron gas (2DEG) to enable high current density transport. The amount of 2DEG can be readily calculated by the spontaneous and piezoelectric polarization charges in a specific layer structure design. Using advanced III-N materials systems such as AlInN, it is common to see 2DEG density of greater than 10^{13} cm^{-2} in GaN HEMTs these days.

It is noted that the m-planes in a Wurtzite crystalline are non-polar planes. There is no spontaneous polarization field lined up in this direction. Devices built on non-polar planes are immune from the spontaneous polarization effect. It is a highly desired property for nitride-based optoelectronic devices to avoid the light-emission efficiency droop in solid-state lighting. In the case of power electronic device applications, however, GaN HEMTs built on non-polar planes have not seen much significant advantages as of today. The availability of high-quality large-size non-polar GaN wafers are also still under development.

2.2 Material and Electronic Properties of GaN

Shown in Table 2.1 is a summary of some key material properties for binary III-N materials. Compared to typical cubic semiconductors (e.g. Si, Ge, InP, and GaAs), III-N crystalline typically is in the form of the Würtzite structure. They are known to have direct energy bandgap, comparable thermal conductivity, chemical inertness, and high electron mobility, among a few interesting properties in these materials. A radar chart of normalized characteristics for semiconductor materials systems for power electronic devices are shown in Fig. 2.3 for a comparison [4]. Some of the key electronic and material properties pertaining to power transistors are discussed in detail below.

As discussed in the previous section, WBG materials have unique carrier transport and design considerations. The lattice constants for WBG materials are significantly smaller from the conventional cubic semiconductor materials. A lack of low-cost native substrates over the years of GaN technology also create significant challenges: all commercially available substrates have lattice mismatch to GaN

Table 2.1 The comparison table for various compound semiconductors [2, 3]

Material	Si	GaAs	InP	4H-SiC	GaN
Crystalline structure	Diamond	Zincblende	Zincblende	Wurtzite	Wurtzite
Bandgap structure	Indirect	Direct	Direct	Indirect	Direct
Bandgap energy (eV)	1.1	1.42	1.35	3.26	3.49
Mobility (cm^2/Vs)	1500	8500	5400	700	2000
Peak electron velocity ($x10^7 cm/s$)	1	2.1	2.2	2	2.7
2D Electron gas (cm^{-2})	–	$<4\times10^{12}$	–	–	$>1\times10^{13}$
Breakdown field (MV/cm)	0.3	0.4	0.5	2	3.3
Thermal conductivity (W/cm-K)	1.5	0.5	0.7	4.5	>1.7
Dielectric constant (ε_r)	11.8	12.8	10.8	10	9

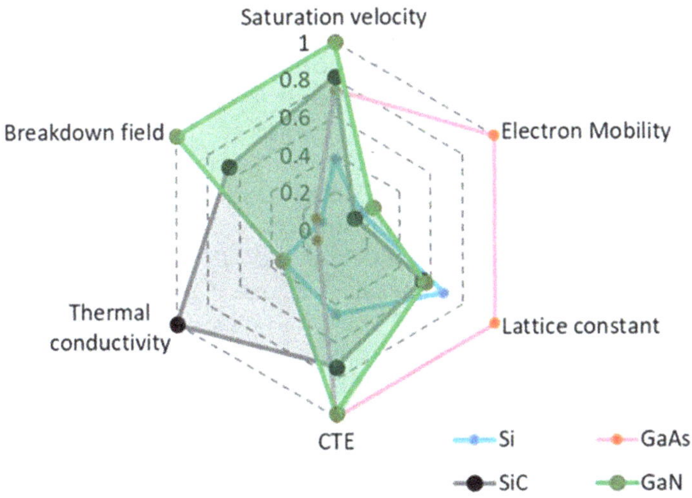

Fig. 2.3 The normalized characteristics of compound semiconductors [4]. The strength of wide band gap ones, such as GaN and SiC, can be seen in green and gray colors

materials, and the inevitable heteroepitaxy of III-N materials requires significant engineering optimization. Strain management and buffer layer designs are key to the success of today's GaN technology development. For example. GaN-on-Si technology is typically built on (111)-oriented silicon wafers. It is because the interatomic distance on the (111) plane of silicon is 3.84 Angstroms, while the lattice constant of (100) silicon is 5.43 angstroms. When a GaN epi-layer is grown on top of the (111) silicon surface, it presents a 17% of lattice mismatch. Careful strain management and engineering will be needed. The choices of substrates will be discussed in further detail later.

III-N materials have the bandgap energy range from 3.4 eV (for GaN) to 6.2 eV (AlN). The wide bandgap property extends the operable temperature range for WBG electronic devices because the amount of thermal excitation is exponentially decreased with an increase of the energy bandgap. It is common to see that GaN

HEMTs or SiC MOSFET operating at elevated temperature beyond 250 °C. In addition to GaN, ternary compounds such as AlGaN and InAlN, as well as the quaternary compound (AlInGaN) typically have a critical electric field of greater than 3MV/cm. This value is at least 10 times higher than that for GaAs and Si. The high breakdown field can allow for a shorter carrier drift region in a transistor design, while maintaining the same voltage breakdown ratings. A shorter drift region design translates to a less resistive channel, and a much reduced on-state resistance for power electronic switches.

In power electronic device design, the energy loss during the switching transient is critical for energy efficiency enhancement. In contrast to Si and SiC, GaN and related materials are "direct bandgap" semiconductors. Direct-bandgap semiconductors are inherently a better choice for electronic switches in two folds. First, the recombination of the electron and hole pairs in direct semiconductors are in the form of photons, not phonons (or heat). Second, the carrier recombination lifetime is in the order of sub-nanoseconds. Because of ultra-short carrier lifetime in III-N materials, high-power GaN transistors can easily operate at a switching frequency of up to MHz range with minimum off-state transient. This feature stands out from SiC and Si power electronic devices. Si-based devices typically operate at a switching frequency of less than 20 kHz, and SiC power devices are optimally operating at 100 kHz, all because of the relatively long carrier lifetime of indirect semiconductors. The high-switching frequency nature of GaN transistors allows for new opportunities to miniaturize power modules and overall system implementations.

Another interesting property of III-N materials is the electron saturation velocity and electron mobility. III-N materials share similar electron drift velocity overshoot characteristics as seen in GaAs, and reaches a value of >$2x10^7$ cm/s. The high electron drift velocity combined with high electron mobility has made GaN HEMTs a favorable choice for mm-wave power amplifiers. This feature also offers low-switching loss GaN-based power switches in terms of specific on-state resistance. It is common to see reports of the electron mobility of greater than 1800 cm^2/V-s in the 2DEG of AlGaN/GaN HEMT structures.

The flexibility of bandgap engineering in III-N materials systems and the associated lattice mismatch in heterostructures add another source of electronic modulation through the polarization engineering, as shown in Fig. 2.4. For example, a strained AlGaN layer grown on top of a Ga-faced GaN layer experiences tensile strain. The strain will induce additional piezoelectric polarization field that adds to the existing spontaneous polarization, enhancing the 2DEG density in the GaN HEMT channel. Conversely, if an InGaN layer was grown on top of a Ga-faced GaN layer, the piezoelectric field is lined up in the opposite direction to the spontaneous polarization field. The beauty of compound semiconductor engineering allows researchers and engineers to exploit their creativity with new designs in III-N power electronics.

As a transistor switches, the dissipation of charges during the switching transient inevitably creates ohmic loss and creates undesired heat. Excessive heat will create device degradation and shortens product lifetime. III-N materials have comparable thermal conductivity compared to that of silicon; it is however inferior to

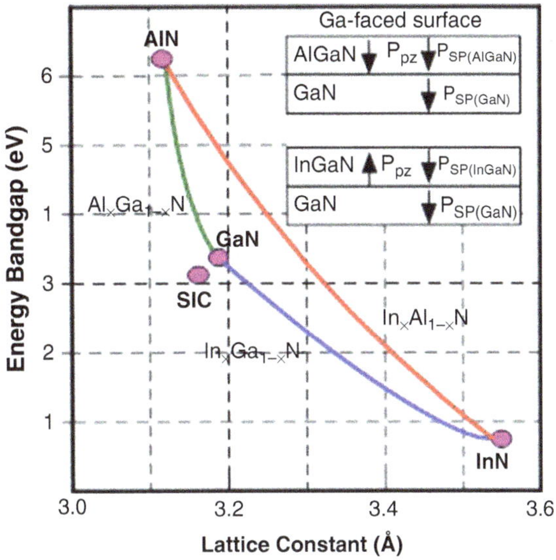

Fig. 2.4 A chart of the bandgap energy vs. the lattice constant for III-N materials systems [5]

SiC. Thermal management in power GaN transistors will be an integral part of the device technology development. Much effort has been focused on materials studies of heat spreaders such as diamond films [6]. The associated high-power package technology is also actively developed in WBG industries in recent years.

2.3 Substrates for GaN

Due to high cost of native substrates as of 2024, GaN has been grown on foreign substrates for commercial products. The growth of GaN and related III-N materials on foreign substrates have been challenges, and in a way, hampered the progress of high-quality GaN device development for power electronics. The growth of III-N epitaxy was usually carried out by inserting a buffer layer in between the underlying substrate and high-quality epitaxial layer for transistors. The concept of the buffer layer was first implemented in GaN on sapphire substrates by Yoshida et al. in 1983 and Amano et al. in 1986. This buffer layers is a set of strained hetero-epitaxial layers to stop the propagation of lattice-mismatch-induced defects and cracks. The buffer layer engineering is considered as one of the most important technology know-hows in the GaN device development.

Commercially available native substrates are GaN substrates and AlN substrates. The development of native substrates involved specialized crystal growth methods such as HVPE and Ammon thermal growth. These native substrates have high crystal quality but relatively low wafer throughput. As a result, the cost of native substrates have been limited to applications that require ultra-low defect density such as lasers or other optoelectronic devices. It may take a few more years before the use of native GaN or AlN substrates can be used in power electronics as a cost-effective solution.

On the other hand, there are a handful of commercially available foreign substrates to choose from for III-N material growth. These foreign substrates are sapphire, SiC, Si, and engineered substrates. These foreign substrates come with different wafer sizes that can support up to 8-inch of device processing lines as of 2024. Si substrates also enjoy the mature development in the silicon industry and are poised to offer 12-inch Si for future GaN technology expansion. A brief discussion of these large-size substrates will be presented below.

The sapphire substrate, unlike the gemstone, is a synthetic single crystalline aluminum oxide (Al2O3.) The electronic-grade sapphire was perfected to create a porous-free and grain-boundary-free crystalline. The sapphire substrate is highly resistive to chemical corrosion with a hardness of 9 and a dielectric constant of 9.3. It is a highly resistive material ($>10^{11}$ ohm-cm at 300 K). The final form of the sapphire substrate is called alpha-alumina (Al_2O_3). Sapphire is essentially a rhombohedral lattice. However, an equivalent hexagonal lattice unit cell can be drawn by including three primitive cells of the sapphire. In this fashion, hexagonal wurtzite GaN can be grown along the c-axis ([0001]), or the basal plane of a sapphire substrate. One of the drawback of sapphire for power electronic devices is the relatively poor thermal conductivity. It is the worst performed heat conductor among the foreign substrates for GaN with a thermal conductivity of <30 W/(m·K).

SiC substrates have been actively developed for at least more than 3 decades and offered a choice of substrate platform for GaN material growth. The growth of GaN on SiC was first explored in the LED industry, and spread across the entire application space of GaN electronics. Today, many high-performance RF power GaN HEMTs are built on the GaN-on-SiC device platform and successfully extended the applications in the millimeter-wave systems. SiC comes with polytype formations. Of many polytypes in SiC, silicon-faced 4H- and 6H-SiC are commonly used for the heteroepitaxy of GaN materials. The 4H- and 6H-SiC are referred to different periodicity of the repeatability of hexagonal stacking of SiC in the polytype formation. The lattice constant for 4H-SiC is 3.073 Angstroms and that for 6H-SiC is 3.081 Angstroms. The lattice constant of SiC exhibits a ~3.5% of lattice mismatch w.r.t. that for GaN. Furthermore, SiC has one of the highest thermal conductivity (450 W/(m·K)) available among the foreign substrates for GaN. One of the major disadvantages of SiC substrates is its high substrate cost. In recent years, 8″ SiC substrates have become available and hopefully this substrate technology may eventually reach a scale of economy to the WBG communities.

Si substrates, as mentioned in the previous section, has been one of the most economical solutions to serve as a substrate for GaN epitaxy. In spite of the high-level of lattice mismatch (~17%), Si has enabled the scale of economy in the substrate supply, and is ready for upgrading the wafer size from 6″ to 8″ or larger in the near future. Many GaN applications up to 650 V have been using GaN-on-Si platform for commercial productions.

One of the parameters to consider in the choice of the foreign substrates for GaN is the coefficient of thermal expansion (CTE). It is because the GaN epitaxy usually requires high-temperature thermal cycles to grow high-quality semiconductor films for electronic device implementation. CTE mismatched substrates could result in undesired cracks on the epitaxial films during the cooling process. Special care in the cooling steps and proper thermal budget management during the epitaxial growth could help mitigate the occurrence of the CTE-mismatch-induced epitaxial film failure. Another solution to thermal-expansion-induced epi film failure is to find CTE-matched substrates. To this end, engineered substrates with nearly CTE-matched characteristics may offer new opportunities for future GaN device development. For example, QST™ (Qromis substrate technology) substrates are designed to have a nearly identical CTE at typical growth temperature of the GaN epitaxy. The CTE-matched property enabled a crack-free thick film GaN epitaxy with a thickness greater than 10 microns. Such substrate technology may be highly desired for GaN-based power electronics for beyond -650 V applications in the future.

2.4 Growth Technologies for GaN and Related Materials

2.4.1 MBE

Molecular beam epitaxy (MBE) was first proposed in 1958 by K. G. Günther [7] and developed in Bell Laboratory by A. Cho and J. Arthur and other labs during 1960s and 1970s [8–10]. It uses sublimed atomic or molecular beams to deposit epitaxy layer onto a heated substrate. The chamber requires ultra-high vacuum, usually in the 10^{-10} Torr range, to ensure the long mean free path of the atoms or molecules.

2.4.1.1 Basic MBE Chambers

Figure 2.5 shows a regular MBE system (main chamber only). It usually has a load lock, a major chamber, and several vacuum chambers for material analysis, if needed [11]. The main chamber needs to be kept under ultra-high vacuum, usually in the 10^{-10} torr range, in the standby mode. During the growth, the effusion cells and the cracker (for As or other sources) are heated up to evaporate the elemental

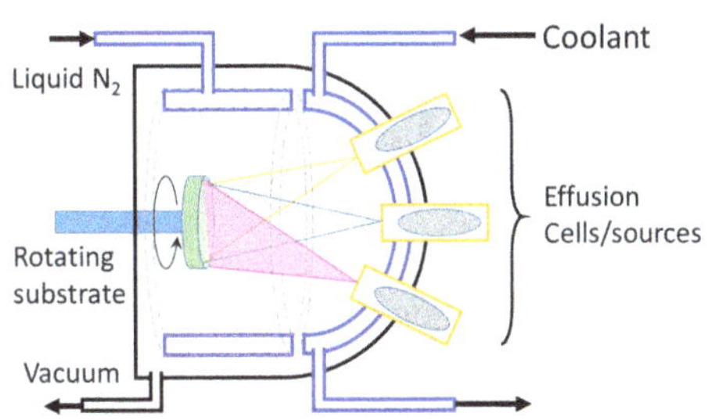

Fig. 2.5 A schematic diagram of a MBE growth system

sources (such as Al, Ga). There are shutters before these cells to control the open and close of the evaporated molecular flow. When the shutter is open, the atoms/molecules can go directly towards a rotating substrate in the chamber. With the right condition (temperature and pressure), the atoms can be placed layer by layer and form a uniform crystal. In some cases, a plasma or RF source will be necessary for the growth due to the difficulties to obtain pure solid materials. The reason to keep the chamber at ultra-high vacuum is to ensure the mean free path of the molecular beam can be long enough to reach the substrate. An additional particle absorption can be provided by the water cooling panel and cryogenic shroud (with liquid nitrogen). A mechanical transfer arm (not shown) is used to move the wafer on the rotating substrate before and after the growth.

Figure 2.6 shows a generic growth model in a MBE system. Usually we expect to form a stabilized surface by metallic atoms. Then the nitrogen family atoms (or oxygen, and/or carbon family) can be added to the surface via physisorption or chemisorption. In physisorption, the van der Waals force pulls the atoms to the right location. On the other hand, the chemical bond reacts between atoms and the surface in the chemisorption. The surface temperature is also important to make the elements move around on the surface in order to find the suitable location. The desorption process could happen if the adsorption does not happen and the reacting atoms/molecules would leave the substrate surface.

MBE is particularly beneficial for InN or high indium content nitride growth because it is capable of delivering high-quality film at low deposition temperature which is critical for the material [12]. The metallic elements, such as gallium, are easy to implement in the regular MBE. However, for nitrogen, it is far too stable in the ordinary condition such that different sources are found to be applied in the MBE system. Ammonia, electron-cyclotron resonance (ECR), radio-frequency sources, or ion sources have been used in the past. Because nitrogen incorporation is low at the low substrate temperature, plasma sources are popular among the MBE systems [12, 13].

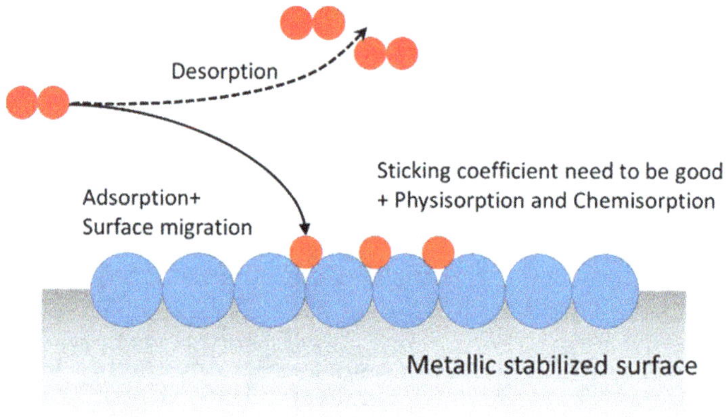

Fig. 2.6 The schematic diagram for growth model in MBE (using GaAs as an example) [11]

The substrates we used for the nitride growth can vary among different labs. Liu and Edgar collected an extensive list for the GaN-based growth [14]. From sapphire, silicon, to oxides or even metal nitrides, we found many choices for this heteroepitaxy. Sapphire is the most popular substrate we used today, and the nitridation of the sapphire is the critical step in the MBE growth of GaN [15, 16]. In the early days of this technology, Yoshida et al. found an extra AlN buffer layer was actually helpful to enhance the Hall mobility and increase the cathodoluminescence by an order [17]. The oxygen atoms on the surface are replaced with the nitrogen ones to form a thin layer of AlN or AlNO, which can be a homogeneous layer for the subsequent GaN growth. Nitridation at low temperature (around 200 °C) is preferred for a uniform AlN layer and thus good for GaN quality [18, 19].

Compared to Sapphire, SiC is another choice for substrate materials in MBE systems. The SiC has a better thermal conductivity, and lower mismatch to GaN and AlN (sapphire is much different in the lattice constant than SiC's to the GaN materials). However, SiC wafer surface is usually rougher than sapphire's, which is not very favorable in the MBE growth. To handle this situation, an etch with HCl or high temperature hydrogen etch could be used to smooth the surface [12].

When the substrate is considered for GaN growth, Si is an important candidate that we cannot neglect. Silicon has the advantages of low-cost, large substrate with high quality, and the easy integration with mature IC fabrication; therefore, it is very attractive to be applied in the GaN epitaxial growth as a substrate. However, it is difficult to grow a Wurtzite GaN on a cubic Si wafer. We usually use (111) Si wafer because the atomic arrangement of a diamond crystal structure in this orientation is close to the Wurtzite's. The choice of buffer layer is AlN, whose binding energy is stronger than that of silicon and nitrogen's to prevent the formation of insulating Si_xN_y layer before the real epitaxy growth [12]. Sanchez-Garcia et al. reported a detailed exploration of growth condition to improve the quality of this AlN layer. They achieved a much better subsequent GaN growth via the surface analysis, Al coverage, and a low growth rate [20].

Figure 2.7 shows a generic growth procedure of a MBE system for GaN-related materials. An initial high-temperature (900 °C) cleaning is followed by a low-temperature nitridation step. A short GaN buffer growth at a relatively low temperature (around 600 to 700 °C) can proceed before we really raise the substrate temperature to 800 °C for the epitaxy layers [12].

2.4.2 MOCVD

The MOCVD is the short term of Metal-Organic Chemical Vapor Deposition. The concept was developed in the 1960s [21] and became more practical in the 1970s [22]. This technology followed its predecessor: vapor phase epitaxy (VPE) closely, as it needs two or more chemical vapors mixed and heated up together in order to obtain the III-V crystal. These chemical reactions in the vapor or at the surface of the substrate can produce the crystallized material onto the substrate and form a thin

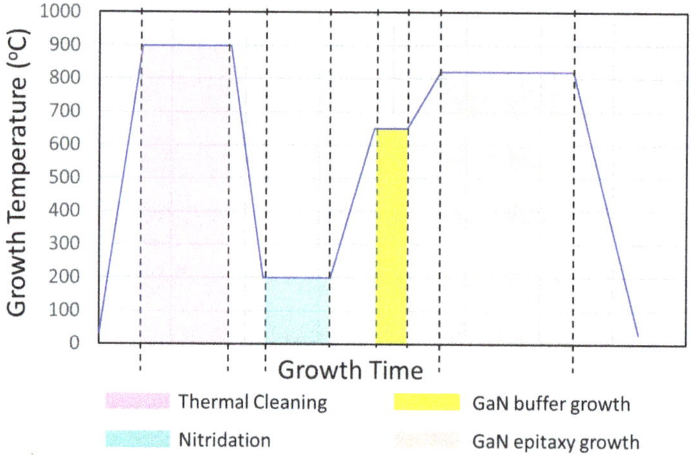

Fig. 2.7 A generic growth process of a MBE system [12]

film [22]. In today's world, the wide market penetration of the MOCVD system has made it the critical ingredient in the photonic, energy, and high-speed electronics industries. The MOCVD system provides high throughput and steady wafer growth when compared to other types of epitaxial technology. From InP, GaAs, to GaN, almost all compound semiconductors need the MOCVD technology to commercialize. High-speed transistors, semiconductor lasers, solid-state lightings, and micro LEDs are just some of its great products [23–25].

In a MOCVD system, the metal-organic compounds are the sources of group 3a elements (such as Ga, Al, and In) while the hydrides or the mixture of the hydride and metal-organic chemicals are the sources of group Va elements (such as As, P, and N). Trinethylgallium (TMGa or TMG) and ammonia (NH_3) are usually used for the sources of Ga and N in GaN growth while Trimethylindium (TMIn) or Triethylindium (TEIn) can be used for the indium source. In the past, the traditional chamber design together with the rotating susceptor, the incoming gas/vapor flow can develop a constant supply of reactants on the surface of the wafer. The heated environment, either by regular heater or RF heater, provide the energy to the chemicals. Combined with proper design of the gas flow, substrate rotation, and growth temperature, we can expect a uniform growth rate on the substrate. However, at the beginning of the GaN-based epitaxy, it is very difficult to obtain materials with good quality. Due to the difficulties to obtain GaN substrates, people instead used sapphire from the beginning. The large lattice mismatch (>13%) and thermal expansion coefficient difference causes the cracked surface after the GaN epitaxial growth. There would not be any device development having such poor material quality. Amano et al. used AlN buffer layer during the growth and showed significant improvement of GaN layer smoothness, and only the pits instead of the cracks remained [26]. As the material's basic quality improves, people started to pay attention to the potential device application of the GaN material which requires implementing both p-type and n-type dopants. The n-type dopant for GaN-based material was easily accomplished by silicon. Meanwhile, the p-type dopants posted the most difficulties task for the scientific community. Zn and Mg were popular choice for p-type dopants back in the 1980s. To incorporate them into GaN, we need chemicals like Bis-cyclopentadienyl magnesium (Cp_2Mg) or diethylzinc ($(C_2H_5)_2Zn$)/dimethylzinc($(CH_3)_2Zn$) [27]. Both zinc and magnesium were considered for the p-type dopant candidates from the beginning. Although zinc is quite popular in the traditional compound semiconductor, its sticking coefficient is not as high as magnesium in the case for GaN growth [28], which makes zinc a less effective dopant. A relative high activation energy for both Mg and Zn [29, 30] led to poor ionization of these incorporated dopants and thus very bad conductivity. Without effective carrier concentration, a basic p-n junction diode of GaN was impossible. Several important breakthrough efforts helped the community to alleviate this critical issue. H. Amano et al. used a low-energy electron beam to irradiate (LEEBI) the wafer [31, 32], and a remarkable reduction of resistivity was observed (from $>10^8$ Ω-cm down to 35 Ω -cm) [32, 33].

Another breakthrough was brought by Dr. Nakamura's thermal annealing method, which was achieved by an ambient of N_2 and an elevated temperature over 700 °C [34], and the resistivity of the p-type GaN thin film dramatically dropped for more than six orders (10^6 to 10^0 Ω -cm). An equally important modification was the two-flow growth of MOCVD that separates the inactive gas flow from the reactive species as shown in Fig. 2.8b, provided a uniform result across the wafer [36, 37]. In Fig. 2.9, a generic growth temperature process flow is shown, combining all the results demonstrated in the past few decades, which include the initial nitridation and AlN deposition, a proper temperature ramps up to over 1000 °C for GaN buffer, followed by a low-temperature growth (around 800 °C) for InGaN, and finalized by an over-1000 °C GaN cap growth [38]. All these efforts and others set up a healthy environment for the GaN/InGaN/AlN-based materials to play a critical role in photonics, power electronics, and consumer electronics at present and in the future.

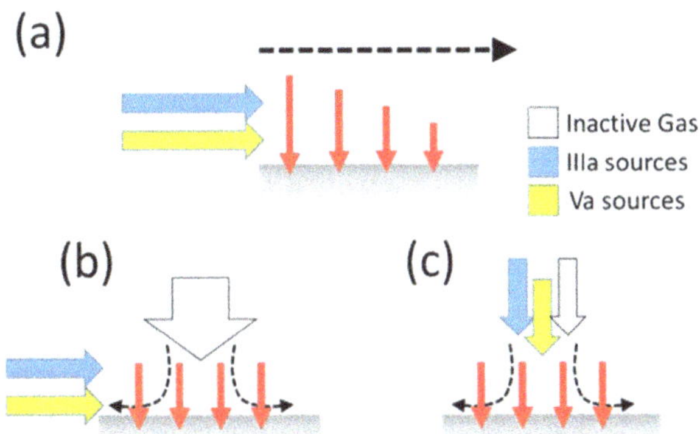

Fig. 2.8 Different MOCVD growth designs: (**a**) The traditional MOCVD gas flow and the substrate diagram [35]. (**b**) Using an inactive gas perpendicular to the normal gas flow [36]. (**c**) All gas flows are perpendicular to the substrate [35]

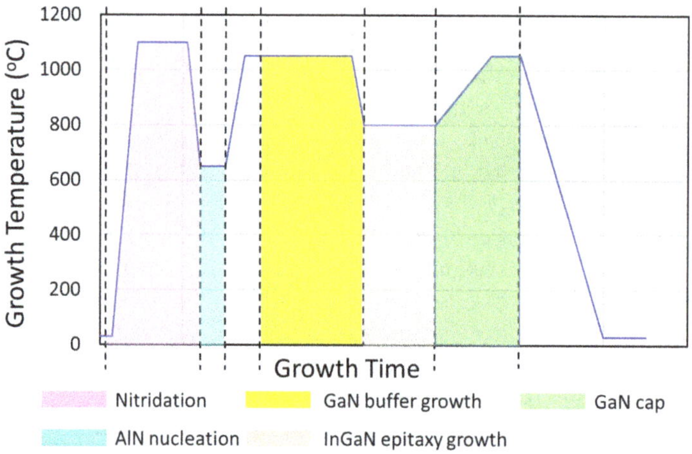

Fig. 2.9 A generic growth process of a MOCVD system [38]

References

1. C. Wood, D. Jena, *Polarization Effects in Semiconductors: From Ab Initio Theory to Device Applications*, 1st edn. (Springer, Berlin, 2007)
2. A.R. Powell, L.B. Rowland, SiC materials-progress, status, and potential roadblocks. Proc. IEEE **90**(6), 942–955 (2002). https://doi.org/10.1109/JPROC.2002.1021560
3. R.T. Kemerley, H.B. Wallace, M.N. Yoder, Impact of wide bandgap microwave devices on DoD systems. Proc. IEEE **90**(6), 1059–1064 (2002). https://doi.org/10.1109/JPROC.2002.1021570
4. S.S.H. Hsu, C.W. Tsou, Y.W. Lian, Y.S. Lin, GaN-on-silicon devices and technologies for RF and microwave applications, in *2016 IEEE International Symposium on Radio-Frequency Integration Technology (RFIT)*, 24–26 Aug. 2016 2016, pp. 1–3, https://doi.org/10.1109/RFIT.2016.7578151
5. S.-C. Shen, III-Nitride Platforms, in *Extreme Environment Electronics*, ed. by J.D. Cressler, H.A. Mantooth, 1st edn., (CRC Press, Boca Raton, 2013), pp. 263–273, p. 1041
6. J.D. Blevins et al., Recent Progress in GaN-on-Diamond Device Technology, Presented at the CS MANTECH, Denver, Colorado, USA, 2014
7. K.G. Günther, Aufdampfschidhten aus halbleitenden III-V-Verbindungen. Zeitschrift für Naturforschung A **13**(12), 1081–1089 (1958). https://doi.org/10.1515/zna-1958-1210
8. A.Y. Cho, J.R. Arthur, Molecular beam epitaxy. Prog. Solid State Chem. **10**, 157–191 (1975). https://doi.org/10.1016/0079-6786(75)90005-9
9. J.E. Davey, T. Pankey, Epitaxial GaAs films deposited by vacuum evaporation. J. Appl. Phys. **39**(4), 1941–1948 (1968). https://doi.org/10.1063/1.1656467
10. M.B. Panish, Molecular Beam Epitaxy. Science **208**(4446), 916–922 (1980). https://doi.org/10.1126/science.208.4446.916
11. M.A. Herman, H. Sitter, *Molecular Beam Epitaxy: Fundamentals and Current Status (The Springer Series in Materials Science)* (Springer, Berlin, 1996)
12. X. Wang, A. Yoshikawa, Molecular beam epitaxy growth of GaN, AlN and InN. Prog. Cryst. Growth Charact. Mater. **48-49**, 42–103 (2004). https://doi.org/10.1016/j.pcrysgrow.2005.03.002
13. B. Heying, R. Averbeck, L.F. Chen, E. Haus, H. Riechert, J.S. Speck, Control of GaN surface morphologies using plasma-assisted molecular beam epitaxy. J. Appl. Phys. **88**(4), 1855–1860 (2000). https://doi.org/10.1063/1.1305830

14. L. Liu, J.H. Edgar, Substrates for gallium nitride epitaxy. Mater. Sci. Eng. R Rep. **37**(3), 61–127 (2002). https://doi.org/10.1016/S0927-796X(02)00008-6
15. N. Grandjean, J. Massies, M. Leroux, Nitridation of sapphire. Effect on the optical properties of GaN epitaxial overlayers. Appl. Phys. Lett. **69**(14), 2071–2073 (1996). https://doi.org/10.1063/1.116883
16. C. Heinlein, J. Grepstad, T. Berge, H. Riechert, Preconditioning of c-plane sapphire for GaN epitaxy by radio frequency plasma nitridation. Appl. Phys. Lett. **71**(3), 341–343 (1997). https://doi.org/10.1063/1.119532
17. S. Yoshida, S. Misawa, S. Gonda, Improvements on the electrical and luminescent properties of reactive molecular beam epitaxially grown GaN films by using AlN-coated sapphire substrates. Appl. Phys. Lett. **42**(5), 427–429 (1983). https://doi.org/10.1063/1.93952
18. G. Namkoong, W.A. Doolittle, A.S. Brown, M. Losurdo, P. Capezzuto, G. Bruno, Role of sapphire nitridation temperature on GaN growth by plasma assisted molecular beam epitaxy: Part I. Impact of the nitridation chemistry on material characteristics. J. Appl. Phys. **91**(4), 2499–2507 (2002). https://doi.org/10.1063/1.1435834
19. M. Losurdo, P. Capezzuto, G. Bruno, G. Namkoong, W.A. Doolittle, A.S. Brown, Role of sapphire nitridation temperature on GaN growth by plasma assisted molecular beam epitaxy: Part II. Interplay between chemistry and structure of layers. J. Appl. Phys. **91**(4), 2508–2518 (2002). https://doi.org/10.1063/1.1435835
20. M.A. Sánchez-García et al., Growth optimization and doping with Si and Be of high quality GaN on Si(111) by molecular beam epitaxy. J. Electron. Mater. **27**(4), 276–281 (1998). https://doi.org/10.1007/s11664-998-0399-2
21. H.M. Manasevit, Single-crystal gallium arsenide on insulating substrates. Appl. Phys. Lett. **12**(4), 156–159 (1968). https://doi.org/10.1063/1.1651934
22. R.D. Dupuis, Metalorganic chemical vapor deposition of III-V semiconductors. Science **226**(4675), 623–629 (1984). https://doi.org/10.1126/science.226.4675.623
23. C.-C. Lin et al., The micro-LED roadmap: Status quo and prospects. J Phys Photonics **5**(4), 042502 (2023). https://doi.org/10.1088/2515-7647/acf972
24. F. Ren et al., GaN/AlGaN HBT fabrication. Solid State Electron. **44**(2), 239–244 (2000). https://doi.org/10.1016/S0038-1101(99)00229-4
25. M.R. Krames et al., Status and future of high-power light-emitting diodes for solid-state lighting. J Disp. Technol. **3**(2), 160–175 (2007). https://doi.org/10.1109/JDT.2007.895339
26. H. Amano, N. Sawaki, I. Akasaki, Y. Toyoda, Metalorganic vapor phase epitaxial growth of a high quality GaN film using an AlN buffer layer. Appl. Phys. Lett. **48**(5), 353–355 (1986). https://doi.org/10.1063/1.96549
27. R.D. Dupuis, Metalorganic Chemical Vapor Deposition (MOCVD), in *Encyclopedia of Physical Science and Technology (Third Edition)*, ed. by R.A. Meyers, (Academic Press, New York, 2003), pp. 495–511
28. H. Amano, M. Kitoh, K. Hiramatsu, I. Akasaki, Growth and luminescence properties of Mg-doped GaN prepared by MOVPE. J. Electrochem. Soc. **137**(5), 1639 (1990). https://doi.org/10.1149/1.2086742
29. M. Ilegems, R. Dingle, Luminescence of Be- and Mg-doped GaN. J. Appl. Phys. **44**(9), 4234–4235 (1973). https://doi.org/10.1063/1.1662930
30. M. Ilegems, R. Dingle, R.A. Logan, Luminescence of Zn- and Cd-doped GaN. J. Appl. Phys. **43**(9), 3797–3800 (1972). https://doi.org/10.1063/1.1661813
31. H. Amano et al., Electron beam effects on blue luminescence of zinc-doped GaN. J. Lumin. **40-41**, 121–122 (1988). https://doi.org/10.1016/0022-2313(88)90117-2
32. H. Amano, M. Kito, K. Hiramatsu, I. Akasaki, P-type conduction in mg-doped GaN treated with Low-Energy Electron Beam Irradiation (LEEBI). Jpn. J. Appl. Phys. **28**(12A), L2112 (1989). https://doi.org/10.1143/JJAP.28.L2112
33. S. Yamasaki et al., p-type conduction in Mg-doped Ga0.91In0.09N grown by metal-organic vapor-phase epitaxy. Appl. Phys. Lett. **66**(9), 1112–1113 (1995). https://doi.org/10.1063/1.113829

34. S. Nakamura, T. Mukai, M. Senoh, N. Iwasa, Thermal annealing effects on P-type Mg-doped GaN films. Jpn. J. Appl. Phys. **31**(2B), L139 (1992). https://doi.org/10.1143/JJAP.31.L139
35. M. Dauelsberg, E.J. Thrush, B. Schineller, J. Kaeppeler, Chapter 4 – Technology of MOVPE production tools, in *Optoelectronic devices: III Nitrides*, ed. by M. Razeghi, M. Henini, (Elsevier, Oxford, 2005), pp. 39–68
36. S. Nakamura, Y. Harada, M. Seno, Novel metalorganic chemical vapor deposition system for GaN growth. Appl. Phys. Lett. **58**(18), 2021–2023 (1991). https://doi.org/10.1063/1.105239
37. S. Nakamura, GaN growth using GaN buffer layer. Jpn. J. Appl. Phys. **30**(10A), L1705 (1991). https://doi.org/10.1143/JJAP.30.L1705
38. O. Ambacher, Growth and applications of Group III-nitrides. J. Phys. D. Appl. Phys. **31**(20), 2653 (1998). https://doi.org/10.1088/0022-3727/31/20/001

Chien-Chung Lin received the B.S. degree in electrical engineering from the National Taiwan University in 1993, and the M.S. and Ph.D. degrees in electrical engineering from Stanford University, USA. He joined National Taiwan University in August 2021 as professor of Graduate Institute of Photonics and Optoelectronics and Department of Electrical Engineering. From 2009 to 2021, he was with National Chiao Tung University (NCTU), Tainan, Taiwan, where he started as an assistant professor and became a full professor. The major research efforts in his group are in design and fabrication of novel semiconductor optoelectronic devices, including LEDs, solar cells, and lasers. Since 2016, he has been jointly appointed by the Industrial Technology Research Institute (ITRI), where he participates and leads several key projects in the micro LEDs, 3DIC, and silicon photonics. From 2015 to 2017, he was the recipient of the Young Investigator Research Grant by Ministry of Science and Technology of Taiwan. From 2019 to 2020, he served as the chairperson of the Taipei chapter of the IEEE Photonics Society. He also served as the Optoelectronic Devices Committee member of the IEEE Electron Device Society between 2020 and 2021. He has more than 280 journal and conference publications and is a Fellow of the OPTICA (formerly the Optical Society of America) and a senior member of the IEEE.

Shyh-Chiang Shen is a Full Professor in the School of Electrical and Computer Engineering at Georgia Institute of Technology in Atlanta, GA, USA. His research is focused on wide bandgap semiconductor-based microelectronics and optoelectronic devices and manufacturing. His research has yielded 8 awarded U.S. patents and 200+ publications in refereed journals and technical conferences. In 2022, he took a leave of absence from Georgia Tech and joined Vanguard International Semiconductors, Corp. (VIS) in Taiwan as a Director of GaN Program Development. Dr. Shen is a senior member of the IEEE and a Fellow of Optica (formerly known as the Optical Society of America).

Chapter 3
Manufacturing Processes

Sonu Daryanani, Aurore Constant, Cristina Tringali, and Ferdinando Iucolano

In Sects. 3.1, 3.2, and 3.3, semiconductor expert Sonu Daryanani explores the intricate world of Gallium Nitride (GaN) device fabrication, focusing on epitaxy, defect characterization methods, and the challenges and applications of ion implantation in GaN power devices. Daryanani delves into the complexities of epitaxial growth techniques, discusses various defects inherent in GaN wafers and their characterization, and elucidates the nuances of ion implantation as a crucial process in optimizing GaN power devices. In the last Sect. 3.4, authored by Aurore Constant, Cristina Tringali, and Ferdinando Iucolano from STMicroelectronics, we delve into the intricate process of Gallium Nitride (GaN) device processing. The authors provide a comprehensive overview of the techniques, methodologies, and challenges involved in fabricating GaN-based devices.

S. Daryanani (✉)
Technical Writer, Las Palmas de Gran Canaria, Spain

A. Constant
Research and Development Engineer, STMicroelectronics, Tours, France

C. Tringali
Responsible for GaN Power Low V and RF Technology, STMicroelectronics, Catania, Italy

F. Iucolano
Senior Design Architecture ENG II, STMicroelectronics, Catania, Italy

© The Author(s), under exclusive license to Springer Nature Switzerland AG 2024

M. Di Paolo Emilio (ed.), *GaN Technology*,
https://doi.org/10.1007/978-3-031-63238-9_3

3.1 GaN Device Epitaxy

Sonu Daryanani

A subset of CVD is MOCVD, commonly known as Metal-Organic Vapor Phase Epitaxy (MOVPE). Chemical reactions from provided precursors occur in the vapor phase during CVD, resulting in the production of a solid substance placed on a substrate. The group III element, in this case Ga, is supplied by the metal-organic complexes. Heat from an external source is required to fracture the precursors. Early inventions and patents in the late 1960s are attributed to Manasevit and his colleagues at Rockwell [1]. Before MOCVD, the two main methods used to grow III/V and II/VI compounds were liquid-phase epitaxy (LPE) and vapor-phase epitaxy (VPE). Even while these methods offer the benefit of a quick deposition rate, they lack the control necessary to create abrupt interfaces and thin, homogeneous layers. An alternate method for creating superior III/V and other compound materials is called molecular beam epitaxy (MBE); however, it requires extremely high vacuum systems and has difficulties when scaling up to mass manufacturing.

The widespread usage of MOCVD for GaN did not start till the 1990s with some key breakthroughs in the MOCVD and p-doping of GaN that led to development of high brightness blue-green and white LEDs. The development of low-temperature buffer layers on sapphire substrates and the breaking up Mg-H bonds for improved p doping were essential advances. GaN MOCVD faced three key challenges:

- No lattice matched substrate available, requiring a lot of work on creating a nucleation scheme that could grow the epitaxial layers.
- The Ga mass transport to the surface is inhibited by complex reactions that take place in the vapor phase.
- No good p-dopants available.

These challenges have now been largely mitigated and MOCVD is the primary growth method used in GaN power device manufacturing.

3.1.1 GaN Epitaxy

Tri-methyl Gallium (TMGa: $Ga(CH_3)_3$) and ammonia (NH_3) are used as the source chemicals for GaN, with Trimethyl-Aluminum (TMAl: $Al_2(CH_3)_6$) added to create the AlN and AlGaN heterostructure layers needed for the HEMTs. The liquid metal-organics are typically delivered through stainless steel bubblers, which are heated to deliver a reasonable vapor pressure. A carrier gas, such as a mix of nitrogen and hydrogen, is used to deliver the material to the growth chamber. Two basic types of MOCVD chambers can be used:

- Cold-wall: here the substrate is the hot surface while the chamber is kept cold. These are more common, and have advantages of a simpler design with greater

temperature control. The graphite/carbide-coated substrate holder (susceptor) is typically resistively heated.
- Hot-wall: the entire chamber and substrate is heated. The advantages here are more efficient cracking of the pre-cursor chemical and potentially less wafer bowing during growth due a reduced thermal gradient in the growth zone. Here, the quartz chamber can be induction heated.

The deposition is typically carried out at pressures of 50–100 mbars. The growth should be performed under a laminar gas flow for high uniformity, and a lower pressure reduces the possibility of a turbulent flow. The growth temperature can typically vary from 500 °C to 1300 °C. A lower temperature results in a reaction limited regime where growth rates are more sensitive to temperature. A higher temperature results in the preferred mass-transport limited region where growth is limited by diffusion of reactants through the boundary layer. An even higher temperature can lead to lower growth rates, from effects of desorption and decomposition. The V/III ratios during for GaN are typically much higher (> 500) than that used for other III/V materials like gallium arsenide (GaAs) [1].

3.1.2 Substrates

Sapphire (Al_2O_3) substrates are commonly used for GaN LED manufacturing, with advantages of being relatively cheap and optically transparent. Silicon carbide (SiC) offers advantages of better lattice matching and thermal conductivity than Si, but is more expensive. Si (111) is the commonly used substrate for power HEMT applications.

3.1.2.1 Nucleation Layer (NL)

A NL is essential to accommodate the large lattice mismatch between GaN and Si (˜17%) and provide a two-dimensional surface for GaN growth. Surface treatments prior to the NL growth are needed, such as spraying some Al or NH_3 to create a rough SiN_x surface. This helps mitigate the lattice mismatch-induced stress. The NL can be grown at temperatures between 1100 °C–1300 °C, with typical thicknesses of around 100–200 nm.

3.1.2.2 Buffer layer (BL)

The GaN BLs needs to be highly resistive to prevent DC leakage and parallel conduction. Undoped GaN is typically n-type, hence needs a p-dopant to counter this. Iron (Fe) and carbon (C) are candidates for this. Carbon doping can be achieved from the source TMGa by adjusting the deposition conditions such as the flow rate, pressure, and temperature, and can have advantages of not having a strong memory effect, which Fe has. The level of C doping can be modulated for a given breakdown voltage, approximately 1e19 cm^3 needed to achieve breakdowns over 700 V. This buffer layer can typically be grown at around 980 °C–1080 °C.

Control of dislocations such as threading defects (TDs) are a key function of the BLs. Several methods have been tried. A stepped doping of the C content can also be done in parallel. The authors [2] showed that the SL and IL approaches can reduce the density of slip lines and cracks, allowing for a thicker net buffer layer to be used to meet the breakdown voltage requirements while minimizing wafer bow. Vertically propagating defects can be terminated at the SL or IL interfaces. Many other studies, such as from [3], have demonstrated that low temperature (LT: grown at ˜ 600 °C) ILs of ˜ 10 nm AlN placed in the buffer layer can reduce stress and improve the crystal quality of the GaN channel layer. BL thicknesses of several microns are typically needed.

3.1.3 Barrier Layer

The AlGaN barrier layer composition and thickness determine key device performance metrics like the threshold voltage (Vth) and the two-dimensional electron gas (2 DEG) channel density and mobility. The Al percentage can vary from 15–40%. In an e-mode p-GaN HEMT, the AlGaN layer can typically be 10–20 nm. High Al composition can create difficulty in forming ohmic contacts, with the tradeoff of improved channel charge density. The use of a thin (˜ 1 nm) AlN layer at the channel/barrier interface serves to create a barrier and prevents the 2 DEG wavefunction from penetrating into the barrier. This also allows for high Al composition barriers.

Another tradeoff to consider is in its thickness. To increase the Vth, a thin barrier over the gate is typically desirable, trading off with the 2 DEG charge density. A partial barrier recess in the gate region, followed by re-growth can balance this by creating a thicker barrier in the 2 DEG access regions, and a thin barrier under the gate.

3.1.4 Selective Area Growth (SAG)

A key use of MOCVD is in the growth of films in select regions of the device. A masking layer, such as silicon dioxide (SiO_2), can be used to prevent the growth in other areas. Some examples of SAG in GaN HEMT development are:

- SAG of the ohmic contact regions [4]: here n-GaN was regrown in the source/drain regions of the device to reduce source resistance.
- One of the key challenges in e-mode GaN metal-insulator-semiconductor devices such as the MISFET and MOSFET is the plasma damage created by the dry etch of the barrier layer at the gate formation steps. This creates gate leakage and reliability issues. A solution, exemplified in [5], is a recessed gate GaN MOSFET

created by selectively masking off the gate area and using SAG to regrow the epitaxial layers. An Aluminum oxide (Al_2O_3)/AlGaN/GaN MISFET was demonstrated with improved performance due to the avoidance of etch damage.

- In e-mode p-GaN HEMT, the etch damage of this layer along the side-walls and the channel access regions next to gate can severely degrade device performance. SAG of the Mg-doped p-GaN layer in the gate was done [6] with improved gate control.

High-performance, high-productivity MOCVD tools are now available with built-in process monitoring/control and automated wafer handling, chamber cleaning.

3.2 Wafer and Defects and Their Characterization Methods in GaN Power HEMT Fabrication

Sonu Daryanani

Numerous inherent material benefits, including a large bandgap (WBG) and strong electron mobility, are provided by gallium nitride (GaN). Utilizing this as a lateral high electron mobility transistor (HEMT) device with negligible capacitances and no reverse recovery loss allows for improved power conversion performance. Since this technology is being used in a larger range of applications, it is important to have a thorough understanding of the mechanics underlying yield and reliability enhancement. We list some of the flaws seen during the production of GaN wafers in this section, along with characterization methods that may be applied to find them.

3.2.1 GaN Crystal Structure and Base Substrates

The homoepitaxial growth of GaN substrates is challenging due to the high melting temperature and dissociation pressure. Heteroepitaxial growth is thus commonly used, with Si (111) and sapphire the preferred candidates for cost (larger wafer sizes), wide availability, high electrical insulation (in the case of sapphire), and the ability to use existing CMOS Fabs and potentially integrate with CMOS devices (in the case of Si). The Ga-facing wurtzite crystalline structure (Fig. 3.1) is typically grown on these substrates as it is thermodynamically stable.

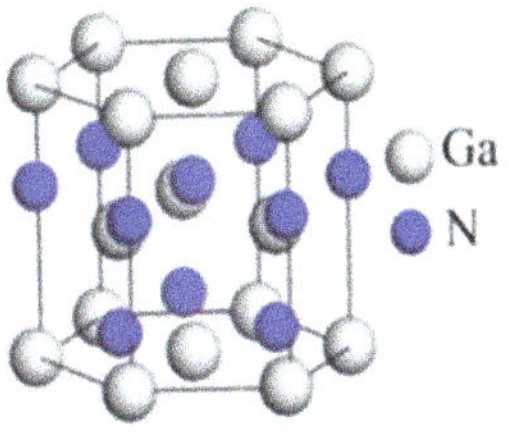

Fig. 3.1 Wurtzite crystal structure of GaN. (Source: Ref. [7])

Table 3.1 Some crystal properties of GaN and common substrate materials

Mismatch	Si	SiC	Sapphire	AlN	GaN
Crystal structure	FCC	HCP	HCP	HCP	HCP
Lattice constant (Å)	5.43	3.08	4.758	3.112	3.189
Lattice mismatch (%)	−16.0	3.5	16.08	2.4	–
Thermal expansion (10^{-6} K)	3.59	4.3	7.3	4.15	5.59
Thermal mismatch (%)	55	30	−23	34	–

Source: Ref. [8]

Although silicon carbide (SiC) has better lattice and thermal expansion coefficient (TEC) matching to GaN, the higher cost of these substrates limits it use to RF and other niche applications. Table 3.1 lists some of the lattice properties of GaN and commonly used substrates. GaN-on-Si is today the dominant technology used for the creation of GaN-based power electronic devices, with high volume production in several 6-inch and 8-inch Fabs.

The polarization effects of the AlGaN/GaN heterostructure leads to the formation of a two-dimensional electron gas (2-DEG) used as the conduction channel, which is the basis of the HEMT device operation.

3.2.2 *Manufacturing of Semiconductor Wafers in GaN Material*

Semiconductor wafers are the unsung heroes powering our modern electronics. From smartphones to cars, these tiny marvels made of materials like silicon navigate the realm between conductors and insulators, adapting their current flow to fit various applications.

Silicon stands as the reigning champion among semiconductor materials, inhabiting nearly every electronic device with its versatile capabilities. Yet, there are moments when silicon requires a sidekick, like gallium nitride (GaN), especially in high-temperature or high-frequency scenarios. While newer materials occasionally vie for the spotlight, silicon remains the cornerstone, thanks to its cost efficiency. Thus, it often serves as the foundational material for next-gen semiconductor materials.

The journey begins with silicon ingots, elongated logs sliced into wafer-thin disks, barely a millimeter thick. These semiconductor wafers become the canvas upon which electronic circuits are etched or serve as the substrate for other semiconductor materials, such as GaN.

Step 1: Setting the Stage—Preparing the Substrate

In the chemistry lab's controlled environment, the substrate undergoes meticulous cleaning. Whether destined for pattern growth or Metal-Organic Chemical Vapor Deposition (MOCVD), the base plates receive expert care. In the case of sapphire substrates, a sustainable approach reuses cleaned disks. Immersed in a heated potassium hydroxide bath, the disks shed their GaN remnants, emerging pristine and ready for the next phase.

Step 2: Crafting the Canvas—Pattern Making and Characterization

Within the sterile confines of the cleanroom, the cleaned disks embark on a journey of transformation. Through deposition, spinning, and etching processes, nanometer-scale cavities emerge, mapping out the blueprint for nanostructure growth. Scanning electron microscopy becomes the lens through which these patterns are scrutinized for their size, distribution, and purity, crucial for epitaxial growth.

Step 3: The Art of Epitaxy—Growing Semiconductor Wafers with Precision

Within an epitaxial growth reactor, a delicate dance unfolds. Gases mingle under stringent conditions of pressure, temperature, and gas composition, decomposing upon the substrate's surface. Guided by the substrate's pattern and crystal plane, epitaxial growth commences, ushering forth high-quality films or nanostructures. Amidst a nitrogen-filled chamber, substrates rest close together, ensuring uniform conditions vital for crystal growth.

Step 4: Refining the Masterpiece—Further Characterization with Precision

Once epitaxial growth concludes, scrutiny intensifies. Scanning electron microscopy returns, delving into the intricacies of nanostructures. Every defect, or dislocation, within the semiconductor material is scrutinized, for they dictate energy efficiency. Collaborative efforts yield unconventional growth methods, like nanowire coalescing, promising lower defect densities in GaN material. Through electron beams and meticulous observation, defects are quantified, paving the way for optimization.

Step 5: Breathing Life into Innovation—Manufacturing Components and Testing

Having analyzed the GaN crystals, experts embark on crafting electronic components. Within cleanroom confines, lithographic methods, etching, and metal deposition fashion these components atop the epitaxially grown material. With precision, electrical conductivity, elemental composition, and surface smoothness undergo rigorous testing. Capacities such as current flow, capacitance, breakdown voltage, and leakage current are measured, evaluating each component's performance.

3.2.3 GaN Defects

3.2.3.1 Threading and Other Dislocations During Growth

Threading dislocations (TDs), on the order of 1×10^{10} cm^2, are created at a very high level due to the large lattice mismatch between GaN and Si. These flaws, which may be categorized as one-dimensional or line defects, start at the substrate and move towards the epilayers. The optimization of buffer layers positioned between the substrate and the GaN channel has been a laborious task. These can now consist of a sequence of low-temperature GaN or AlN interlayers and a combination of superlattices (e.g., GaN/AlN).

Additional dislocations [9] include point-defects like nano-pipes, two-dimensional defects like stacking faults that can happen between crystal planes, and three-dimensional volume defects like interstitials connected to a single atomic site (these can be formed, for example, during ion implantation).

3.2.3.2 Particles

Particles can be deposited on the wafer front or back surface during any of the many steps related to the fabrication. Some of these could be metallic, such as from the metal-organic chemical vapor deposition (MOCVD) chamber used to deposit the buffer and epi-layers. Wafer handling can also be a dominant cause for these. Standard chemical cleans used in Si wafer processing can be used with GaN to reduce these.

3.2.3.3 Wafer Bowing and Breakage

The large lattice and TEC mismatch between GaN and Si can lead to epilayer cracks, wafer warp, and or breakage, especially for thicker films that are needed for higher voltage-rated devices. Stress management in the buffer layer is critical. Other steps during wafer fabrication would typically need to be optimized such as the wafer clamping force in tools, temperature ramp up and down rates during anneals, adjusting spin speeds in a scrubber, etc. [10].

3.2.4 Defect Characterization Techniques

In general, there are three classifications for defect characterization:

- Non-destructive: such as optical and X-ray based measurements.
- Destructive: such as Atomic Force Microscopy (AFM), secondary ion-mass spectroscopy (SIMS), and transmission electron microscope (TEM).
- Electrical measurements on either full devices or test structures.

Let us look at examples of each of these:

Optical and X-Ray Defect Inspection and Characterization
An example of an advanced inspection tool that can be used for GaN defect detection is the Candela® 8520 from KLA Instruments. This tool uses proprietary optics

to simultaneously measure scattering intensity from normal and oblique lasers incident on the wafer, as well as surface reflectivity, topographic variations, phase shift, and photoluminescence (PL). Automated full wafer inspections using muti-channel detection enables a wide range of defects to be detected and classified [11]. As an example, the Candela® platform can help detect hillocks and micropits formed during MOCVD growth and flag improper reactor conditions.

In contrast to lasers used for PL, cathodoluminescence (CL) uses high-energy electrons to stimulate the material and then analyze generated photons. GaN epitaxial TD defects can be detected on Horiba Scientifics CL tool when imaging at the band edge emission wavelength for GaN (362 nm).

High-resolution X-Ray diffraction (HR-XRD) rocking curves are also commonly used to characterize crystalline quality in epitaxial growth. The authors in [12] showcased an example of this technique to compare AlGaN/GaN epilayers grown on sapphire and silicon substrates, with the sapphire substrate exhibiting a tighter peak in the detected waveform. This can be attributed to the better lattice and TCE matching between GaN and sapphire.

EMMI and TEM

When electrical leakage is a failure signature, semiconductor product and device failure analysis frequently uses emission microscopy (EMMI). If the leakage is confined in space, it can be highly beneficial. The product has an electrical bias at a specific leakage level, and the failure point may be located via infrared detection. The EMMI spot sizes, which are usually in the region of 1 μm, are usually too big to identify individual crystal flaws. The scientists in [13] identified individual dislocations in a GaN HEMT device generated during high-temperature reverse bias stress testing (RBSTs) using a novel combination of EMMI with SEM and Scanning TEM (STEM) imaging. The gadget was equipped with an inbuilt series p-GaN resistor to control current and guard against catastrophic damage. Initially, the leakage increase during RBST was correlated to 9 places along the device width of 2 distinct devices using EMMI. The focused ion beam (FIB) holes produced close to the EMMI locations provide as reference points for the TEM investigation that follows. The AlGaN barrier layer's structural deterioration was then correlated with these spots via planar STEM. High-angle annular dark-field (HAADF) and annular bright field (ABF) detection were combined. While minor fractures and structural faults are identified by the HAADF, TDs are highlighted as lines by the ABF. These faults were then seen and further correlated to a specific dislocation inside the volume using cross-sectional TEM imaging. Therefore, at defect sites, a three-dimensional route of dislocations can be formed. Using this combination of analyses, it was concluded that leakage-causing flaws observed under RBST were not necessarily caused by pre-existing TDs in the AlGaN/GaN structure. It is possible that high fields at the p-GaN/AlGaN interface was the reason for the RBST leakage in this study since the device did not employ field plates to decrease the electric field at the gate.

3.3 Ion Implantation in GaN Power Devices

Sonu Daryanani

Power electronics devices are anticipated to utilize a growing amount of silicon carbide (SiC) and gallium nitride (GaN), two wide Band Gap (WBG) semiconductors. When compared to older silicon (Si) based devices, they provide a number of important advantages, including increased switching frequency, efficiency, and power density. The most common technique for achieving selective doping in silicon devices is ion implantation. There are many obstacles when using it for WBG device processing. We will discuss a few of them in this section along with a summary of some of its possible uses in GaN power devices.

3.3.1 Dopant Species

Several factors determine the practical usefulness of ion-implanted dopant species in semiconductor device fabrication:

- The existence of low ionization energy in the occupied lattice sites. Si is blessed with having shallow donor (for n-type doping) and acceptor (for p-type doping) elements that can be ionized. Deeper levels within the bandgap lead to poor ionization, especially at room temperature, and hence poor conductivity for a given implanted dose.
- Source material that can be ionized and implanted in commercial ion-implanters. Both solid and gas source material compounds may be used, and the practical usefulness of these depends on stability with temperature, safety, the efficiency of ion generation, ability to create unique ions that can be mass separated, the energy needed to achieve the given implantation depth, etc.
- Diffusivity within the implanted material. High diffusion rates under normal post-implant process conditions can result in hard to control junctions and dopant movement into unwanted regions of the device, hence degrading device performance.
- Activation and damage recovery. Dopant activation involves the creation of vacancies at high temperatures that allow for movement of the implanted ions from interstitial to substitutional lattice sites. Damage recovery is key for the healing of amorphization and crystalline damage created during the implantation process.

Table 3.2 lists some of dopants commonly used in SiC and GaN device fabrication, along with their ionization energies.

Relatively shallow doping species exist for n-type doping in both SiC and GaN; however, a key challenge in the creation of p-type doping with ion-implantation is the high ionization energies of the elements available.

Table 3.2 Dopant species commonly used in SiC and GaN power devices

Material	N-type dopant	P-type dopant
SiC	N, P (about 60 meV)	Al (about 200 meV); B (280–300 meV)
GaN	Si (12–62 meV)	Mg (170–245 meV)

Source: Ref. [14]

3.3.2 *Implants in GaN Power Devices*

Key implantation and annealing characteristics in Gallium Nitride (GaN) are crucial for optimizing the performance of GaN power devices. Unlike Silicon Carbide (SiC), hot implantation does not confer a significant advantage over room temperature implantation in GaN.

One notable feature of GaN is the behavior of common n-type dopant Silicon (Si), which can exhibit amphoteric properties, acting as either an n- or p-type dopant depending on its site occupancy. This behavior is influenced by GaN growth conditions and can lead to partial compensation effects.

P-doping in GaN presents additional challenges due to the high background concentration of electrons. Achieving p-type conductivity with Magnesium (Mg) doping requires high levels of Mg dopant, but excessive doses can introduce defects, trapping carriers at deeper energy levels, and reducing dopant activation.

GaN is susceptible to decomposition at temperatures exceeding 840 °C under atmospheric pressure, resulting in loss of nitrogen and formation of gallium droplets on the surface. To mitigate this, capping layers like SiO2 and various annealing techniques such as rapid thermal annealing (RTA) are employed. Annealing temperatures for GaN are typically lower (< 1500 °C) compared to SiC, with methods including high-pressure annealing, multi-cycle RTA, microwave, and laser annealing being explored. Despite these efforts, achieving a p + implanted junction remains a significant challenge in GaN devices.

3.3.2.1 Use of Mg and N Co-implantation in Vertical GaN Diodes

In a vertical Si and SiC power device, a common method for edge termination is the use of rings of p-type doping created by ion-implantation. The formation of vertical GaN devices is likewise made easier if such selective doping can be achieved. Mg-doped ion-implanted junctions have many challenges, some of which can be listed below.

- The high ionization potential (as shown in Table 3.2)
- Defects created during implantation can create permanent clusters, leading to deactivation
- Relatively high temperatures (>1300 °C) required to achieve activation. This is higher than the decomposition temperature of GaN, hence requiring special methods. A successful example of this is the use of ultra-high-pressure annealing (UHPA) with a N_2 pressure of 1 GPa [8]. Annealing at 1300–1480 °C resulted in >70% activation with good surface carrier mobility

- Mg diffusion at these elevated temperatures, as it interacts with point defects in the damaged region, can result in a graded junction. The control of the Mg profile in the p-GaN e-mode HEMTs, even with the MOCVD or MBE growth processes, is a key challenge

The use of a nitrogen (N) co-implant with the Mg has been shown to improve Mg dopant activation as well as suppress diffusion [15, 16]. The authors attributed the improved activation to the suppression of vacancy clustering from the N implant, which in effect was responsible for recombination with these vacancies at annealing temperatures above 1200 °C. In addition, the diffusion is restricted by the trapping of Mg by vacancies created by the N implant, leading to steeper junctions. This concept was used in the fabrication of a vertical planar GaN MOSFET through an all-ion implantation process [17]. This 1200 V device achieved an impressive specific on-state resistance R_{DSon} of 0.14 Ohms-mm^2. This flow, if it can be developed for mass-manufacturing, can be cost-effective and follow the commonly used process-flows in the manufacture of Si and SiC planar vertical power MOSFETs.

3.3.2.2 Isolation Implants in HEMTs

Because of the aforementioned issues, Mg-doped p-GaN may usually be grown rather than implanted in p-GaN e-mode High Electron Mobility Transistors (HEMTs). Ion-implant technology can be applied in HEMTs for the isolation of the lateral device. Several implant species have been tested here, including oxygen (O), argon (Ar), hydrogen (H), nitrogen (N), iron (Fe), etc. The creation of damage-related traps is the main process. Device planarity is a benefit of this approach over a mesa-etched isolation flow.

3.3.2.3 Si Implants to Reduce HEMT Source Resistance

While several studies have been conducted on the use of Si implantation for the creation of n + Ohmic contacts in GaN layers, the practical implementation can be challenging due to the high impurity concentration and lattice damage created. One motivation for the use of Si implants is the achievement of low resistance contacts with a Si CMOS compatible process flow that does not use gold (Au), or the subsequent post-metallization alloy process.

3.3.2.4 Use of Fluorine Implants

In HEMTs, low-dose Fluorine (F) implants have been used, taking advantage of the strong electronegativity of F to improve the breakdown voltage (BV) of the device. A negatively charged region at the backside of the 2-DEG electron gas is created, which suppresses source-injected electrons from reaching the high field region [18].

Another interesting application of ion-implantation in GaN is the use of F implants in vertical Schottky Barrier Diodes (SBDs) [19]. Here, F implantation placed at the surface next to the top anode contact was used to create a high-resistance edge termination region. The reverse current was decreased by 5 orders of magnitude, while also boosting the BV.

3.4 GaN Device Processing

Aurore Constant, Cristina Tringali, and Ferdinando Iucolano

Gallium Nitride (GaN) is a wide-bandgap semiconductor material that has revolutionized various industries due to its exceptional properties. The inherent strengths of GaN technology make it a game-changer in fields such as power electronics, telecommunications, and optoelectronics.

GaN technology boasts superior electron mobility, enabling faster, smaller, and more efficient electronic devices. This translates to reduced power consumption, and in the creation of energy-efficient solutions that can contribute to a sustainable future. GaN technology is envisioned to enable efficient power conversion within battery chargers, computers, servers, automotive, lighting systems, and photovoltaics.

In recent years, numerous GaN power products have been introduced to the market. To compete in terms of cost with Si trench MOSFETs (<200 V rated) and super-junctions (650 V rated), the industry has focused on hetero-epitaxy (GaN-on-Si) and addressing the hurdles associated with processing a III-V material in CMOS Si fabs.

Choosing GaN-on-Si over GaN-on-SiC is not just about saving costs. While Si substrates are indeed cheaper, they also support CMOS-compatible processes, facilitating large-scale production. Moreover, Si is widely adopted due to its outstanding quality and availability in sizes up to 300 mm. Currently, hetero-epitaxy of III-nitrides using MOCVD stands out as the preferred method for producing industry-scale AlGaN/GaN high-electron-mobility transistors (HEMTs).

Regarding the processing aspect, integration challenges such as cross-contamination, non-standard wafer thickness, high risk of wafer breakage, and the adoption of Au-free metallization are of outmost importance. Processing III-V materials in a CMOS process line presents several challenges.

Regarding cross-contamination, Ga and Al are life-time killers for Si devices. To mitigate this, a specific contamination control protocol must be developed. This involves limiting the number of tools dedicated to GaN processing, with the majority being shared with the Si production line to ensure rapid fab cycle times. To prevent cross-contamination, dedicated cleans and wafer boxes should be used after processing in GaN-dedicated tools, particularly when AlGaN and GaN surfaces are exposed. Additionally, controlling the wafer bevel-edge is necessary to minimize cross-contamination through clamping or inadvertent etching of GaN. One of the biggest concerns is related to GaN wafer breakage during processing. Because of the high-stress formation during the hetero-epitaxy growth, the GaN-on-Si wafers are prone to plastic deformation. Special care has to be taken for all wafer handling, clamping, and (de)-chucking of the wafers.

A high-electron-mobility transistor (HEMT) relies on a two-dimensional electron gas (2DEG) that is formed by the spontaneous polarization at the interface between AlGaN and GAN layers. Thanks to the presence of this high conduction layer and related high electron mobility, it is easy to fabricate a transistor structure without the need for material doping via implantation and annealing. The device is by nature normally on and named depletion mode. This type of condition is useful in the case of RF application, instead for power application a normally off device is preferred, also called enhanced mode. Various approaches have been developed to achieve this:

- Depletion mode MISHEMT in cascode configuration with a Si MOSFET to switch the HEMT.
- Enhanced mode pGaN gate HEMT: a depletion region under the gate is formed and controlled by a pGaN gate contacted through a Schottky contact.
- Enhanced mode recessed-MIS gate HEMT: the gate recess technique consists in etching the heterostructure under the gate and forming a MIS channel instead.

Among the different e-mode device concepts, only the p-GaN gate structure has been successfully commercialized.

The process flow is schematically reported in the different sections of Figs. 3.2, 3.3 and 3.4. It begins with the application of a photoresist to facilitate the processing of the zero-layer mask, which is crucial for aligning subsequent masks. Following the zero-layer mask, the pGaN is defined and etched with high selectivity towards AlGaN while minimizing plasma damage. This etching is a crucial process step in the fabrication of HEMT devices because the removal of the pGaN layer from the active area outside the gate creates the 2DEG channel. In order to selectively etch GaN over AlGaN, oxygen or fluorine (e.g. sulfur hexafluoride (SF6)) is added to the chlorine-based chemistry (Cl2/BCl3). The mechanism of achieving selective etching of GaN towards AlGaN can be explained by the formation of a thin (Al, Ga)Ox or (Al, Ga)Fx surface passivation layer on AlGaN surface in an atmosphere containing oxygen or fluorine, which is strongly resistant to etching and thus leads to low etch rate of AlGaN. The selectivity of etching depends on the aluminum content in the AlGaN layer. The lower the Al content, the lower the etch selectivity of GaN over AlGaN, which is associated with the formation of less etching-resistant passivation.

Subsequent to the pGaN gate formation, passivation layers are deposited on top of the pGaN and AlGaN surface, starting with a thin dielectric layer via ALD, followed by PECVD, defining the initial interlayer.

Afterward, the devices must be electrically isolated from each other when integrating GaN HEMTs. For GaN HEMTs, the ion implantation technique has already proven to be an attractive isolation approach over another isolation technique such as mesa etching, providing lower leakage and higher breakdown voltage of the isolation regions. Ion implantation introduces several defects into the GaN

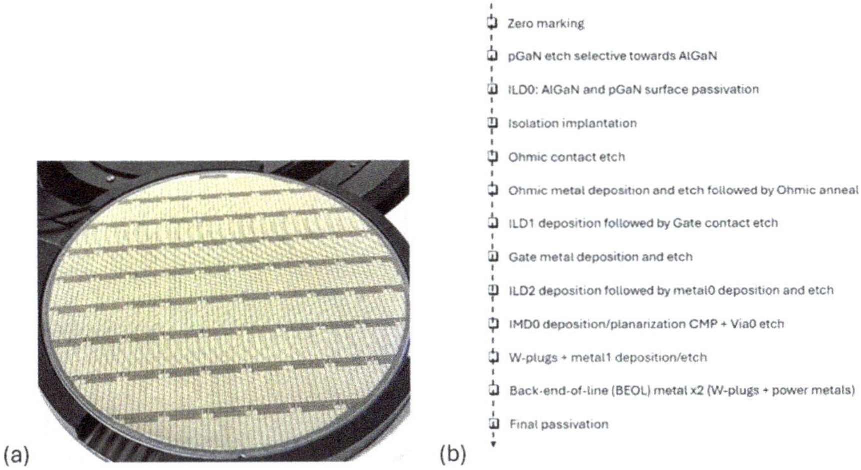

Fig. 3.2 (**a**) Fully processed normally off p-GaN gate HEMT devices on a 200 mm GaN-on-Si wafer, and (**b**) the corresponding process flow description

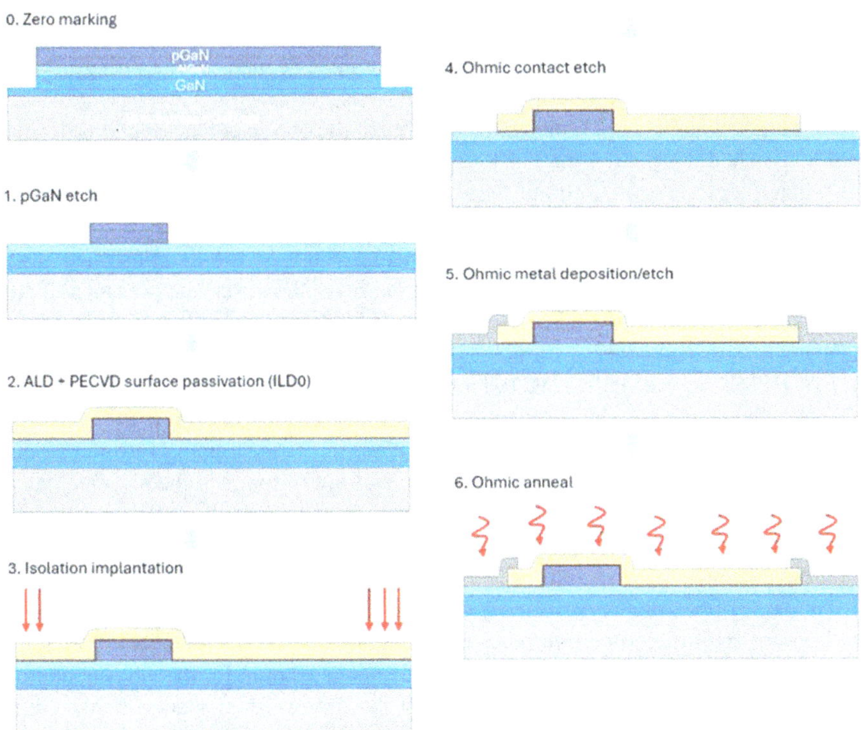

Fig. 3.3 Schematics of the process flow used for the p-GaN gate HEMT devices up to ohmic contact module

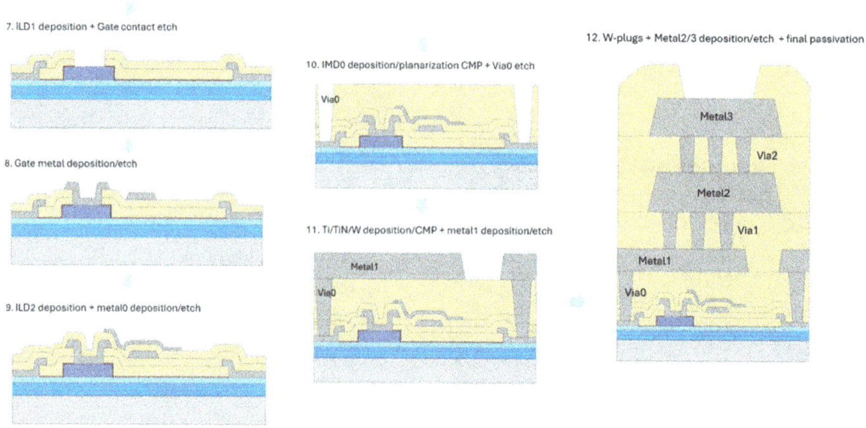

Fig. 3.4 Schematics of the process flow used for the p-GaN gate HEMT devices from gate module till final passivation

heterostructure that damages the crystal lattice and suppresses the 2DEG formation. The ion implantation-induced damages disappear after annealing at high temperatures, typically above 600 °C, thereby compromising isolation quality. Therefore, a low post-implantation thermal budget must guarantee in the manufacturing flow to guarantee high-quality isolation of HEMT devices.

Once the drain/source ohmic contacts defined, the ohmic contact metal is deposited and etched. Currently, the most commonly reported contact stacks on AlGaN/GaN heterostructures incorporate elements with a low metal work function, such as Ti, Ta, or Al, often with a protective capping layer to prevent out-diffusion and undesired oxidation from the top of the stack. Many of the demonstrated low-resistance contacts utilize an Au-based capping layer; however, the presence of Au is undesirable in an industrial Si-CMOS fab. Therefore, developing Au-free ohmic contacts is a critical concern for achieving high-volume and cost-effective production of HEMT devices in Si-CMOS fabs.

Once the source and drain regions are formed, a second interlayer is deposited. Next, the gate contact is etched, with the etching process stopping into the pGaN layer. This is followed by the deposition and etching of the gate metal, which also defines the first field-plate level. Subsequently, a third interlayer is deposited to cover the first field plate. Metal is then deposited and etched to define the second field-plate level.

An oxide deposition is performed, followed by planarization. The via0 is crucial in this step, as it directly connects to metal0, gate, and ohmic metals. The etching process for via0 is particularly critical, as it needs to account for the varying depths depending on the landing metals. Following the formation of via0 with W-plugs, metal 1 is sputtered and defined. Finally, as in conventional CMOS processing, the back-end-of-line (BEOL) process is completed up to metal3, including the final passivation step.

References

1. G.B. Stringfellow, *Organometallic Vapor-Phase Epitaxy: Theory and Practice* (Academic Press, Boston, 1999)
2. D. Marcon et al., 200mm GaN-on-Si epitaxy and e-mode device technology. IEDM (2015)
3. G. Cong et al., Design of low-temperature AlN interlayer for GaN grown on Si (111) substrate. J. Cryst. Growth **276**, 381 (2005). https://doi.org/10.1016/j.jcrysgro.2004.11.419
4. C.-H. Chen et al., High-transconductance self-aligned AlGaN/GaN modulation field effect transistors with regrown ohmic contacts. Appl. Phys. Lett. **73**, 3147 (1998). https://doi.org/10.1063/1.122701
5. J. Zhang et al., High-mobility normally OFF Al2O3/AlGaN/GaN MISFET with Damage free recessed gate structure. IEEE Electron Device Letters **39**, 1720 (2018). https://doi.org/10.1109/LED.2018.2872637
6. H. Yuliang et al., AlGaN/GaN high electron mobility transistors with selective area grown p-GaN gates. J. Semicond. **37**, 114002 (2016). https://doi.org/10.1088/1674-4926/37/11/114002
7. H. Qin et al., *Influence of Pressure on the Mechanical and Electronic Properties of Wurtzite and Zinc-Blende GaN Crystals*, vol 8 (Crystals, 2018)
8. D.G. Zhao et al., Stress and its effect on optical properties of GaN epilayers grown on Si(111), 6-H SiC(001), and c-plane sapphire. Appl. Phys. Lett. **83**, 677 (2003)
9. F.K. Yam et al., *Gallium Nitride: An Overview of Structural Defects* (n.d.). https://cdn.intechopen.com/pdfs/20503/InTech-Gallium_nitride_an_overview_of_structural_defects.pdf
10. S. Chowdhury et al., *5000+ Wafers of 650 V Highly Reliable GaN HEMTs on Si Substrates: Wafer Breakage and Backside Contamination Results* (ASMC, 2020). https://doi.org/10.1109/ASMC49169.2020.9185385
11. M Raghunathan, et al., *Defect Inspection for Compound Semiconductor Wafers* (n.d.). https://csmantech.org/paper/defect-inspection-for-compound-semiconductor-wafers/
12. S. Pharkphoumy et al., Correlation of crystal defects with device performance of AlGaN/GaN high-electron mobility transistors fabricated on silicon and sapphire substrates. Electronics **12** (2023). https://doi.org/10.3390/electronics12041049
13. M. Stabentheiner et al., On the insignificance of dislocations in reverse bias degradation of lateral GaN-on-Si devices. J. Appl. Phys. **135** (2024). https://doi.org/10.1063/5.0178743
14. F. Roccaforte, F. Giannazzo, G. Greco, Ion implantation doping in silicon carbide and gallium nitride electronic devices. Micro **2**, 23 (2022). https://doi.org/10.3390/micro2010002
15. H. Sakurai et al., Highly effective activation of Mg-implanted p-type GaN by ultra-high pressure annealing. Appl. Phys. Lett. **115** (2019). https://doi.org/10.1063/1.5116866
16. A. Uedono et al., Dopant-activation process in Mg-implanted GaN studied by monoenergetic positron beam. Sci. Rep. **11**, 20660 (2021). https://doi.org/10.1038/s41598-021-00102-2
17. R. Tanaka et al., Demonstration of 1200 V/1.4 mOhm-cm2 vertical planar MOSFET fabricated by an all ion implantation process. Jpn. J. Appl. Phys. **59**, SGGD02 (2020)
18. M. Wang, K. Chen, Improvements in the Off-State Breakdown Voltage with Fluorine Ion Implantation in AlGaN/GaN HEMTs. IEEE Trans. Electron Devices **58**, 460 (2011). https://doi.org/10.1109/TED.2010.2091958
19. Z. Liu et al., High-voltage vertical GaN-on-GaN Schottky barrier diode using fluorine ion implantation treatment. AIP Advances **201**. https://doi.org/10.1063/1.5100251

Sonu Daryanani received his Ph.D. in Electronics Engineering in 1993 from the University of Bradford, England, while doing research at AT&T Bell Labs, NJ, on GaAs optoelectronics. Over the last 25 years, he was a Device/Process Integration Engineer at Microchip Technology, AZ, working on microcontroller and discrete devices. He is now a consultant and writer for PowerElectronics News. He can be reached at sldaryanani@gmail.com

Aurore Constant has Ph.D. degree in Electrical Engineering in 2011 from the Polytechnic University of Catalonia, Barcelona, Spain, where she worked on SiC gate oxide process design and characterization for power devices. Then in 2011, she joined Onsemi's corporate R&D department in Belgium, where she was involved in the development of power devices on GaN. Since 2021, she has worked with ST on R&D where she is currently responsible for the technology development of GaN power devices.

Cristina Tringali graduated in Industrial Chemistry from the University of Catania in 2000 before joining ST as an etching process engineer. In 2006, she became section Head for R&D Etching Group, in 2014, responsible for R&D Process Development Group and in 2017, responsible for GaN Integration in the R&D GaN Device Group. Since 2020, she is in charge of GaN Technology Development for both RF and low-voltage power GaN applications, including process, Device Physics and Simulations.

Ferdinando Iucolano graduated in Physics from the University of Catania in 2004. In 2008, he received a Ph.D. degree in Physics on Nanoscale approach to charge transport phenomena in GaN-based structures. From 2011, he joined STMicroelectronics as power device designer for the development of GaN technology in the R&D department. In 2014, he became responsible for device physics group for wide bandgap materials and since 2017, he oversees technological development and design of product for GaN-based power devices including RF-power, Power Switches devices and GaN ICs. He is author of more than 140 papers in scientific journals and proceeding of National and International conference with more 2500 citations. Moreover, he is a Member of the Standardization Committee for GaN JC-70 Wide Bandgap Power Electronic Conversion Semiconductors with an active role in the identification of failure modes on GaN TDDB.

Chapter 4
GaN Technology

Florin Udrea, Kengo Ohmori, Ming Su, Michael de Rooij, Sam Abdel-Rahman, Walter Balzarotti, Doug Bailey, Andrew Smith, Oliver Häberlen, Tim McDonald, Mark Davidson, Troy Baker, Ryo Takeda, Takamasa Arai, and Stephen Oliver

Chapter 4 of this volume provides a comprehensive exploration of Gallium Nitride (GaN) technology's pivotal role in revolutionizing power electronics. Florin Udrea, CTO of Cambridge GaN Devices, lays the foundation by delving into essential aspects and the dynamic landscape of GaN devices, emphasizing material properties, device configurations, and integration levels. Following this, Kengo Ohmori and Dr. Ming Su from Rohm Semiconductor offer a detailed examination of GaN transistor evolution and performance characteristics, showcasing its dominance in high-power, high-frequency applications.

The chapter further delves into layout considerations with Dr. Michael de Rooij from Efficient Power Conversion addressing parasitic inductance challenges. Sam Abdel-Rahman, System Architect at Infineon Technologies, underscores the

F. Udrea
Cambridge GaN Devices, Cambridge, UK

K. Ohmori · M. Su
Rohm Semiconductor, Santa Clara, CA, USA

M. de Rooij
EPC, El Segundo, CA, USA

S. Abdel-Rahman
Infineon Technologies Americas Corp., Morrisville, NC, USA

W. Balzarotti
Infineon Technologies AG, Neubiberg, Germany

D. Bailey · A. Smith
Power Integrations, San Jose, CA, USA

O. Häberlen
Infineon Technologies Austria AG, Villach, Austria

T. McDonald
Infineon Technologies, Americas Corp, El Segundo, CA, USA

© The Author(s), under exclusive license to Springer Nature Switzerland AG 2024
M. Di Paolo Emilio (ed.), *GaN Technology*,
https://doi.org/10.1007/978-3-031-63238-9_4

importance of efficiency and compactness in data centers, with a focus on packaging and thermal solutions. Walter Balzarotti of Infineon Technologies discusses efficient gate-driver circuit designs for GaN HEMT applications, followed by Doug Bailey and Andy Smith from Power Integrations exploring GaN technology's robustness and reliability through extensive testing procedures.

Oliver Häberlen and Tim McDonald from Infineon elaborate on ensuring industrial reliability for GaN transistors, while Mark Davidson and Troy Baker analyze the landscape of Vertical GaN technology. The chapter culminates with insights from Ryo Takeda and Takamasa Arai from Keysight Technologies addressing challenges and enabling technologies for accurate dynamic characterization of GaN power semiconductor devices. Additionally, Stephen Oliver, VP Corporate Marketing & Investor Relations at Navitas Semiconductor, provides an exploration into the realm of Gallium Nitride Integrated Circuits (ICs), offering insights into their evolution, technological advancements, and comparative analysis between discrete and monolithic integration.

4.1 Structure and Design

Florin Udrea

Gallium Nitride (GaN) is arguably the most exciting material in the field of power electronics today, enabling the development of high-voltage devices with increased power density, reduced on-resistance, and very high-frequency response [1]. The wide bandgap of the material (E_g = 3.4 eV) results in a high critical electric field (E_c = 3.3MV/cm) which can lead to designs of devices with shorter drift region, and therefore lower on-state resistance and faster switching when compared to silicon-based counterparts for the same voltage rating.

Table 4.1 shows the physical properties of different materials vs those of silicon. GaN offers concomitantly higher carrier mobility (for the two-dimensional electron gas—2DEG), and higher critical electric field than both Silicon and Silicon Carbide (SiC). The availability of relatively inexpensive GaN-on-Si wafers makes GaN a favorite choice among the wide bandgap materials, especially in medium- to-high-end applications where cost remains a powerful consideration. GaN-on-Si lateral devices are more suited for relatively lower voltage (40 V to 900 V) and lower

M. Davidson (✉)
Semiconductor Executive, Downington, PA, USA

T. Baker
CTO, Bolt Semiconductor, Hummenlstown, PA, USA

R. Takeda · T. Arai
Keysight Technologies, Tokyo, Japan

S. Oliver
Navitas Semiconductor, Torrance, CA, USA

Table 4.1 Material properties of GaN in comparison with Silicon and other wide bandgap materials

Physical property	Si	4H-SiC	GaN	SC CVD diamond	β-Ga_2O_3
Bandgap (eV)	1.1	3.2	3.4	5.5	4.9
Relative permittivity	11.9	10	9.5	5.7	10
Breakdown field (MV/cm)	0.3	3	3.3	5	8
Thermal conductivity (W/K/cm) *[note: Reference 28 for details about it]*	1.48	3.30	1.30 (GaN substrate) 1.5(Si) 0.35 (Sapphire) 3.3 (SiC)	24.00	0.13
Mobility ($cm^2/(Vs)$)	1350	700	1200 (bulk) 2000 (2DEG)	3800 for holes 4500 for electrons (intrinsic doping)	200–300
Saturation velocity (10^7cm/s)	1	2	2.5	2	2
Available wafer diameters	8–12″	6″	6–8″ Si 6″ SiC 6″ Sapphire 3–4″ GaN	<1″	<1″

Table 4.2 Figures of merit (FOM) for GaN and other wide bandgap materials relative to Silicon

Figure of merit (FOM)	Si	4H-SiC	GaN (HEMT)	SC CVD Diamond	β-Ga_2O_3
Baliga's FOM	1	134	537	141,635	3571
Johnson's FOM	1	20	27.5	74	27

power applications (<10 kW) while SiC vertical devices are best for voltages above 1.2 kV with power levels in excess of 1 kW. Nevertheless, it is expected that GaN devices will also penetrate the 1.2 kV territory and their applications will scale up from tens of kWs to hundreds of kWs by using parallel chips. GaO and Diamond offer interesting material properties as their bandgaps are wider than those of SiC and GaN, but GaO suffers from the absence of a reliably p-type dopant, poor electron mobility, and very low thermal conductivity while Diamond suffers from a lack of effective n-type dopants, fast degradation of mobility at high temperatures, and the lack of shallow dopants (even for p-type). While GaO and diamond remain an option for the future in specific applications, it is expected that GaN and SiC would make the stronger market penetration in the next 10 years.

Both Baliga FOM and Johnson FOM for GaN and SiC, respectively, and shown in Table 4.2 [2], are orders of magnitude higher than those of Silicon making a strong case for the use of these materials in power semiconductor devices in order to improve the efficiency of power systems and/or reduce their form factor by operating at higher frequencies. Recently, the FOMs for different materials have been adjusted to take into account a 2D and 3D superjunction design and the parasitic JFET effect within the superjunction pillars [4, 5]. This led to a slightly different dependence of the specific on-state resistance, Ron with the critical electric field, Ec ($R_{on} \sim 1/Ec^{2.5}$ rather than $1/Ec^3$).

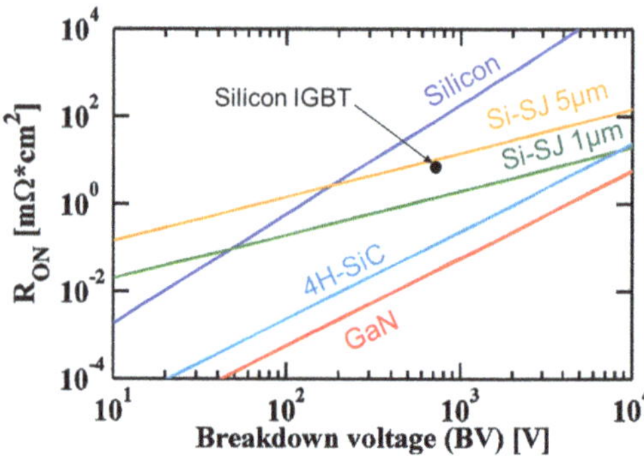

Fig. 4.1 Specific Ron vs breakdown voltage limits for different materials and technologies

A graph displaying the on-state resistance, Ron vs breakdown limits is a powerful and well-established way to assess different technologies and materials (Fig. 4.1). These limits are closely related to the static figures of merit. The original limit of Silicon is based on a one-dimensional Power MOSFET design and no longer represents an "ideal" case as this has been invalidated by the theory of superjunction. The original limits for wide bandgap materials are shown in Fig. 4.1.

The superjunction Silicon limits for a 5 and 1 micron cell pitch are also plotted showing the ability of silicon to outperform its one-dimensional limit. However, one can note that even when pushing the superjunction concept to the absolute limits (with very aggressive narrow pillars), there is still plenty of ground to justify the development of wide bandgap materials such as SiC and GaN. Besides this, both SiC and GaN can further improve their corresponding limits by employing various multi-dimensional forms. The graph in Fig. 4.1 also shows that GaN offers a better static trade-off than its SiC rival. Nevertheless, for a more detailed picture, one should consider other factors such as maturity of technology (in favor of SiC), thermal conductivity (in favor of SiC), channel resistance (in favor of GaN) and wafer size, and CMOS infrastructure (in favor of GaN for lateral devices) [1–5].

4.1.1 Lateral vs Vertical Configuration

The use of an AlGaN/GaN heterostructure in lateral GaN devices allows the formation of a two-dimensional electron gas (2DEG) at the hetero-interface where carriers can reach very high mobility (μ = 2000 $cm^2/(Vs)$) values. This is because the 2DEG acts as a quantum layer where electrons can move fast without scattering from neighboring ion impurities. Transistors in this technology are named High Electron Mobility Transistors (HEMT). In addition, the piezo-polarization charge present at the AlGaN/GaN heterostructure results in a high electron density in the 2DEG layer (e.g., 1×10^{13} cm^{-2}). These properties allow the development of transistors with unique and very competitive performance.

The lateral configuration has also some attractive features compared to the vertical configuration:

- It allows monolithic integration of sensing, protection, and drive circuits with relatively simple isolation and efficient shielding between different components.
- It allows easy access to all terminals (including the high-voltage terminal).
- It allows integration of half-bridge devices and multiple power devices operating on a common substrate.
- In some cases it allows a physical connection of the substrate to a heat sink.
- It allows simple hybrid integration with silicon devices via surface wire bonding.
- It allows simpler and cheaper packaging.
- It allows devices to be self-terminated (i.e., do not require an additional termination region to shape the electric field at the edge of the device as in the case of vertical structures).

The vertical configuration remains however a preferred option for higher power levels. At higher voltage levels (>1.2 kV) and higher power levels, the competition from SiC should not be underestimated. GaN indeed can offer some advantages in terms of its high critical electric field and higher channel mobility, but SiC presents a native gate oxide and its dopants are shallow (in the bandgap) and very well controlled. A good review of vertical GaN technologies and prospects can be found in [6, 7].

4.1.2 GaN Devices and Levels of Integration

Now and over the next decade, lateral GaN devices will exist in different voltage ratings from 40 V to 900 V. In particular, the classes of 30-40 V, 100 V, 200 V, 600-700 V, and the 900 V are very popular.

There are several types of HEMT-based discrete power devices and power integrated circuits in the market (or close to entering the market) today. Examples of devices and technologies from a number of companies are given below.

Discrete or Discrete-like devices

- Cascode technology based on a high-voltage depletion HEMT on GaN on Si and an enhancement low-voltage MOSFET—Transphorm, Nexperia.
- Series connection of Si MOSFET with high-voltage depletion HEMT with direct access to the HEMT gate and additional Si circuitry—Visic.
- Enhancement mode p-GaN HEMT on Si substrates—Infineon, GaN-Systems, EPC, Innoscince.

Power IC and System on Chip devices

- Hybrid integration of Silicon drives/controller and enhancement HEMTs on GaN on Si—Texas Instruments.
- Hybrid integration of Si drives and controllers with GaN on Si or on Sapphire—Power Integrations, ST Microelectronics.

- Monolithic integration of smart interfaces for sensing/protection, ease of use, and enhanced gate reliability—Cambridge GaN Devices (CGD).
- Monolithic integration of gate drive and sensing circuits—Navitas, EPC.

Table 4.3 shows some of these solutions with features and challenges. The first 4 rows present the schematic cross-sections of different GaN HEMTs used today.

Table 4.3 GaN power devices and architectures

<table>
<tr><td>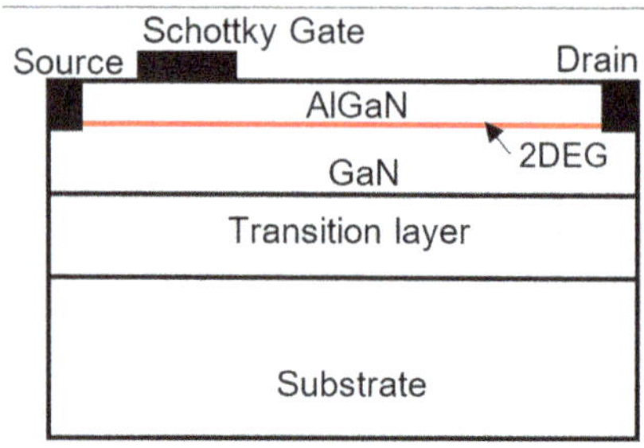
</td><td>Depletion mode HEMT
Features: (i) V_{th}—negative, normally-on operation (ii) low Ron.
Challenges: (i) Limited use as stand-alone device - used mainly in a Cascode configuration. (ii) Schottky gate voltage cannot go beyond +0.8 V. (iii) Requires relatively large negative voltages for turn-off</td></tr>
<tr><td>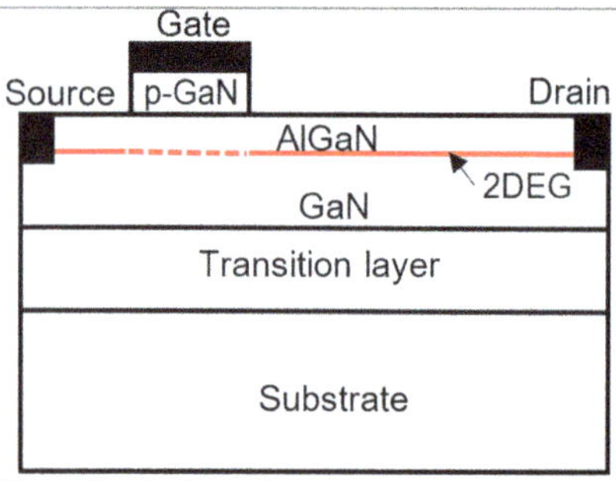
</td><td>Enhancement mode p-GaN gate HEMT
Features: (i) normally off operation, (ii) p-GaN gate using Mg doping, (iii) choice of gate Ohmic contact on the p-GaN (for enhanced gate reliability) or Schottky contact (for reduced leakage current).
Challenges: (i) low threshold voltage V_{th} ~ 1.5 V, (ii) fragility of the gate with max voltage ~7 V, (iii) requires negative voltages for safe and reliable turn-off</td></tr>
<tr><td>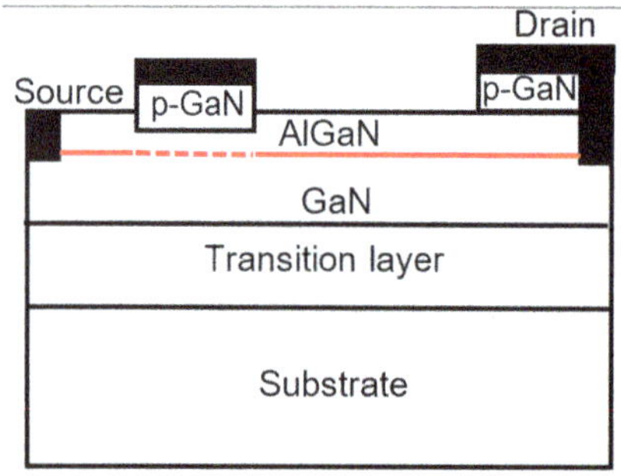
</td><td>Gate injection HEMT with additional p-GaN hole injector connected to drain
Features: (i) similar to the p- GaN HEMT, (ii) enhanced reliability provided by the p-GaN hole injector connected to the drain, (iii) thicker AlGaN under the hole injector compared to that under the gate, to avoid depletion of the 2DEG.
<u>Challenges</u>: Same as p- GaN HEMT</td></tr>
<tr><td>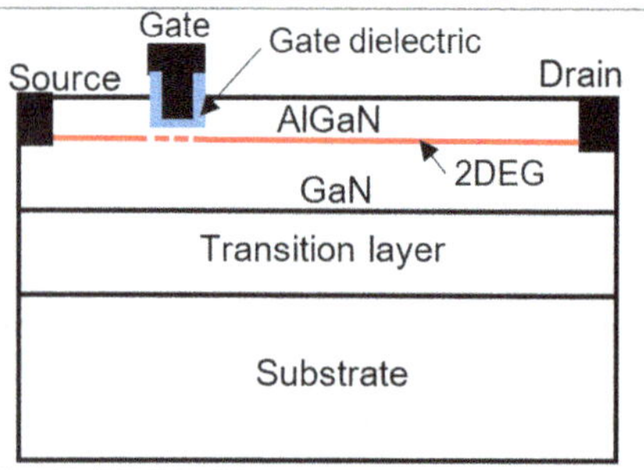
</td><td>Enhancement mode MIS HEMT
Features: (i) recessed insulated (MIS) gate for positive V_{th}, (ii) low gate leakage, (iii) extended gate voltage range
Challenges: (i) Reliability of the gate given traps and interface states, (ii) reproducibility and stability of the threshold voltage, (iii) hot carrier injection from AlGaN into the gate dielectric, (iv) high electric fields at the corner of the trench gate dielectric</td></tr>
</table>

(continued)

Table 4.3 (continued)

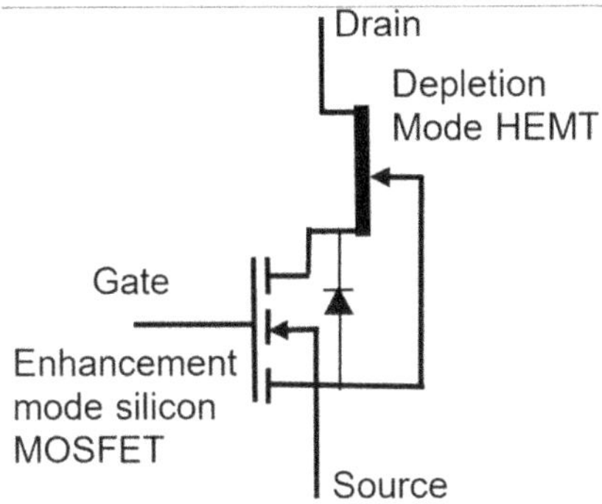	*Cascode device* *Features:* (i) depletion mode HEMT in series with a low-voltage enhancement mode silicon MOSFET, (ii) stable, reliable, and easy to use drive provided by the silicon MOS gate, (iii) good reliability, (iv) low on-state forward drop in the reverse conduction (diode) mode (third quadrant) *Challenges:* (i) co-packaging of two or multiple components, (ii) voltage sharing, (iii) some reverse recovery losses due to the anti-parallel bipolar diode, (iv) no possibility of adjustment of slew rate on the HEMT gate
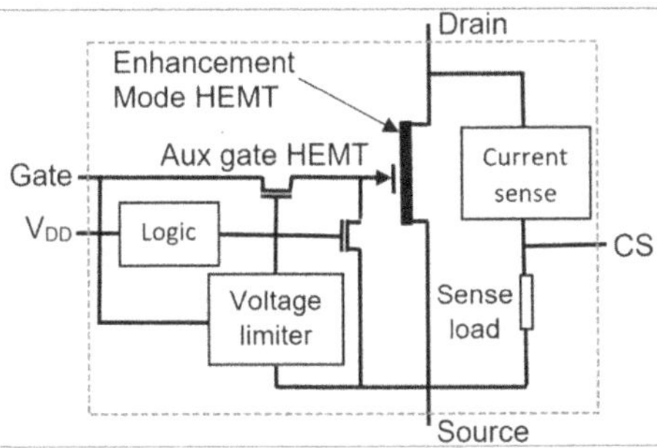	*Smart HEMT (ICeGaN HEMT)* *Features:* (i) monolithically integrated smart sensing and protection circuits, (ii) stable, reliable, and easy to use drive provided through a GaN interface attached to the gate, (iii) high V_{th} ~ 2.5 V, (iv) extended voltage range, up to 20 V, (v) no negative voltage requirement, (vi) current sensing *Challenges:* (i) extra GaN area for the smart interface.
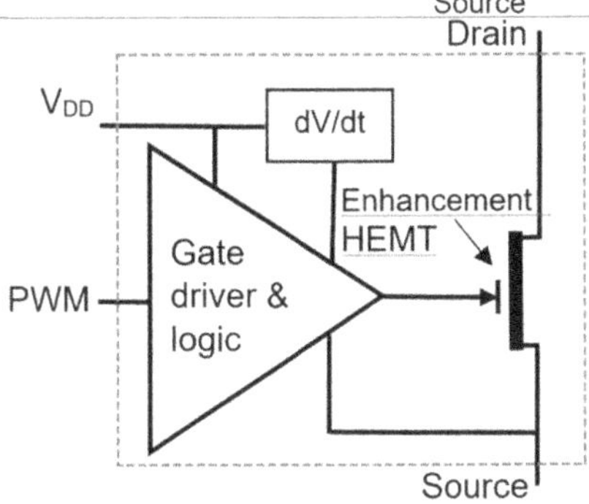	*Power IC with HEMT and integrated drive* *Features:* (i) monolithically integrated drive & logic circuits with slew rate control, (ii) stable & reliable operation, (iii) compact solution *Challenges:* (i) extra GaN area for the drive circuit, (ii) thermal consideration, as the drive & logic circuits can get hot during HEMT operation, (iii) limited performance of the drive and logic circuits—availability of only n-channel low-voltage GaN transistors (no p-channel devices).
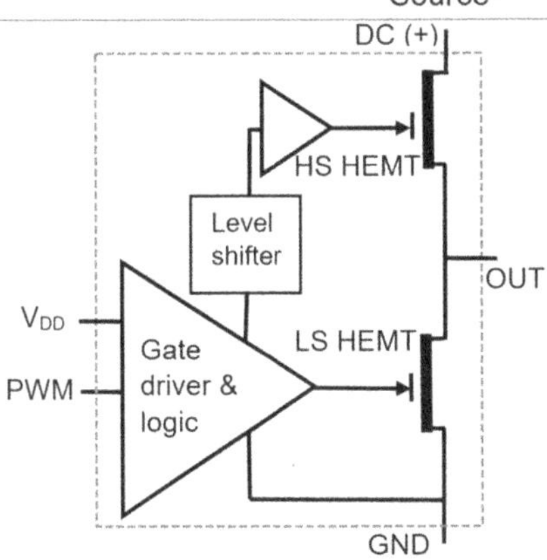	*Power IC with half-bridge integration* *Features:* (i) low-side HEMT integrated with high-side HEMT in half-bridge, (ii) driver, logic and level shifter circuits monolithically integrated, (iii) compact solution and low parasitics *Challenges:* (i) extra GaN area for the circuits, (ii) thermal consideration, as the drive and logic circuit can get hot during operation of the half-bridge, (iii) limited performance of the drive and logic circuits—availability of only n-channel low-voltage GaN transistors (no –p channel), (iv) interference between LS and HS HEMTs

The next 4 rows present configurations where such devices are used to enhance their performance, safe operating area, and ease of use. What transpires is the diversity that GaN offers today to provide tailored solutions for different applications in power electronics. From the simplicity of enhancement, GaN discrete transistors to the complex half-bridge monolithic integration with logic and drive circuits, GaN offers several layers of complexity to maximize its benefits against Silicon solutions.

4.1.3 ICeGaN Technology

ICeGaN™ can be driven in a similar manner to a silicon MOSFET and unlike all other enhancement GaN solutions, it is compatible with any silicon-based drivers. Its zero reverse recovery losses and very low output charges make it an ideal choice for high-frequency, high-efficiency applications. ICeGaN™ devices have a higher threshold voltage, Vth > 2.5 V in order to suppress dV/dt-related spurious turn-on events and as a result, allow safer operation. No negative voltages are needed to drive IcEGaN and for this reason, leading to lower on-state losses when the device is in reverse conduction. Moreover, the ICeGaN™ devices can be driven in the forward conduction with gate voltages of up to 20 V (well in excess of the standard 7 V for p-GaN HEMT) without any compromise in the device transconductance or dynamic performance. The presence of the Miller clamp transistor ensures very fast turn-off and very high dV/dt immunity. Finally, there is no need for RC drive, or any clamping diodes to be attached to the gate reducing the BOM. A block description of an ICeGaN™ device which includes the main HEMT, the sense HEMT, the sensing load, the Auxiliary HEMT, the Miller clamp, and other analogue/logic blocks is shown in Fig. 4.2 and its schematic drive scheme and transfer characteristics shown in Fig. 4.3.

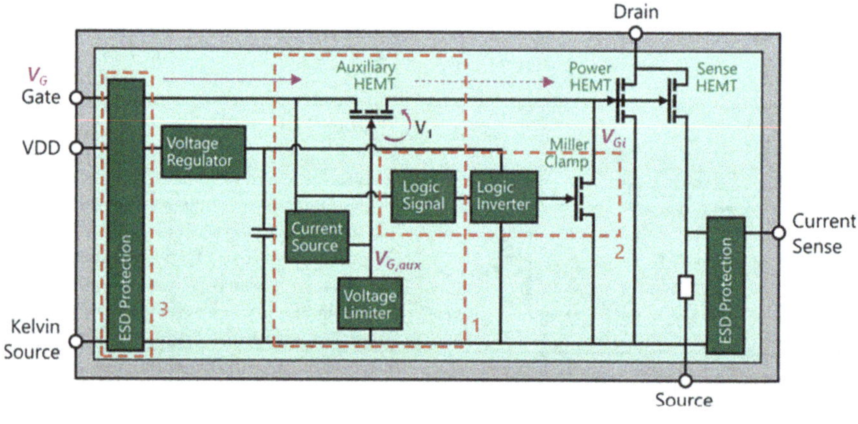

Fig. 4.2 ICeGaN technology platform integrating several sensing and protection functions

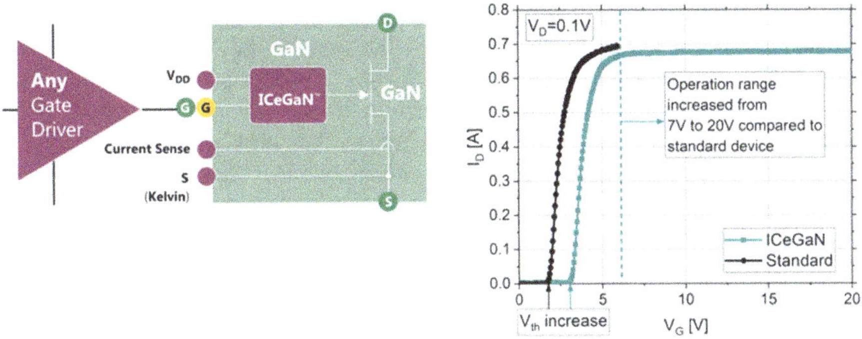

Fig. 4.3 ICeGaN drive and transfer characteristics. One can note that ICEGaN™ has a higher threshold and wide voltage range, but similar transconductance to a standard discrete HEMT

4.1.4 *Reliability of GaN HEMTs*

The field of power electronics is experiencing a major and exciting change with the adoption of wide bandgap semiconductors. GaN is currently seeing an exponential increase in demand. Its attractiveness in the market comes from the excellent performance benefits as a result of its enhanced physical properties compared to those of silicon. For system designers, GaN devices have the ability to operate at higher frequencies, which can cut the size, volume, and ultimately the cost of passives (transformers, inductors, and capacitors), while still benefiting from higher system efficiency, thus reducing the cost of heat sink or, in some specific cases, denying the need for thermally enhanced PCBs. However, one area where silicon still has the edge is reliability.

There were several potential concerns for reliability of GaN, and most of which have now been largely addressed or are in the process of being overcome.

4.1.4.1 Vertical Leakage

The vertical leakage current from the drain of a lateral GaN HEMT to the semiconductor substrate used to be a major concern in the past. The quality of the epi-layers (the nucleation layer, the transition layer, and the buffer) has been tremendously improved, and the vertical leakage (though still dominant at very high blocking voltages) has been reduced by orders of magnitude and is no longer a major problem.

4.1.4.2 Dynamic Ron

2DEG, which inherently exists at the AlGaN/GaN hetero-interface, has been found to be unstable and loose electron charge when a prior high voltage stress was applied. This phenomenon is known as dynamic Ron or current collapse. When the device is under high voltage stress in the off-state, a fraction of the electrons in the 2DEG is lost by trapping mechanisms in either traps in the bulk of the GaN, the transition layer (i.e., the layer placed between the substrate and the GaN buffer), or at the surface of the device. Hot carrier injection in the passivation layer, next to the gate (the control terminal), also may play a role, though this is not believed to be the major contribution to the dynamic Ron phenomenon. This loss of charge from the 2DEG layer to the surrounding traps leads to a subsequent decrease in the conductivity during on-state and thus an increase in the on-state resistance. The effect could be seen also during the switching or when the device is in operation in a real system.

The hot carrier injection and the trapping effect of the electrons in the bulk of the GaN or at the surface of the device in the vicinity of the gate not only lead to a loss in the 2DEG layer conductivity but are also associated with a shift in the threshold voltage. The trapping and the hot carrier injection are significantly more prominent at higher electric fields. One of the peaks of the electric field occurs at the drain side of the edge of the gate structure.

There are two possible ways to minimize the effect of the dynamic increase in the on-state resistance: (i) use of field plates in order to reduce the field peaks in the structure, in particular around the gate structure, to minimize the effect of electron trapping from the 2DEG and/or hot carrier injection, and (ii) the use of a hole injector layer to inject holes into the bulk and actively passivate the bulk traps. There are several theories describing the effect of the injected holes. One possible mechanism is that the injected holes change the electric field distribution from the 2DEG to the substrate, by moving the high fields away from the 2DEG vicinity, toward the substrate. As a result, the 2DEG is exposed to lower electric fields, which slow down the mechanism of electron trapping in the vicinity of the 2DEG [1].

The dynamic Ron has been greatly reduced today, with most GaN chip manufacturers guaranteeing a maximum of 20–30% increase in Ron, while others claim a negligible increase (below 10%). In the next five years, it is entirely possible that this problem will be completely eliminated.

4.1.4.3 Fragility of the Gate

The p-GaN gate using magnesium doping is considered today the most promising solution for an enhancement HEMT. It delivers a normally-off operation with a relatively low, positive threshold voltage of ~1.3 V. The gate cannot be biased above 7 V with respect to the source as the gate current increases exponentially. The Cascode or the smart interface solutions (ICeGaN™) have been introduced to address exactly these problems, achieving a higher threshold voltage and a gate range similar to those offered by silicon devices (e.g., 3 V and up to 20 V, respectively).

The threshold stability has also been found to be affected by traps in the p-GaN, AlGaN, and at the interface between the two. In general, an Ohmic gate placed on the p-GaN offers a potentially more robust solution, but a Schottky contact has the benefit of a lower gate leakage current. The latter is particularly important at high temperatures.

The fragility of the gate can also be addressed by integrating the driver or adding protection (clamping) circuits. ICeGaN™ technology is particularly special in this respect. The auxiliary HEMT provided between the external gate terminal and the internal gate terminal can absorb high voltages. The maximum static voltage applied to ICeGaN is 30 V, while the maximum dynamic voltage is in excess of 70 V, beating specifications of insulated gates in silicon and silicon carbide devices.

4.1.5 *Future of Lateral GaN*

The development of GaN will continue at an accelerated rate in the next decade. Several foundries and vertically integrated power device businesses are now considering GaN as a key technology in their portfolio. 6 inch GaN wafers are the standard today with 8 inch wafers already available from some foundries. 12 inch GaN on Silicon is not too far out with other substrates being considered for higher rated voltages (e.g., 1.2 kV sapphire).

In the long term, we expect concepts such as FINFET (Tri-Gate), Superjunction and Multi-channel GaN to be employed in lateral HEMTs [8]. These concepts can be used in isolation or in combination to maximize the benefits. For example, the Tri-Gate technology can be used to form three-dimensional gate structures around the gate to deplete multi-channels within the gate region. The superjunction can employ p-GaN-type stripes displaced in the third dimension of the HEMT to distribute the field more uniformly between the source and the drain, and thus further reduce the drift region dimensions (for the same rated breakdown). Finally the multi-channel technology is based on forming a plurality of GaN/AlGaN layers with stacked heterojunctions where multiple 2DEG layers are formed. This results in a structure featuring parallel conduction paths between drain and source which result in a lower specific on-state resistance.

In the medium to long term, we expect GaN to be used in more complex Power ICs. Emergence of p-channel-type low-voltage GaN using for example FINFET technology could allow integration of more efficient circuit blocks for sensing/protection and gate drive circuitry. Monolithic integration of full-bridge, 3-phase inverter chips and/or basic control circuits will also be on the roadmap of many companies.

4.2 Device Performance

Kengo Ohmori and Ming Su

Since the first GaN enhancement mode transistors became generally available, GaN devices have continued to mature and become a dominant player in high-power, high-frequency applications. Advanced manufacturing of substrate materials and GaN crystal growth, along with new techniques for P-type doping and efficient device construction, has brought the price of GaN transistors within the range of their silicon counterparts. When the enhanced specifications of GaN switches are considered with respect to energy efficiency and speed, the overall system cost for many applications has gone down while performance has increased. As a result, GaN devices are overtaking many other technologies like SiC MOSFETs as the preferred solution.

Periodic table group 3A elements, commonly indium and gallium, can be alloyed with group 5A elements such as nitrogen, phosphorus, and arsenic to create direct bandgap semiconductors. Gallium nitride (GaN) in particular must be chemically synthesized by subjecting gallium and ammonia to high temperature and pressure to form a crystal exhibiting a robust Wurtzite structure. The complexity of this structure makes GaN semiconductors limited in size and purity, but the resulting electrical properties offer many performance benefits in the fabrication of electronic components.

As with all semiconductors, GaN requires doping to control its conductivity. For N-type devices, this can be achieved with silicon or germanium dopant atoms. P-type devices, on the other hand, can only be created through a complex process using magnesium and high-temperature annealing. This fact has contributed to the slow adoption of GaN devices where both P- and N-type regions are required.

4.2.1 GaN Transistor

To build a transistor from a GaN crystal, a heterojunction (different bandgap semiconductors) must be formed using molecular beam epitaxy or chemical vapor deposition on top of the base crystal. Using a 3A-5A alloy of aluminum nitride and gallium nitride for this layer allows precise control of the bandgap delta. The strain created at the boundary between these two layers results in the formation of a two-dimensional electron gas and is the foundation of the high electron mobility transistor (HEMT).

The HEMT is a relative newcomer in the world of mass-produced power switching devices. First demonstrated in the late 1990s and early 2000s, HEMTs make use of a unique band diagram that results in a conceptual gas of majority electron carriers within the undoped low bandgap material. The resulting structure enables operation at extremely high frequencies due to the lack of scattering dopant atoms. A comparison between a silicon-based vertical MOSFET and a GaN HEMT is shown in the cross section below. In the silicon power MOSFET, current flows vertically from the bottom side drain to the source contacts through the bulk doped silicon, resulting in a relatively high on-state resistance. In the HEMT device, current flows laterally through the very thin electron gas layer—where high mobility and lack of scattering atoms yields very low on-state resistance and very high switching speeds (Fig. 4.4).

The switching mechanism of the HEMT device is enabled by the inclusion of a gate terminal atop a small doped GaN region to modulate the electron gas layer. This results in enhancement mode operation wherein a grounded or negatively biased gate depletes the electron gas and effectively turns off the conduction

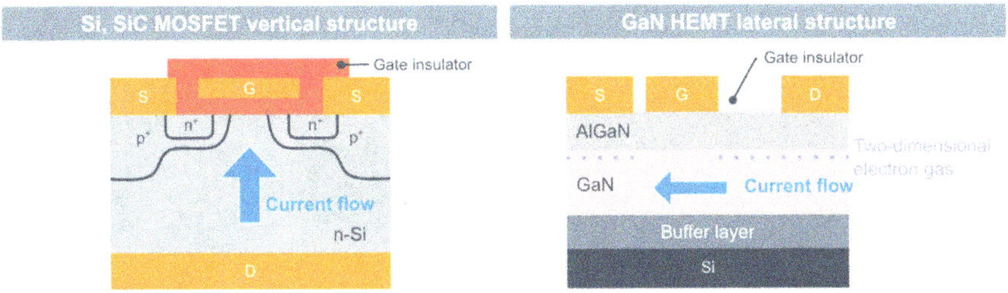

Fig. 4.4 Doped silicon transistor cross section compared to GaN HEMT

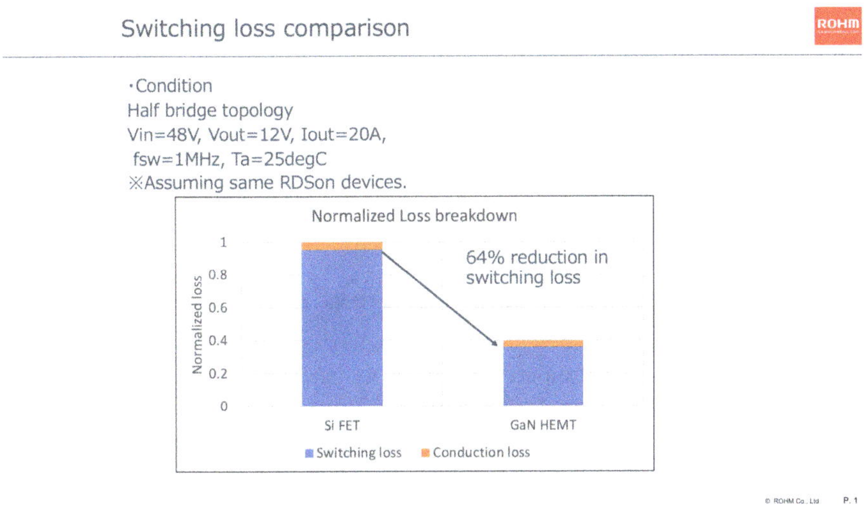

Fig. 4.5 Switching loss comparison

channel. Application of a positive gate-source potential reestablishes the electron gas layer and enables efficient conduction from drain to source.

Combining the high-mobility electron gas layer, horizontal conduction structure, and fundamental material properties of the GaN semiconductor yields a tremendous performance advantage, especially in high-power, high-frequency switching applications. A common figure of merit for switching transistors multiplies the on-resistance of the channel (Rdson) with the charge stored in the gate (Qg). This captures the energy loss required to activate the gate and the energy loss in the channel itself. Comparing switching loss between a GaN HEMT and a comparable Si MOSFET illustrates this performance advantage clearly. Shown in the figure below, the GaN HEMT exhibits a 64% reduction in switching loss when compared to a silicon MOSFET with similar RDS(ON) (Fig. 4.5).

From the efficiency perspective, the total switching loss can be compared between GaN HEMTs and Si MOSFETs as a function of switching frequency, with the GaN device outperforming the Si device switching power across the entire frequency range. Such a drastic difference directly translates into simpler cooling strategies, smaller form factor, and ultimately an overall reduction in cost (Fig. 4.6).

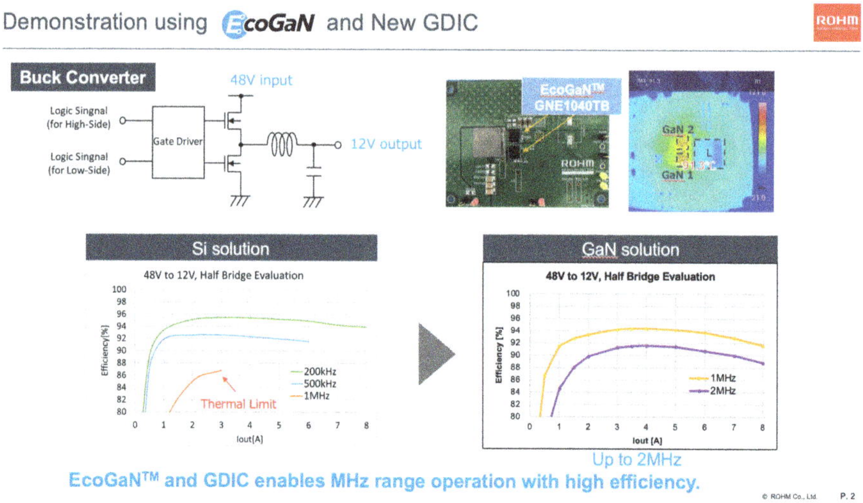

Fig. 4.6 demonstration using GaN solution

4.2.2 *Supporting Circuity for GaN Transistors*

As with any discrete electronic component, its performance cannot be considered in isolation, but rather with regard to any unusual or unique supporting devices that might be required to make it operate. In the case of HEMT transistors, the parasitic model looks very much like a traditional silicon MOSFET model, and in fact, most silicon MOSFET gate drivers can be used interchangeably with HEMTs.

GaN HEMT gate drivers can be categorized into isolated and non-isolated topologies, and then further into those with zero Vgs off voltage and those with negative Vgs off voltage. More complex versions of each of these also exist for driving half- or full-bridge circuits, and for driving GaN HEMTs in parallel.

An example is shown in the figure below using ROHM's BD2311NVX-LB ultrafast gate driver in a LiDAR system. This non-isolated driver is capable of driving a GaN HEMT with pulse widths as small as 1.24 ns in a miniscule 2.0 × 2.0 × 0.6 mm package. The separate low and high driver pins are optimized for switching performance to eliminate overshoot, while also providing undervoltage lockout protection (Fig. 4.7).

Since most GaN HEMTs can only tolerate 6 V driving voltage on their gate, overshoot during high-speed switching events can be critically important for reliability. This is demonstrated in the figure below, where the use of a dedicated GaN driver IC provides an additional 600 mV of headroom in a 5 V signal environment (Fig. 4.8).

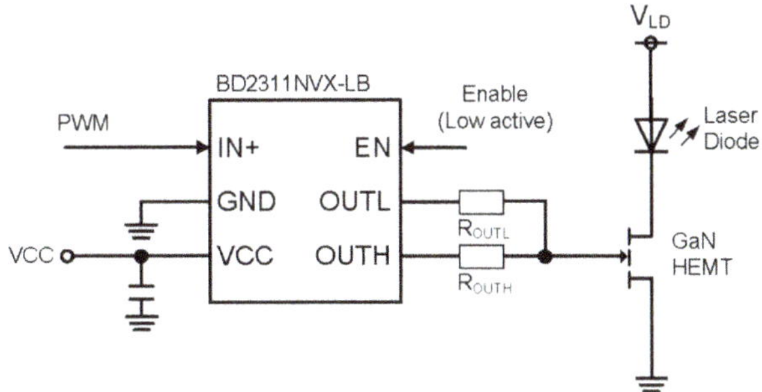

Fig. 4.7 Example gate drive circuit with negative Vgs off and isolated supplies

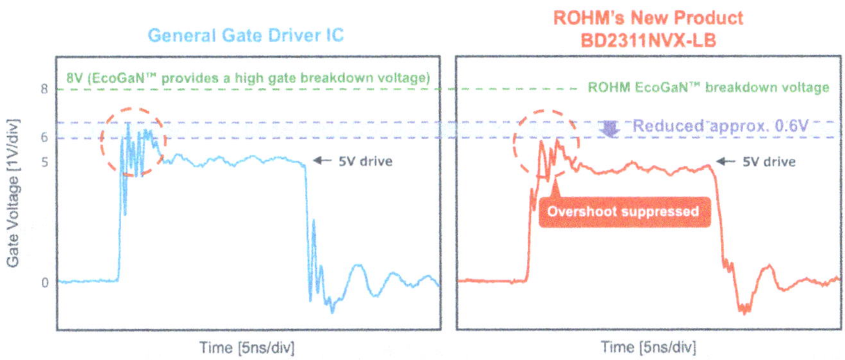

Fig. 4.8 Overshoot of traditional gate driver compared to specialized GaN driver IC

4.2.3 Applications

The sweet spot of GaN HEMT operation is characterized by very high-frequency switching and moderately high power levels. This combination leads to a very specific set of application spaces that are growing more important every day, and are gaining commensurate market share of the electronics industry. Three of the most common examples are motor power circuits, LiDAR pulsed laser drivers, and niche RF output stages.

Motor drives are typically built from a series of voltage converters. An AC source is first converted to high-voltage DC, and then converted back to a variable frequency or variable voltage set of AC waveforms to drive the motor itself. In an electric vehicle, the high-voltage DC bus could be used directly.

Another GaN HEMT application gaining broad market visibility is the laser driver stage for LiDAR systems, particularly those used in autonomous driving applications. To create pulses of light for LiDAR, these lasers must be driven by a semiconductor switch. The quality of the switch directly affects the resolution of the LiDAR image, and is typically characterized by turn-on time, peak current, and switching loss. Compared to traditional silicon devices, GaN HEMT switches exhibit extremely fast speeds and up to 65% lower power loss. This yields enhanced LiDAR performance in terms of image quality while simultaneously improving efficiency and thermal complexity.

The market for GaN devices in RF applications continues to grow. Much of this is driven by radar applications, along with the development of shared electronics and antennas, such as electronically scanned array systems. Radar is commonly implemented using a high-power, wide band traveling wave tube (TWT). GaN HEMTs are quickly replacing this unreliable, short operational life device through the application of pulse RF signals. As with the previous examples, the GaN HEMT solution yields higher reliability, improved efficiency, and better high-frequency response.

4.3 Layout Considerations

Michael de Rooij

The capability of GaN transistors to operate at higher frequencies than the aging power MOSFET has put a spotlight on the degrading effects of parasitic inductance in a power conversion circuit [8]. This inductance can hinder the extraction of the full benefit of GaN technology's fast switching capabilities.

For a half-bridge configuration, which is the primary building block of most power converters, the two main sources of parasitic inductance are (a) the high-frequency power loop formed by the two power switching devices along with the bus capacitor, and (b) the gate-drive loop formed by the gate driver, power device, and driver supply capacitor for turn-on. The common source inductance (CSI) is defined by the mutual inductance common to both the gate and power loops. These inductances are shown in Fig. 4.9 with the importance to address given by the label number.

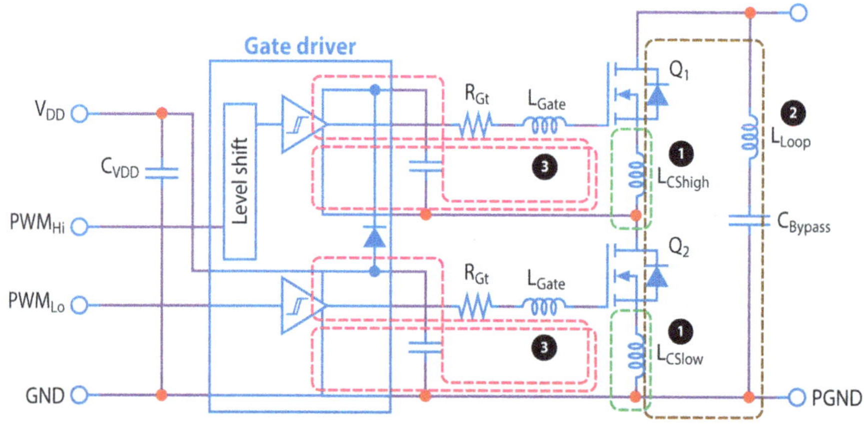

Fig. 4.9 Schematic of a half-bridge power stage showing power (2) and gate drive (3) loops with common source inductance (1)

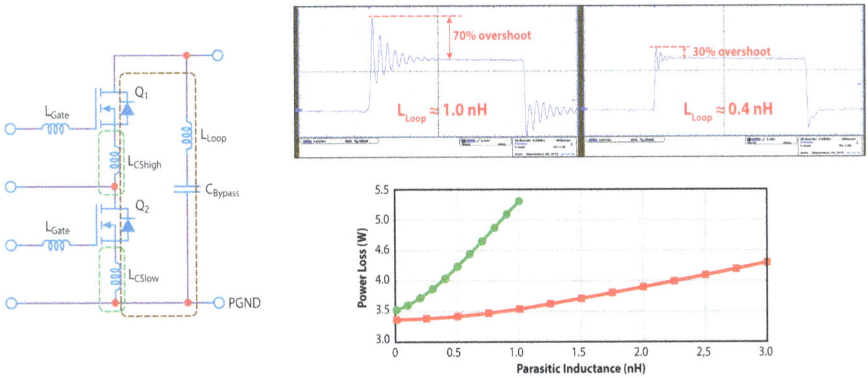

Fig. 4.10 Impact of common source and power loop inductance on the performance of a half-bridge circuit

Fig. 4.11 Effect of gate-loop inductance on the gate voltage switching waveform

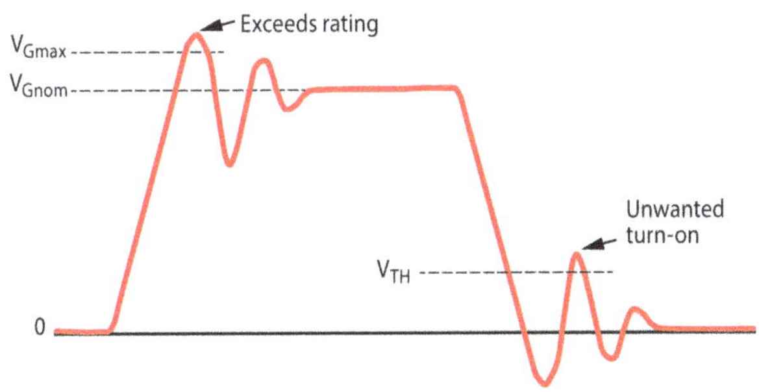

The impact of common-source and power-loop inductance is shown in Fig. 4.10 using an experimental example [9]. Higher loop inductances lead to high switch-node ringing voltages, which in turn, lead to higher losses and EMI generation. Higher common-source inductance also leads to higher power loss but at a higher inductance rate than the power-loop inductance, as shown in the graph of Fig. 4.10, illustrating why it is given the highest priority to address in a design.

The impact of gate-loop inductance is shown in Fig. 4.11. There are three aspects of the half-bridge converter that are impacted by gate-loop inductance, namely, (1) turn-on ringing, (2) turn-off ringing, and (3) the impact on performance. At turn-on, gate-loop inductance can lead to ringing magnitudes that can exceed the voltage rating of the gate and lead to early failure. At turn-off, loop inductance can lead to ringing undershoot with a rebound that can exceed the threshold voltage of the gate thus turning it back on, which leads to higher losses.

The primary method to mitigate the effects of gate-loop inductance is to add a damping resistance in the circuit. That resistance (R_{xx}) can be calculated using Eq. 4.1, where L_{Gxx} is the gate-loop inductance, L_{CSI} is the common-source inductance, and C_{GSxx} is the gate-source capacitance.

$$R_{xx} \geq \sqrt{\frac{4\left(L_{Gxx} + L_{CSI}\right)}{C_{GSxx}}} \tag{4.1}$$

Unfortunately, adding gate-loop circuit resistance will slow the FETs switching transient leading to increased power loss. By minimizing the gate-loop inductance in the design will ensure maximum performance from the GaN device by requiring the lowest gate-loop resistance necessary to damp the effect of gate-loop inductance.

4.3.1 Minimizing Parasitic Inductance

The minimization of all parasitic inductances is vital when considering the layout of high-speed power devices. It is not possible to reduce all components of inductance equally, and therefore, they must be addressed in order of importance, starting with common-source inductance, followed by power-loop inductance and, lastly, gate-loop inductance as indicated in Fig. 4.9.

Before inductance can be reduced in a layout, how it manifests in a layout must be understood. There are two main layout methods that result in different inductance in printed circuit layouts, lateral loops and vertical loops, as shown in Fig. 4.12.

Using Eq. 4.2, lateral-loop results in a significantly higher inductance than the vertical-loop approach can be shown.

$$L = \frac{\mu_0 \text{ length thickness}}{\text{Width}} \tag{4.2}$$

With this knowledge, some typical GaN FET layouts can now be reviewed to identify the approach with the lowest inductance. First, the power loop will be analyzed

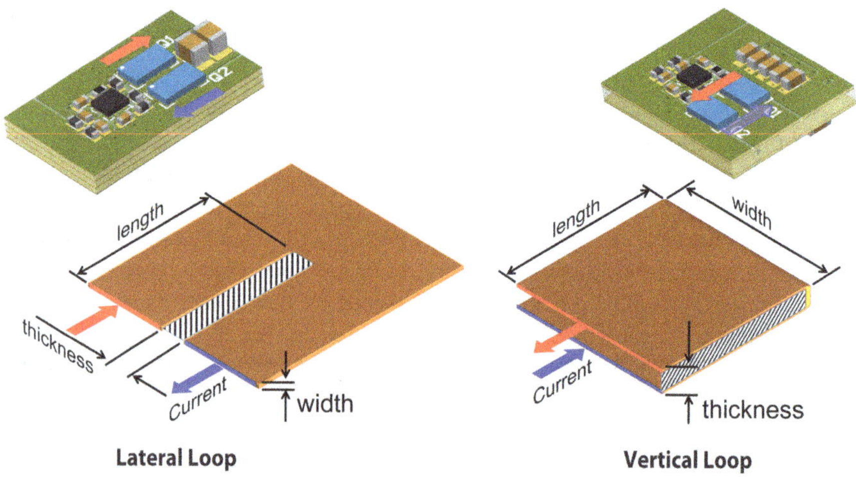

Fig. 4.12 PCB inductance manifestation: lateral loops (left), vertical loops (right)

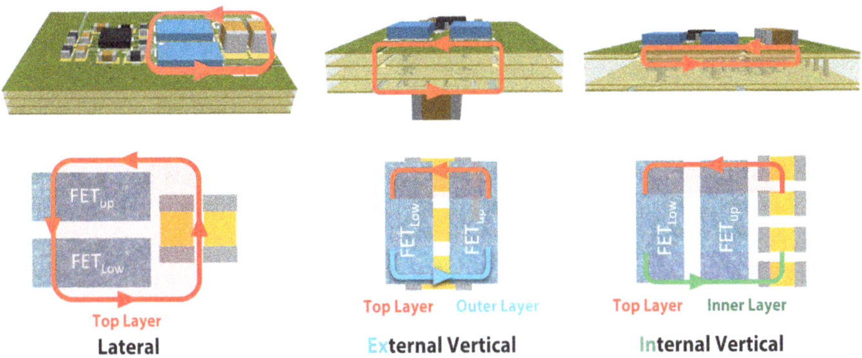

Fig. 4.13 Typical GaN FET half-bridge layouts: lateral (left), external vertical (center), internal vertical (right) [2]

because it serves as a foundation for additional layout techniques. Figure 4.13 shows various power-loop layouts that are popular with power system designers [9], which are (1) a lateral layout, (2) a vertical layout, and (3) an internal vertical layout.

4.3.2 Lateral Power-Loop Design

The lateral layout places the power bus capacitors and devices on the same side of the PCB and are placed in close proximity to each other to minimize the area of the high-frequency power loop. The high-frequency loop for this design is contained on the same side of the PCB and is considered a lateral power loop, since the power loop flows laterally on a single PCB layer. An example of the lateral layout design is shown in Fig. 4.13 (left).

The first inner layer of the lateral layout serves as a "shield layer." This layer plays a critical role in shielding the internal circuits from the fields generated by the high-frequency power loop. The power loop generates a magnetic field that induces a current in the shield layer that flows in the opposite direction to the power loop. The current in the shield layer generates a magnetic field to counteract the original power loop's magnetic field. The end result is a cancellation of magnetic fields that reduces the power-loop inductance. The efficiency of the shield layer to reduce the loop inductance is strongly dependent on the distance between the power-loop layer and shield layer [9], where a smaller distance results in lower inductance.

4.3.3 Vertical Power-Loop Design

In the vertical power-loop layout, shown in Fig. 4.13 (center), the power bus capacitors and transistors are placed on opposite sides of the PCB, with the capacitors located directly beneath the devices to minimize the physical loop size. This is called a vertical-power loop because the loop is connected vertically through the PCB using vias.

The vertical power loop uses a magnetic field self-cancellation method (with currents flowing in opposite directions) to reduce inductance, as opposed to the use of a shield plane.

The board thickness directly affects path length of the current, and hence the inductance decreases, the area of the loop shrinks significantly compared to the lateral-power loop, and the current flowing in opposing directions on the top and bottom layers begins to provide magnetic field self-cancellation. For a vertical power loop to be most effective, the board thickness should be minimized.

4.3.4 Internal Vertical Power-Loop Design

A variation of the vertical power-loop design is the internal vertical power approach. In this layout, the bus capacitors are placed adjacent to the devices and on the same side of the board, similar to the lateral layout, but the current return path uses the first inner layer, similar to the vertical layout and is shown in Fig. 4.13 (right). This positioning achieves the smallest physical loop area combined with magnetic field self-cancellation.

4.3.5 Layout Technique Analysis

Three boards were designed using each of the layout techniques and evaluated [9] with the results shown in Fig. 4.14. From the analysis, it is clear that the internal vertical layout yields the lowest power-loop inductance and can be designed with little impact due to total board thickness.

The same layout techniques apply for the gate-loop circuits. However, given the sensitive nature of the gate circuit, using the internal vertical layout technique offers

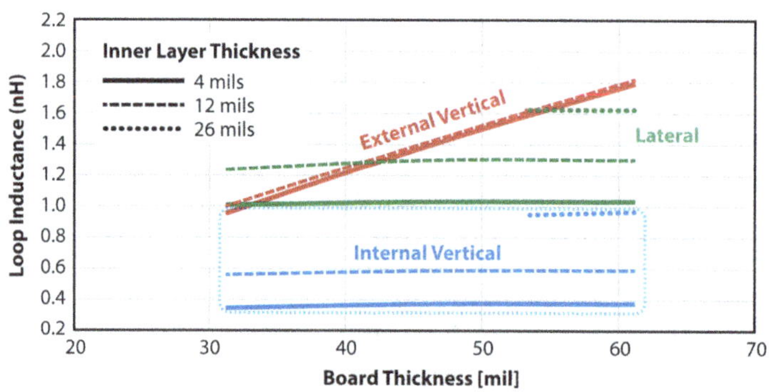

Fig. 4.14 Impact of layout technique on power-loop inductance for various board and inner layer thicknesses

e-field shielding to the gate, particularly with respect to the ground or bus terminals for the high-side driver.

4.3.6 Minimizing Common-Source Inductance

Common-source inductance results from the mutual inductance between the gate and power loops. The challenge becomes how to identify the mutual inductance in a layout. The best method is to determine where current flows in both circuits and any path that is common will result in mutual inductance. The gate and power circuits share the same plane connection to the device, making it more challenging to separate them from each other. To exacerbate the challenge to minimize CSI, most devices do not have a dedicated gate return connection that can push the point of coupling to inside the device [10]. An example of a device with a dedicated gate return connection is the EPC8000 series [11]. Ensuring the gate current flows orthogonally to the power current is one of the best methods to reduce common-source inductance as illustrated in Fig. 4.15.

The position of the connection point is also important. Overlooking the connection point is a common mistake power systems designers make when creating layouts. Figure 4.15 shows the location best suited for the gate return path connection, as opposed to the point further from the gate, which shares a longer path with the power circuit and has higher mutual inductance.

4.3.7 Putting It All Together

Up to now, the parasitic inductances have been considered independently. A practical design requires putting all the techniques together without degrading the performance. Using the internal vertical design technique, shown in Fig. 4.16, is a design

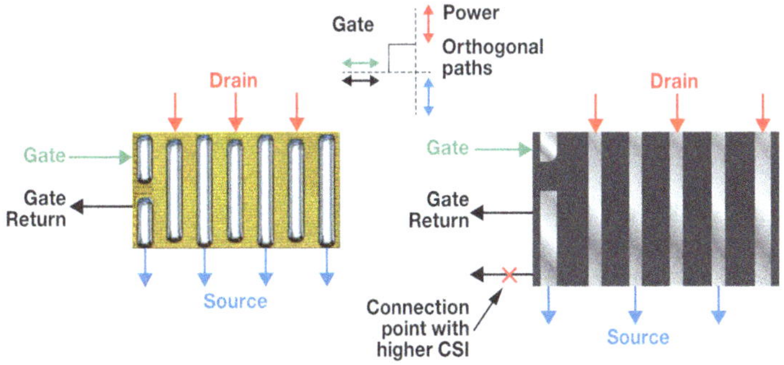

Fig. 4.15 Orthogonality principle applied to minimize common-source inductance to typical GaN FETs

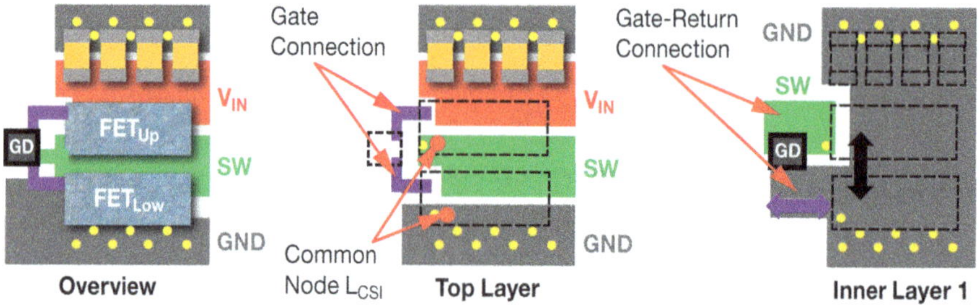

Fig. 4.16 In the overview drawing on the left, the gate driver is located as close as possible to the power devices. The signal loop is treated in a similar manner to the power loop, with the gate signal on the top layer (middle drawing) and the gate return on the inner layer (right drawing). By making the gate signals orthogonal to the power loop, the coupling inductance is minimized

with the gate signal on the top layer and the gate return on the inner layer, together with the power loops using a similar layer assignment. Next, the common coupling point is located close to the device source "closest" to the gate driver, which will result in the lowest possible common-source inductance. To further ensure low common-source inductance, the gate signal is routed orthogonally to that of the power loop as explained earlier.

4.3.8 Variations of the Internal Vertical Layout

There are alternative layout configurations using the internal vertical layout technique that maintain all the positive attributes. These can be useful when faced with layout restrictions or the need to improve the thermal performance of a power system design. Referring to Fig. 4.17, on the left is a layout where the bus capacitors are located next to the lower FET (instead of the upper FET). In this case, the return on the inner layer becomes the main supply bus V_{IN}. This can be useful when faced with layout restrictions such as a 2-phase design. On the right is shown a layout where the bus capacitors are located between the FETs and the return becomes the switch-node. In this case, the switch-node becomes buried between electrically-quiet bus and ground planes and can thus reduce e-field generated EMI [12].

4.3.9 Conclusions

In this section 4.3, various parasitic inductances that are present in a half-bridge topology that affect the performance of GaN FETs were discussed, including evidence of their impact on power system circuit layout design. Various layout techniques, popular amongst designers, were presented such as a lateral, vertical, and internal vertical layout. An analysis of these layout techniques revealed that the

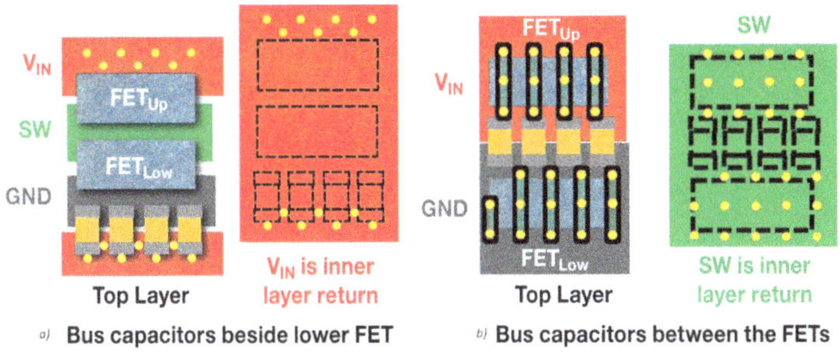

Fig. 4.17 (**a**) An internal vertical layout where the bus capacitors are located next to the lower transistor instead of the upper transistor, and (**b**) a layout where the bus capacitors are located between the transistors and the return becomes the switch-node

internal vertical layout technique offers superior performance through the lowest inductance solution. How to deal with common-source inductance was also presented and culminated in a preferred layout recommendation. Alternative configurations of the layouts were also presented giving designer additional options without loss in performance.

An efficient circuit layout will minimize PCB area, reduce wasteful power dissipation due to slower switching speeds that are limited by parasitic inductances, and improve system reliability, while reducing EMI resulting from reduced voltage overshoot. Layout parasitics that are important when using GaN transistors are the common-source inductance, the high-frequency power loop inductance, and the gate-loop inductance.

4.4 Packaging and Thermal Solutions for Server SMPS

Sam Abdel-Rahman

The rise of cloud-based internet services, artificial intelligence, and cryptocurrency has led to a significant increase in the demand for processing power in datacenters globally. This, in combination with the continued rise of electricity and real estate prices, creates a strong need for highly efficient and compact server power supplies.

In order to achieve the necessary higher power density, it is essential to take into account the following three main contributors:

- *Increased efficiency* in order to maintain an acceptable total power loss in a given volume. This drives the transition to new topologies and technologies such as CoolGaN™ totem-pole PFC.
- *Improved packages and thermal solutions* that can dissipate the power away from the device junction to heatsinks and towards the ambience. This becomes more challenging in smaller SMD packages that are the main enabler for high-density converters.

- *Optimized system design* to achieve maximum density without violating efficiency requirements or temperature-rise limits. This leads to an *increased switching frequency*, driving the transition from conventional to new packages and thermal solutions.

4.4.1 SMD Packages Used in SMPS Topologies

Infineon provides a wide range of bottom-side cooling (BSC) and top-side cooling (TSC) packages that meet the demands for higher power and higher density in server SMPS applications. This section discusses and compares the different packages concerning topics such as assembly, thermal performance, and electrical parasitics.

4.4.1.1 Assembly Implementations of Bottom-Side Cooling and Top-Side Cooling Packages

Figure 4.18 shows the principal cooling concept for bottom and top-side cooling. In both cases, the SMD package is mounted on the PCB, usually via a Reflow Soldering process.

For BSC (Fig. 4.18 left), the main heat flux is directed from the device heatsink (exposed pad) on the bottom of the package through the PCB to an external heatsink mounted on the opposite side of the PCB. Therefore, thermal vias below the package and through the PCB are needed for heat transfer when using an FR4-based PCB. On the backside of the PCB, the external heatsink is mounted on the area with the thermal vias. The heatsink and the PCB are electrically separated via a thermal interface material (TIM). Quite often, a foil with a thickness in the range of 100–500 µm is used as TIM, which has, in the best case, a good thermal admittance λ. This leads ideally to a low thermal impedance Z_{thja} for the overall system.

The PCB has a certain thickness, influenced by the number of necessary copper layers for circuit design and a density limitation for the thermal vias. This reduces

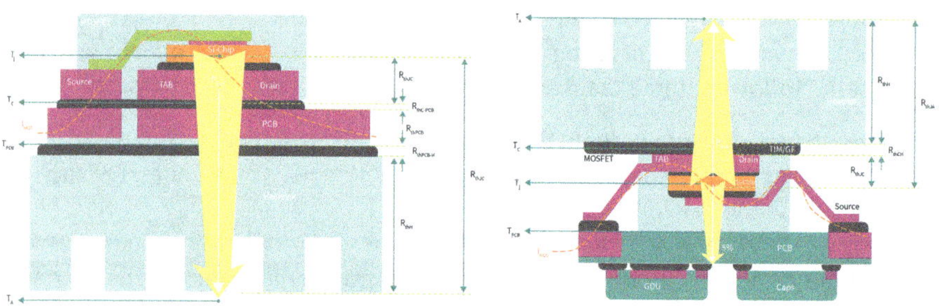

Fig. 4.18 Principal cooling concept for bottom and top-side cooling

the effective cross-sectional area for heat transfer through the PCB when compared with the area for heat transfer that the heatsink offers. This is the first bottleneck. The second bottleneck is the TIM, having a much lower thermal admittance than the device heatsink and the external heatsink.

In TSC packages, the device heatsink on top of the package is interfaced directly to the external heatsink through the TIM (Fig. 4.18 right). In this case, there is no heat passing through the PCB and thermal vias, thus eliminating their thermal impedance from the total thermal impedance. This leads to enhanced thermal conductivity and higher value of the maximum power dissipation for the overall package.

Another advantage of TSC packages is the free area on the opposite side of the PCB that can be used to place other devices such as gate drivers and passive components, as well as the space for signal routing that is available directly below the package body.

For the common heatsink approach, the TIM stack between the device and the heatsink consists of an insulation foil and a gap filler. The gap filler is used for compensating manufacturing tolerances related to the device, the heatsink, and the PCB. Since only a gap filler is used for heat transfer, a reliable electrical insulation between the device and external heatsink must be ensured. Additionally, the gap filler material has to fulfill the necessary breakdown rating. The PCB assembly should be stringent enough to prevent particles from becoming enclosed within the gap filler or blowholes. In general, a clean ambience for PCB assembly can lower the risk of system failures caused by pollution during system manufacturing.

For a further improved Z_{thja} and dynamic power dissipation for TSC, the implementation of an intermediate heat spreader is a good option, as shown in Fig. 4.10. The thermal capacity of this additional heatsink can store for a certain time (some seconds) the additional heat and transfer it over to the common heatsink and the ambience. Depending on the system design, it may also be possible to remove the common heatsink and TIM to improve the system Z_{thja}, where the heat spreader is the primary heatsink and is directly cooled by the fan airflow (Fig. 4.19).

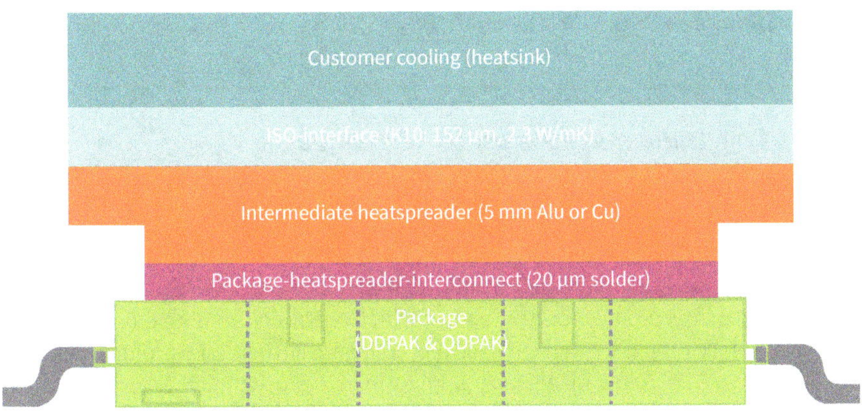

Fig. 4.19 Thermal interface

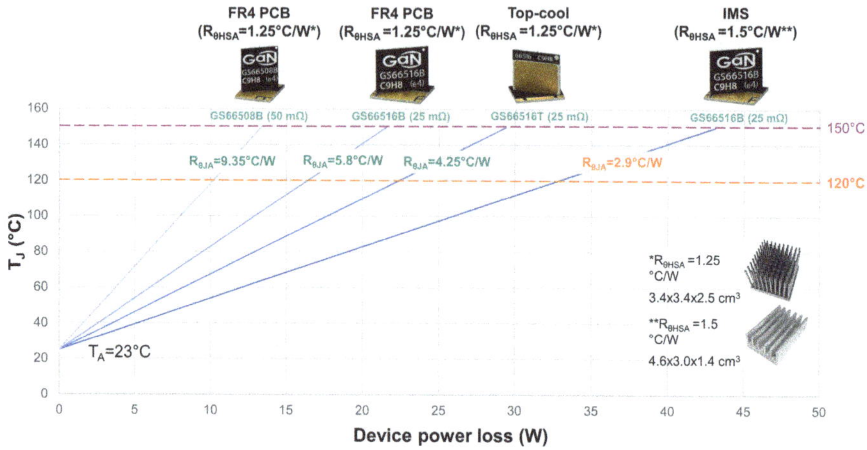

Fig. 4.20 Power dissipation capability of different CoolGaN™ devices and cooling methods

4.4.2 *Thermal Performance*

Figure 4.20 shows the power dissipation capability of different CoolGaN™ devices and cooling methods. We can see that reducing R_{DS} is an obvious step to dissipate more power, since bigger dies lead to smaller $R_{th,js}$. Moving to a top side cooling package provides a significant reduction in total R_{th} and therefore a higher power dissipation capability. Finally, the highest power dissipation can be reached with Insulated Metal Substrate (IMS), in which neither thermal vias nor additional TIM are needed, leading to minimal R_{thja}. Additionally, the external heatsink can be removed as the aluminum core of the IMS board functions as a heatsink. However, the IMS limits the board to a single layer routing and suffers from a reduced number of Temperature Cycles on Board (TCoB), especially for non-leaded SMD packages like TO-leadless (TOLL) or ThinPAK, caused by the rigid IMS-based PCB compared to the more flexible FR4-based PCB.

4.4.3 *Low Inductance Parasitic Advantage in High-Frequency Operation*

Figure 4.21 demonstrates the effect of a package source inductance (L_{Sc}) on the turn-on transient. L_{Sc} is increased from 0 to 4 nH. The rise of drain current (di/dt) causes an inductive voltage drop over L_{Sc}, which subtracts from gate drive voltage and thus reduces the gate current. So, the voltage transient takes longer, and the losses increase. The same mechanism applies in an opposite manner to the turn-off transient.

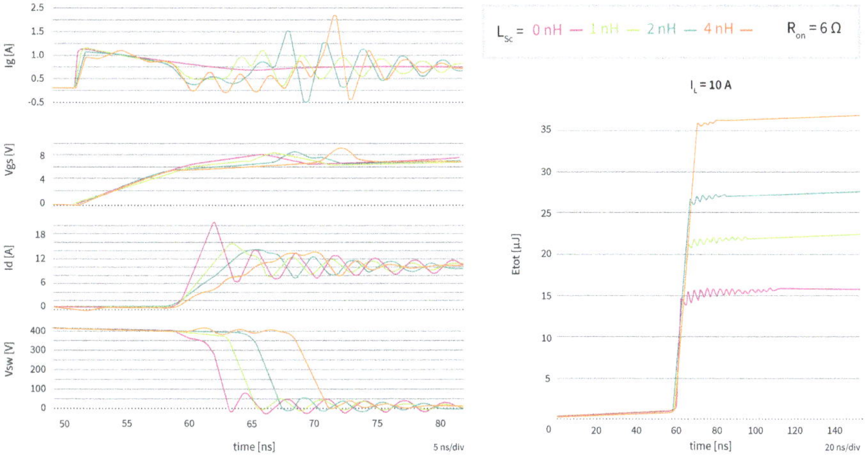

Fig. 4.21 Effect of a package source inductance (L_{Sc}) on the turn-on transient

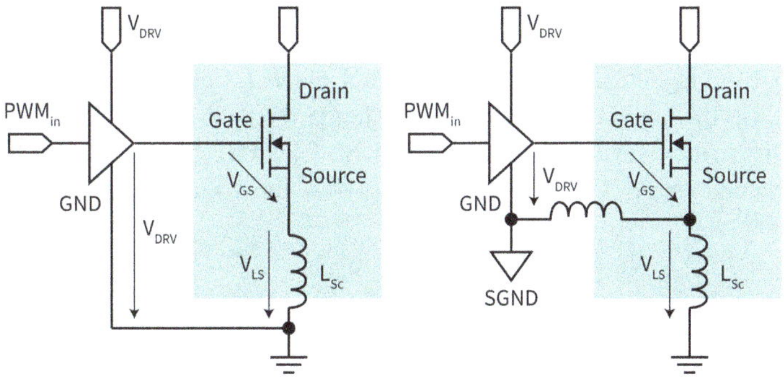

Fig. 4.22 Comparison of gate drive configurations for L_{sc} elimination

The negative effect of L_{Sc} discussed above can be eliminated by using a separate Kelvin-source connection for controlling the gate (Fig. 4.22, right), which effectively reduces switching losses. By using the Kelvin-source connection to drive the gate, the L_{Sc} comes outside the gate drive loop. Therefore, its induced voltage peaks will not feed back into the driving circuit as would happen in the standard configuration (Fig. 4.22, left) that has only a single source connection.

It is important to mention that although the Kelvin source packages nullify the negative effect of the L_{Sc} on the gate drive and switching speed, the L_{Sc} will still add to the total loop inductance. This is a crucial parameter that causes oscillations with the switch capacitance, leading to higher drain voltage overshoot and increase switching losses in fast-switching applications such as server SMPS with CoolGaN™ switches. For that reason, the package L_{Sc} should ideally be the lowest possible value even when using a Kelvin source [13–21].

4.5 Topology and Driving Features

Walter Balzarotti

For any GaN HEMT application, appropriately designing and optimizing the driving circuit is highly important to build an efficient and reliable system. The purpose of the driving circuit is to provide the proper voltage and current to the switch-gate to control the turn-on and turn-off commutations.

The controller is usually not able to generate these voltages and currents and needs another circuit called a gate-driver circuit to transfer and amplify the control signal for proper switch driving. Nowadays, the approach for this application is to integrate these gate-driving circuits in silicon to reach more compactness, ease of implementation, higher reliability, and additional features.

In high-voltage, high-power applications, the use of new alternative materials for switches like silicon carbide (SiC) and gallium nitride (GaN) is increasing, and the gate driver circuit has to adapt to the different features and requirements of these new wide-bandgap devices. For instance, the gate driving voltage for SiC switches is much higher than for silicon devices, because of which specific gate drivers are designed to drive SiC MOSFETs.

For high-voltage GaN HEMTs, dedicated GaN drivers are not yet very well established in the market. Instead, existing standard gate drivers for silicon switches are modified with external components to drive GaN HEMTs as needed.

To select a proper GaN HEMT gate driver, the following features must be considered:

- 4 V UVLO to fit the switch-gate voltage
- Medium/high current output to allow fast transitions
- Short propagation time to allow higher switching frequencies
- Negative turn-off to manage switch low threshold

The gate driver circuit should also be optimized for the chosen GaN technology. The two most common normally off GaN technologies are Schottky gate GaN HEMT and GIT GaN HEMT. They are interchangeable in many applications and need the following adaptations to their driving circuits:

(1) The driving circuit of a Schottky gate GaN HEMT has few additional requirements compared to a standard MOSFET driving circuit. Considering the low threshold voltage of <2 V, in some topologies, precautions are suggested against parasitic turn-ons. Common countermeasures are a strong clamp of the gate to the source or a negative gate-to-source voltage for turn-off. The common way to generate a negative turn-off bias is to set up a dual-isolated supply that provides positive and negative bias for turn-on and turn-off. A different, more economical yet efficient solution to generate a negative voltage is the driving circuit shown in Fig. 4.23 (top)—a charge pump comprising a gate-voltage clamp and a capacitor in series with the gate resistor. During the on-time of the GaN HEMT, the gate driver applies a positive voltage of around 10 V, while a voltage

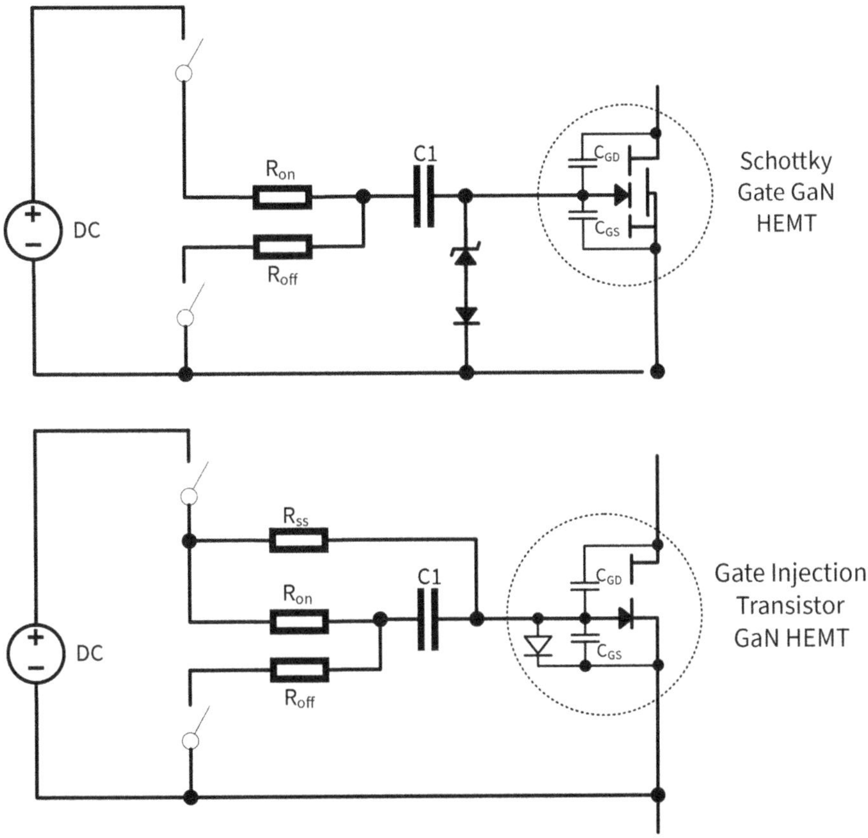

Fig. 4.23 Driving circuits for Schottky gate GaN HEMT (top) and GIT GaN HEMT (bottom)

of 6 V—7 V is clamped directly at the switch gate. This clamp not only protects the GaN gate from overvoltage, but also keeps the capacitor C1 in 4.23 charged to the difference between the driving and the clamping voltage. When the GaN switch is turned off, the gate driver connects the gate resistor to the ground and the series capacitor that is still charged pushes the gate voltage to a negative value. The resulting gate voltage is defined by the charge redistribution between C1 and CGS. This method enables cost-efficient solutions as it is also compatible with bootstrapping, a rather common and inexpensive way of generating a high-side driver supply voltage.

(2) The GIT GaN HEMT driving circuit differs from the Schottky gate circuit in which the gate is not isolated from the transistor's bulk. The gate input characteristic shows a diode-like behavior with a forward voltage drop of approximately 3.5 V. The best way to drive it reliably is by means of a driving source with a high impedance. The resulting stable gate current guarantees constant switch parameters and maintains the gate at a stable and safe voltage level. This

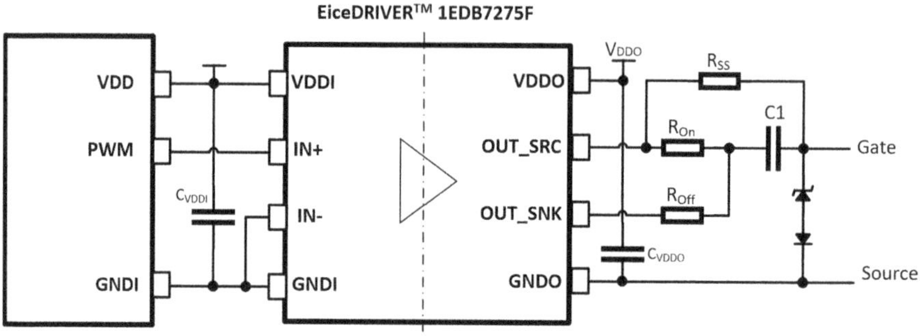

Fig. 4.24 Single-channel, isolated gate driver driving GaN HEMT with universal configuration compatible with SG GaN and GIT GaN, part number 1EDB7275F

on-state constant current is provided by the gate driver through the R_{ss} resistor shown in Fig. 4.23 (bottom). However, during switching transients, a higher gate drive current (up to 1 A) must be available. This high current is passed through the low ohmic R_{on} and R_{off} resistors that are decoupled by the capacitor C1 when the transients are completed.

An isolated gate driver is usually needed to allow independent fluctuation of the gate driver reference compared to the controller voltage, for example, due to the use of a Kelvin source connection. Standard non-isolated gate drivers are usually not suited, except truly differential input (TDI) gate drivers, like, for instance, Infineon's EiceDRIVER™ 1EDN7550B, that allow decoupling of the gate driver and the controller reference. TDI gate drivers provide an even simpler and cost-efficient solution for applications up to 200 V common-mode voltages than isolated gate drivers.

In general, it is possible to find a universal component configuration that allows the control of both Schottky gate and GIT GaN HEMTs despite the slightly different driving requirements. The schematic in Fig. 4.24 depicts the application of a standard, single-channel, isolated gate driver to drive any low-side or high-side GaN HEMT.

Both components needed for driving either of the two GaN switch types are included: R_{ss} for driving GIT and the gate clamp of the Schottky gate driving circuit. To optimize the driving functionality, component values may differ, but the driving circuit layout remains the same.

To test GaN HEMT half-bridge applications, several boards are available with both driver channels isolated or with only high-side isolated and low-side controlled by a TDI driver. For instance, the evaluation board EVAL_2EDB_HB_GaN shown in Fig. 4.25a, b is available for purchase at major catalogue distributors, including guidelines for the selection of component values for both GIT or SG GaN HEMT. Download the application note of the evaluation board for more details [22].

A key driving parameter, particularly for a Schottky gate GaN switch, is the driving voltage. While 5 V is a typical gate-to-source voltage, the maximum rating is often relatively close to the gate-to-source voltage (e.g., 5.6 V). Exceeding the rated voltage can cause reliability issues such as lifetime reduction and device damage.

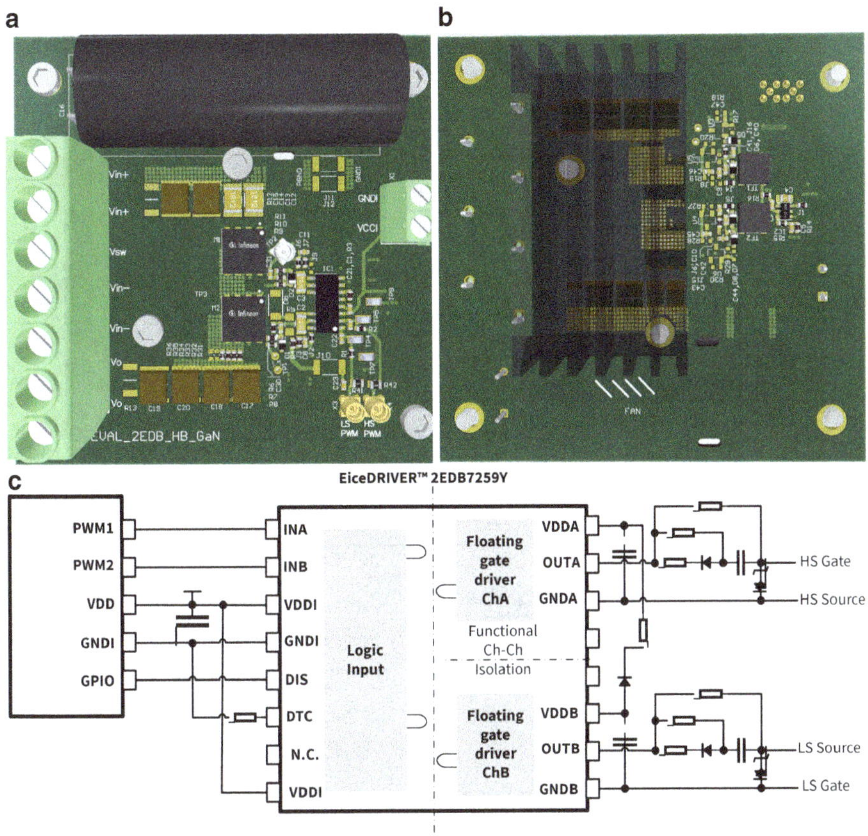

Fig. 4.25 Image of the top of the board (**a**), bottom of the board (**b**), and an application example (**c**) of a half-bridge evaluation board—EVAL_2EDB_HB_GaN

This may pose a problem in half-bridge configurations with bootstrapped high-side supply (Fig. 4.26).

The bootstrap capacitor is normally charged during the low-side switch conduction time. Figure 4.27 shows this charging path, which includes the power supply (VDD), the bootstrap diode, the current-limiting resistor, the bootstrap capacitor, and the low-side switch. When the low-side switch is turned fully on, the voltage applied to the bootstrap capacitor is slightly lower than VDD, due to the drop across the bootstrap diode, which can be partially compensated by the drop across the low-side transistor. However, during dead-time intervals, the low-side GaN HEMT operates in the "body diode" mode with a voltage drop in the range of ~1.5 V–2.5 V, or even higher when a negative VGS is applied in the off-state. This causes overcharging of the bootstrap capacitor during the dead time, resulting in a bootstrap voltage that varies significantly with dead-time duration and operating current, carrying a high risk of exceeding the maximum rating of VGS.

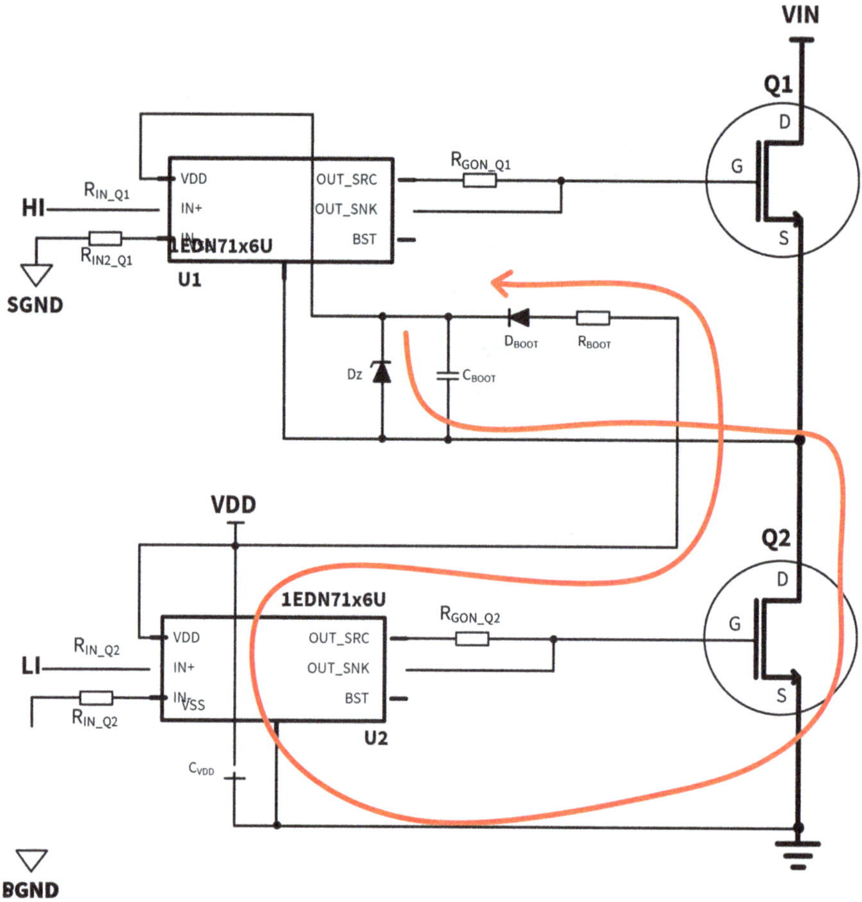

Fig. 4.26 Typical half-bridge configuration using MV GaN HEMT and 1EDN71X6U driver

There are solutions to avoid this overcharging. The EiceDRIVER™ 1EDN71X6U, for example, offers a bootstrap clamping function. Figure 4.27 shows how this solution can be implemented in a half-bridge circuit. The clamp circuit avoids overcharging of the bootstrap capacitance. Rather than connecting the bootstrap diode to the VDD supply rail, it is connected to the BST output of the low-side driver. The BST output operates as a clone of the source and sink output pins, and it is synchronized to the timing of the low-side transistor turning on and off. Therefore, the bootstrap diode can only conduct current when the low-side transistor is turned fully on, but not when it is operating in "body diode" mode during dead time. The active bootstrap clamping scheme is very useful in regulating the high-side driving voltage without the need for additional components. However, in applications where overcharging of the bootstrap capacitor is desired, the bootstrap diode can be connected to the VDD pin instead of the BST pin. A Zener-based regulation scheme is recommended in this case, as shown in Fig. 4.26.

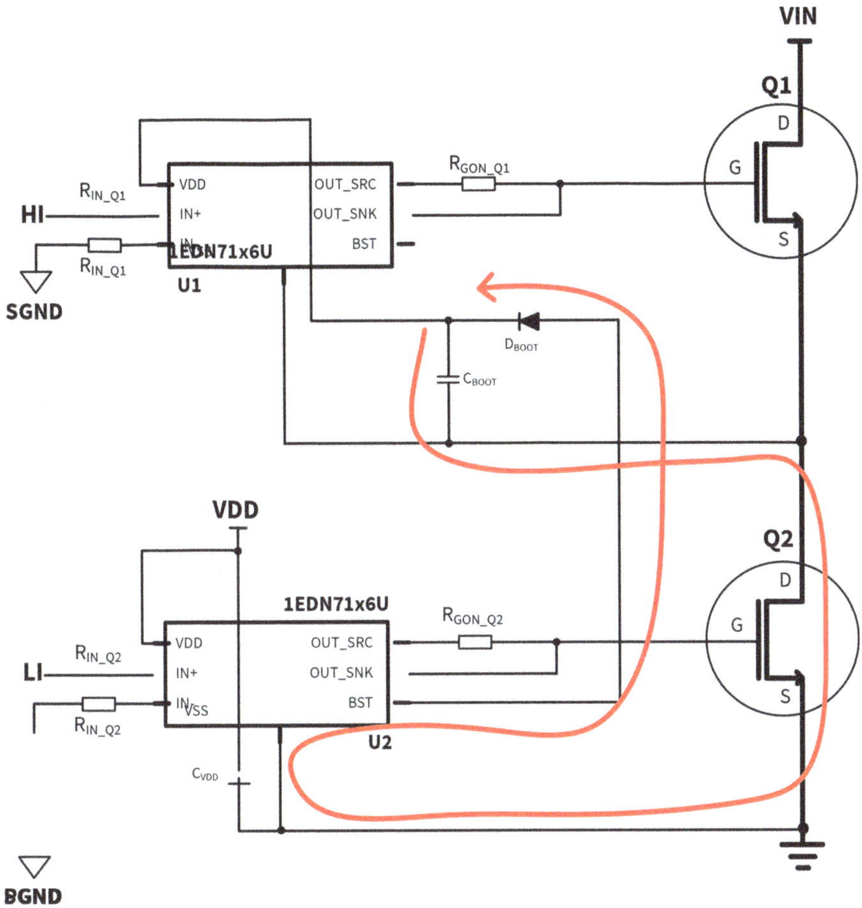

Fig. 4.27 Typical half-bridge configuration using 1EDN71X6U's BST feature

4.6 Quality, Robustness and Reliability—Breaking Down the Bathtub

Doug Bailey

Gallium nitride (GaN) offers significant benefits of increased efficiency and power density, and allows designers to meet far-more challenging power supply specifications than they can with silicon MOSFETs. One reasonable concern about any new technology that offers game-changing benefits is robustness and reliability. To address any doubts that potential users may have, let us discuss the robustness, reliability, and quality of GaN.

GaN, as a power conversion technology, does not have as long a pedigree as silicon. Even so, the bathtub curve as a representation of the failure rate of a device over time is certainly very well established and multiple companies have contributed to

Fig. 4.28 GaN quality and reliability

fleshing out the details as they pertain to the new GaN devices. It is a useful visualization tool for reliability engineering and deterioration modeling, and serves as a clear mechanism to discuss why failures occur and how to predict and prevent them.

"Bathtub" of course, refers to the shape of the curve. The x axis in Fig. 4.28 is the log of time, and the y axis is the probability of failure. The green line—our bathtub curve—is the rate of failure of a typical product over time. In the early phase of a product's life, failures are mainly quality-related due to manufacturing defects. In the center zone, there are devices that were well manufactured, and that have not yet worn out, but they failed anyway, typically because these parts have been overstressed. Then, after some time, parts wear out, so the failure mode is due to lifetime issues.

4.6.1 Lifetime

If we consider failures in the far distant future, the challenge is that users cannot wait for, say, 100,000 h (11 years), or a century, one million hours, for a device to fail before the wear-out mechanisms can be investigated. To accelerate time for the purposes of IC testing, a test called High Temperature Operational Life (HTOL) is employed. This leverages the tendency of physical systems to change faster with temperature, under strong electric fields and in the presence of free ions (humidity and metal migration), to accelerate the effects of time and use on the devices under test. Surprisingly, a relatively few parts (48 for commercial and 77 for automotive) when tested with acceleration over 1000 h provide the information that we need to determine the probability of failure in many years under normal conditions. The test is run on multiple batches of devices to expose any process dependency. Another test, called HALT (highly accelerated life test), applies a similar test under increased humidity but at a slightly lower temperature.

These tests are standard procedure for analog integrated circuits. But because GaN is a new technology, users and manufacturers must make doubly sure that the GaN switch inside the IC is robust and will have a long life, and there are tests specifically designed for discrete transistors called HTRB (high-temperature reverse bias) and HTGB (high-temperature gate bias) which address this. HTRB and HTGB test the device for 1000 h at elevated temperatures under a DC voltage stress to make sure there are no migration issues, or long-term degradation mechanisms associated with high-voltage operation at high temperature. Hot Carrier Tests (HCI) are also performed to find any places where electrons in the channel can get diverted into oxide layers and become trapped. Electro-migration testing makes sure that the metal conductors on the top of the die are appropriately sized for the current that they carry. Gate oxide integrity tests prove the strength of the gate. All these tests are performed on devices intended for industrial and commercial customers.

Of course, GaN is also very suitable for the automotive market, which is even more demanding. So, in addition to the usual tests just described, for automotive devices we run more tests, primarily H3TRB, which increases the humidity and temperature demands above the standard HTRB. Then, we run a series of tests that are designed to switch the part on and off multiple times to prove that the devices are immune to the thermal shock of being switched on. These tests are called Power Temperature Cycling and Intermittent Op Life—the latter was originally a military test. As well as stressing the devices to a higher level, automotive tests are carried out on a larger sample size than for commercial and industrial parts.

4.6.2 Manufacturing Quality

So much for the long term; we also need to address manufacturing quality problems that can cause parts to fail in early life. Power Integrations' quality policy is to prevent any of these parts from entering the product stream. You might think that a robust final test of finished products would catch all of the mis-manufactured devices, and that is certainly an important part of it. However, in order to create an essentially perfect product stream, we work hard to screen out substandard material in prior manufacturing steps—using so-called "process monitors." These monitor tests reduce the number of parts with latent defects that make it into final test and are critical for improving final product quality.

Power Integrations regards epitaxially grown GaN (EPI) as the critical stage in GaN transistor design and production. Because of this, we make our own EPI—we do not buy it, nor finished wafers, from a third party. Therefore, we have complete control over the process, and we can check it continuously. This means that only the best wafers make it through to patterning. After a wafer has been patterned into transistors, we run stress tests to identify parts that are faulty or have a risk of a quality problem.

To validate that our process monitors, our yield improvements and our final test regime yield a perfect product stream, a test called ELFR (Early Life Failure Rate) is used. ELFR is analogous to the High Temperature Op Life (HTOL) test. But instead of doing a small number of parts for a long time, we do a large number of parts for a relatively short time—800 parts for 48 h, across multiple IC lots to ensure that any process dependencies are exposed.

4.6.3 Overstress

We have spoken about reliability and wear-out mechanisms and how to guarantee initial quality. But what happens in the middle section of our bathtub curve? These are parts that have been manufactured perfectly. They have not worn out yet, yet they are dead anyway due to overstress.

To ensure that our devices are robust under the temperature, voltage, and humidity conditions that they will be exposed to during their operational life, we run more tests. These are:

l MSL—moisture sensitivity level
l UHAST—also a moisture test, but under pressure
l TMCL—a temperature cycling test which tests for differential heating and cooling
l HTSL—high-temperature storage life which is the absolute high temperature test

One of the most important attributes of a power semiconductor is its ability to withstand voltage. Power Integrations specifies three types of GaN, with BVs (breakdown voltages) of 750 V, 900 V, and 1250 V. However, even though we talk about BV for convenience, GaN transistors are not specified in the same way as a silicon device. There is a property of GaN called dynamic $R_{DS(ON)}$ which actually increases when high voltage is applied. It recovers later, but under the initial voltage, and for a short while afterwards, it is higher than the typical value.

Power Integrations has chosen 5% as being the limit that we can accept as an increase in $R_{DS(ON)}$. This determines the datasheet "BV" of our GaN devices. We choose the conservative values noted above for the datasheet specification, but actual breakdown does not occur until around 1400 V for the lower voltage families and 2200 V for the 1250 V product line—in all cases, a massive margin with respect to the datasheet BV limits. It is unwise to design systems to use the devices above the datasheet BV, but this headroom is a very useful insurance against infrequent line swells and surges, such as lightning strikes.

With this battery of tests (Fig. 4.29)—process monitors for initial quality, voltage margin to allow for overstressing, wear-out mechanisms that ensure reliability, plus tests related to temperature and moisture sensitivity—whether you are making a smart phone adapter or an automobile, you can have confidence in PowiGaN™, Power Integrations' GaN technology.

Fig. 4.29 GaN devices undergo a battery of tests

4.7 Ensuring Industrial Reliability of GaN Transistors

Oliver Häberlen and Tim McDonald

The future of power semiconductor technology lies in wide-bandgap-based devices such as GaN (Gallium Nitride) as they offer a multitude of advantages over traditional silicon-based ones. These benefits include higher power densities, improved efficiency, and faster switching speeds, making them perfect for industrial applications like motor drives, power supplies, etc.

However, as with any innovative technology, it is critical to ensure that thorough technology development and product qualification procedures are followed to ensure a reliable operation that meets design lifetime and quality requirements in power conversion systems.

Today's state-of-the-art GaN devices are based on the GaN-on-Si technology, where GaN layers are grown hetero-epitaxially on standard silicon wafers for cost and scalability reasons. The concept is of a lateral device based on an AlGaN/GaN heterojunction generating a two-dimensional electron gas (2DEG) of high mobility both in the channel and the drift region of the device close to the surface. The device needs to withstand the required blocking voltage in two directions:

(1) Laterally between the gate and drain terminals, leading to high electric fields at the device surface and the passivation layers,
(2) Vertically across the GaN layers from the drain terminal to the grounded silicon substrate, requiring that the GaN buffer layers also withstand high voltages despite not carrying current in on-state.

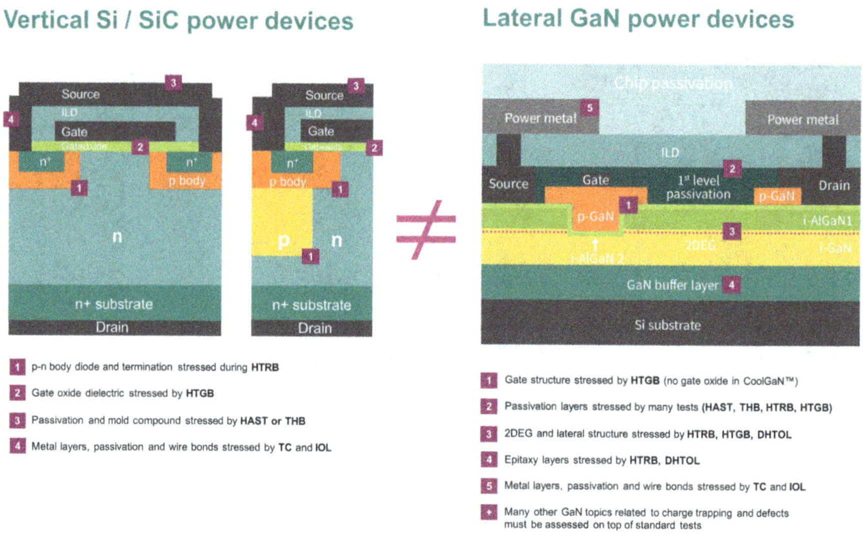

Fig. 4.30 Comparison of vertical Si and SiC with lateral GaN power devices

Also, in contrast to standard Si and SiC MOSFETs, the gate of the normally-off GaN transistor is based on an epitaxially grown p-doped GaN layer without a gate dielectric.

Given these differences, it is clear that new failure modes must be expected in GaN HEMTs that need to be addressed. Figure 4.30 shows how the standard JEDEC qualification tests are different for Si/SiC and lateral GaN technologies, and how additional stress procedures in the case of GaN are called for.

4.7.1 A Proposed Four-Pillar Method for Qualifying GaN Devices

Infineon follows a comprehensive four-pillar methodology based on the application and quality requirement profiles as inputs. By extensive reliability investigations during the development, degradation models for all relevant failure modes have been derived. These allow to set up the correct qualification plan and ensure safe and reliable use of the released CoolGaN™ products, thereby helping our customers to avoid many of the risks they would otherwise encounter (Fig. 4.31).

Assuring reliable operation of GaN HEMTs to planned quality levels and lifespans requires the knowledge of all stresses from which a device can suffer during operation in a specific topology. A comprehensive application profile encompasses pertinent system-level prerequisites such as output power versus time, duration of operation at varying voltages, temperatures, and humidity levels, and whether hard- or soft-switching is involved.

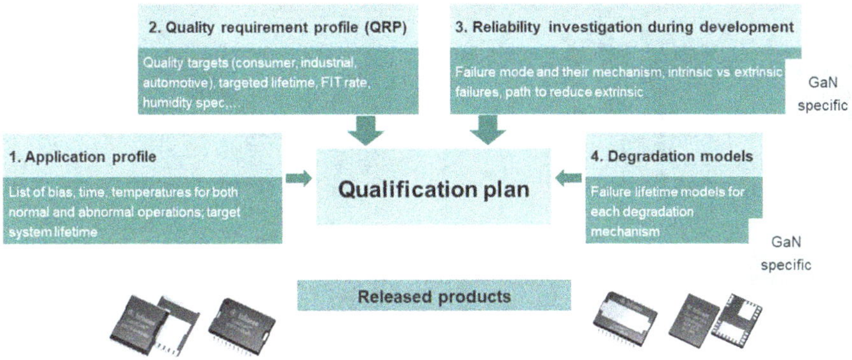

Fig. 4.31 Infineon's four-pillar approach to qualify GaN devices

To complement the application profile, additional quality requirements, such as the customer target lifetime and the maximum allowed cumulative failure rate, must be known. The customer target lifetime can vary widely depending on the application, and the maximum allowed cumulative failure rate is typically different for consumer, industrial, and automotive applications. Moreover, other environmental conditions that could impact performance, including humidity levels and moisture sensitivity level (MSL) ratings, must also be considered.

Infineon has conducted a comprehensive reliability investigation of GaN-specific failure modes while creating CoolGaN™ devices to determine which parameters have the greatest impact on device lifetimes. By creating degradation models for all relevant mechanisms, device lifespans and potential failure rates can be accurately predicted. These predictive models are also essential in determining proper electrical screening conditions necessary to reduce early "infant mortality" failures down to the required ppm levels.

One prominent example to illustrate this procedure is the lifetime modeling for high-voltage stress on the drain terminal of the device in the blocking mode. A typical product-level reliability test for this mode, according to JEDEC and other qualification standards, is the high-temperature reverse bias (HTRB) stress test. It is conducted on a small number of samples (typically 3 × 77 devices) and within datasheet conditions (typically at 80% $V_{DS,max}$ rating), and no part is allowed to fail within the test time (typically 1000 h). However, this is not enough to predict lifetime or failure rates in real-world usage conditions, since due to the strong polar nature of the material, GaN transistors show a time-dependent dielectric-breakdown [23]. The needed time to failure is accelerated with voltage and temperature, and to extract the voltage and temperature acceleration, highly accelerated stress tests with test-to-fail must be conducted in order to generate Weibull plots with failure probabilities over time (see Fig. 4.32).

Steep lines in the Weibull plot indicate good process control as all parts have reproducible times to fail. Out of the different voltage and temperature conditions, it is now possible to create a lifetime model that predicts failure rates over time for

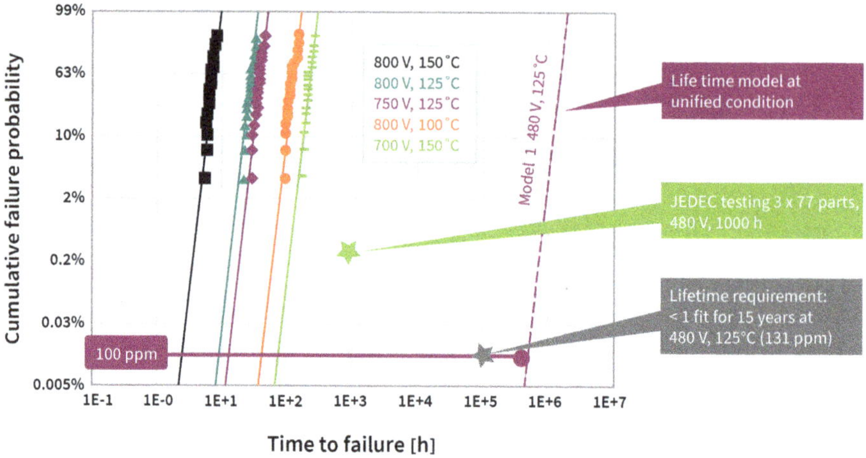

Fig. 4.32 Example of lifetime modeling by making use of highly accelerated stress-to-fail investigations for high-voltage drain biasing [23]

use conditions, where test times would no more be feasible. As shown in Fig. 4.32, data extracted based on typical product-level JEDEC HTRB testing conditions is not enough to provide such a prediction, demonstrating the importance of assessing technology fitness-for-use by lifetime modeling.

4.7.2 Ensuring Reliable Operation of GaN Devices in Applications

All tests mentioned so far have been static. However, for a new device concept (lateral HEMT) combined with a new semiconductor material (GaN-on-Si), it must be ensured that new failure modes are caught, especially when dealing with very fast switching devices with transient dynamic effects. This mandates the design of new stress tests for assessing switching robustness and reliability.

Infineon has accordingly established a setup to run multiple half-bridges in parallel under defined switching modes (hard- and soft-switching) close to application conditions with in situ monitoring of temperature, currents, and voltages of each device under test [24, 25] (see Fig. 4.33). The use of a half-bridge configuration is important to cover failure modes both for low- and high-side switches with their different operating conditions such as the third-quadrant mode conduction, for example. The tests are run in a climate chamber for control of ambient temperature.

The setup described above has been used to run highly accelerated continuous hard-switching tests to extract time-to-fail distributions similar to the previous example for static drain stress (Fig. 4.32), but now for switching stress (Fig. 4.34). As can be seen from Fig. 4.34a, we again obtain steep and controlled distributions that allow extraction of acceleration factors and a hard-switching lifetime model to

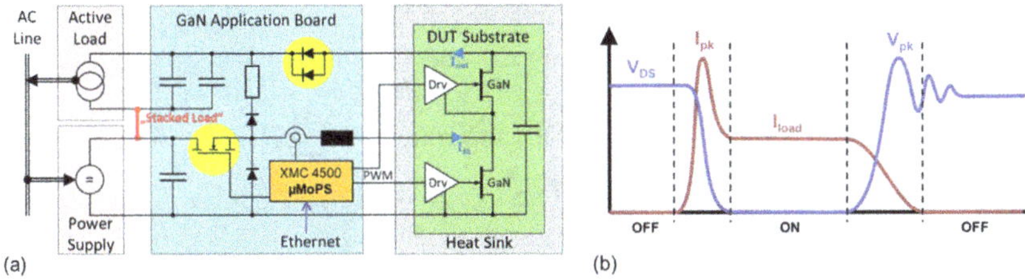

Fig. 4.33 Schematic of (**a**) Infineon's application-related switching stress test system and (**b**) illustrative hard switching turn-on and turn-off transients with current and voltage overshoots seen by the device

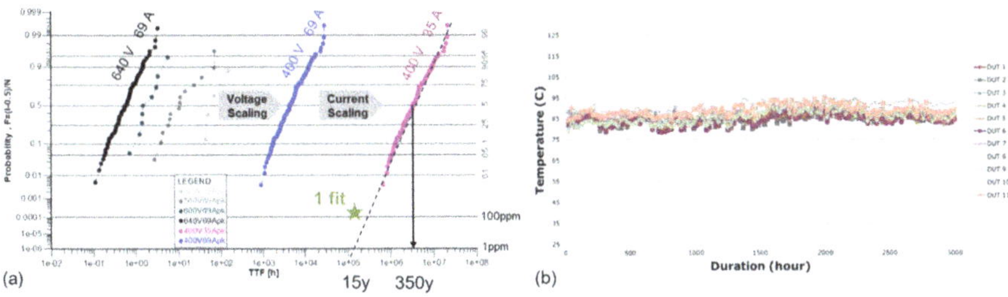

Fig. 4.34 Continuous hard-switching stress test of a CoolGaN™ GIT HEMT in a 100 kHz boost converter with (**a**) test-to-fail conditions under highly accelerated conditions and (**b**) for 3000 h at 500 V DC link voltage (equivalent to >10 years at 400 V DC link)

predict failure rates under typical-use conditions. In the example for a 70 mΩ, 600 V CoolGaN™ GIT running at 400 V DC link and 35 A peak overshoot current, the industrial requirement of 1 FIT (1 failure in 1 billion hours of device operation) has been met. To confirm the extrapolation towards normal-use conditions, runs have been conducted at moderately accelerated conditions (500 V) without any fails or signs of efficiency degradation for up to 3000 h (Fig. 4.34b), which is equivalent to a lifetime of more than 10 years under typical-use conditions of 400 V.

4.7.3 Industry Standards for GaN Device Reliability and Qualification

As detailed in this chapter, Infineon follows a comprehensive qualification regime for CoolGaN™ devices to ensure reliable operation in customers' applications [26]. Nevertheless, the importance for the industry to develop comprehensive standards for device qualification is unquestionable. Industry standards are set to ensure that any quality-related concerns can be addressed simply and quickly, without reference to the details of failure mechanisms for each supplier. Accordingly, Infineon has initiated and contributed to establishing a GaN-focused subcommittee (JC-70.1) within JEDEC [25] to introduce qualification guidelines and standards regarding GaN devices. In this JEDEC effort, Infineon supports the views and approaches described in this publication.

To date, JEDEC has published the following guidelines [27]:

- JEP173—Dynamic On-Resistance Test Method Guidelines for GaN HEMT Base Power Conversion Devices
- JEP180—Guideline for Switching Reliability Evaluation Procedures for Gallium Nitride Power Conversion Devices
- JEP182—Test Method for Continuous Switching Evaluation of Gallium Nitride Power Conversion Devices
- JEP186—Guideline to Specify a Transient Off-State Withstand Voltage Robustness Indicator in Datasheets for Lateral GaN Power Conversion Devices
- JEP198—Guideline for Reverse Bias Reliability Evaluation Procedures for Gallium Nitride Power Conversion Devices

Infineon complies with all currently published JEDEC GaN guidelines and cooperates with and leads efforts of JEDEC committee JC-70 to produce additional standards for wide bandgap devices.

4.8 Vertical GaN

Mark Davidson and Troy Baker

The application of gallium nitride (GaN) technology as vertical conduction transistors represents the potential for significant advancement in high-voltage power conversion with blocking voltages over 1 kV. Vertical GaN power transistors take the full advantage of the material properties of GaN, offering superior efficiency and power density compared to transistors fabricated on traditional materials like silicon and even silicon carbide. Vertical GaN allows for the practical scaling of blocking voltages above 700—900 V since blocking voltage scales with epitaxial properties and not die area. This makes vertical GaN particularly valuable for applications in power conversion and motor drives in applications such as electric vehicle traction inverters and industrial motors. However, despite its benefits, vertical GaN faces challenges such as complex manufacturing processes, high-cost and lack of large-diameter GaN substrates and the need for continued research and development to solve technical hurdles usually encountered during the commercialization of brand new technology. At the time of this writing, there are no commercially available vertical GaN transistors in the market; however, development continues in multiple organizations. This paragraph explores the fundamentals of vertical GaN transistors, their value, the status of GaN substrates, key applications, and the challenges associated with their production.

4.8.1 Fundamentals of Vertical GaN Transistors Vs. High Electron Mobility Transistors

At its core, a vertical GaN transistor is a type of semiconductor device that uses gallium nitride as the principal material. In previous chapters, there have been explanations of why GaN is a superior material for high-voltage and high-performance power transistors. To succinctly re-state, GaN provides high electron mobility and can withstand high electron fields, which allow transistors to work with high-voltage capability and enable extremely high power densities with efficient, ultra-small solutions. GaN allows transistors to be built with low capacitance, meaning switching frequencies can be orders of magnitude higher than silicon and silicon carbide. While this high-frequency operation enables value, it also provides challenges in dealing with high *dv*/*dt*'s in the circuits which create electromagnetic interference (EMI) and large transient voltages. While efforts are underway to build vertical GaN on silicon, to date, the majority of developments have been focused on GaN epitaxial on a GaN substrate. Unlike high-electron mobility transistors (HEMT) which are built on silicon substrates and conduct laterally across the surface, vertical GaN transistors are designed so current flows vertically through the device, including through the substrate.

High Electron Mobility Transistors (HEMTs) utilize surface contacts on the wafer with current flowing through a channel, avoiding the need for impurity doping in GaN. This structure allows for "normally-on" operation, where current flows until a bias is applied to the gate to stop it. However, for safety and compatibility, these devices are often paired with silicon transistors or other design techniques to achieve "normally-off" functionality, suitable for power switching applications.

To build vertical conduction devices, these devices rely on what is known as impurity doping to achieve the area-selective p-type regions employed in the device. This enables devices which are inherently normally-off, meaning no current flows unless a bias is applied to the gate terminal. A vertical-conduction device is favorable as it makes area-efficient use of the substrate, allows for a normally off device, and also provides isolation of high voltages that could be applied between terminals in a die-efficient manner. Unlike HEMTs where increased blocking voltage is achieved through larger die areas which increase cost and degrade performance due to the probability of intersecting with more defects, vertical devices allow for smaller die sizes based solely on Rds-on, thus allowing for an electrically and cost optimized device.

Horizontal-conduction GaN HEMTs are grown on silicon wafers, providing cost optimization thanks to the availability of low-cost, large-diameter silicon wafers. Since the current flows in and out of contacts on the surface, the choice of substrate is not constrained as it is not in the path of current flow. This is an advantage in that silicon is cheap and readily available at high-volumes. The disadvantage is that when growing any semiconductor on top of a dissimilar semiconductor, vast amounts of crystal defects are introduced in the crystal layer grown. For GaN grown on Si, the defect density is ~10^8 to 10^{10} cm-2. While great progress has been made

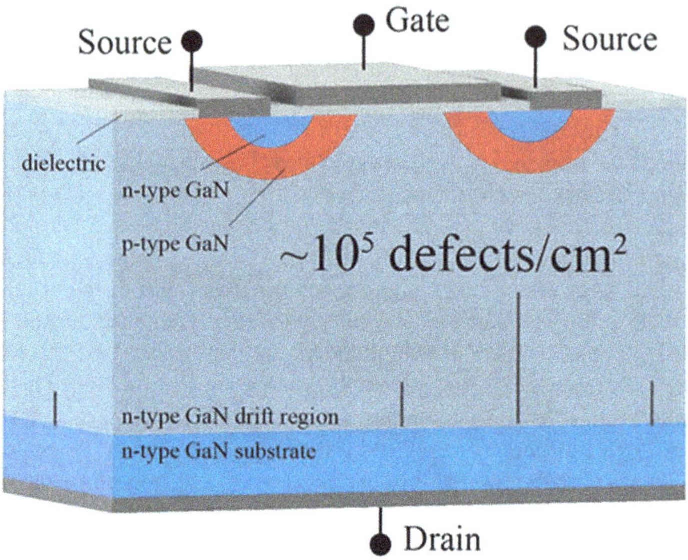

Fig. 4.35 A vertical conduction GaN transistor, relying on low defect density of GaN on GaN. (Source: www.odysseysemi.com/technology)

on ensuring high-quality and long-term reliability, these defects interfere with the high-voltage operational reliability of the devices. As a result, no GaN HEMTs have been commercially released with voltage ratings above 900 volts with the majority being in the 650 volt class (Fig. 4.35).

Vertical-conduction devices require that the substrate be the same material as the drift region. For GaN, this requires a bulk GaN substrate. While more expensive than and not available in the large formats of silicon, the bulk GaN substrates allow vertical-conduction GaN transistors to be made in material with defect densities ~10^3 to 10^5 cm^{-2}. This low defect density allows GaN transistors to have high reliability at voltages above 1000 volts and provide a path to voltages in the multiple thousands of volts.

4.8.2 The Value of Vertical GaN Transistors

The value of GaN can be expressed relative to the other choices available to power supply and motor drive circuit designers. In the above section, we discussed why vertical GaN transistors offer high-voltage advantages versus HEMTs. Relative to silicon and silicon carbide, the advantages are significant. The fundamental value driver for a transistor is the on-resistance of the device, expressed as Rds-on with the units of Ohms. The figure-of-merit of the device can be expressed as specific on-resistance with the units ohm-cm^2. The lower the specific on-resistance, the smaller a die is to provide a nominal Rds-on. GaN offers a theoretical ~100x advantage on specific on-resistance vs. silicon and a theoretical ~10x advantage versus silicon carbide. This means that for a given on-resistance and blocking voltage, a GaN transistor could be as much as 100x smaller than an equivalent silicon transistor and as much as 10x smaller than an equivalent silicon carbide transistor. When a die is smaller for a given Rds-on, additional benefits are enabled. First, the capacitance of the device is lower, making it easier to build a high-frequency and low switching loss circuit. Thus, an overall smaller power circuit thanks to the reduced filter components used in a switch-mode power supply. Additionally, even lower Rds-on can be achieved with minimal increase in die size, which will reduce the conduction loss of a circuit. This figure-of-merit can also be expressed in commercial terms. At the theoretical advantage, a 100 mm GaN substrate can provide up to 4 times as many transistors as a 150 mm silicon carbide transistor of the same Rds-on and voltage rating. The advantage is ~2 times when comparing a 100 mm GaN substrate to a 200 mm silicon carbide. Even with the higher GaN substrate costs, this provides a cost benefit relative to silicon carbide and makes the challenge of providing high-volume shipments easier due to the reduced wafers starts needed. However, as there are no qualified vertical GaN transistors at the time of this writing, it is also noted that there are no wafer fabrication facilities qualified to build vertical GaN transistors at high volume (Fig. 4.36).

4.8.3 GaN Wafers—Technical Status and Commercialization

Free-standing GaN wafers can be used as substrates for epitaxial growth of GaN-based device structures including LEDs, laser diodes, RF HEMTs, and vertical power devices. However, GaN wafers are difficult to grow due to the absence of a liquid phase at normal pressures, which has limited their size, quality, and overall development. Currently, laser diodes are the only semiconductor devices which

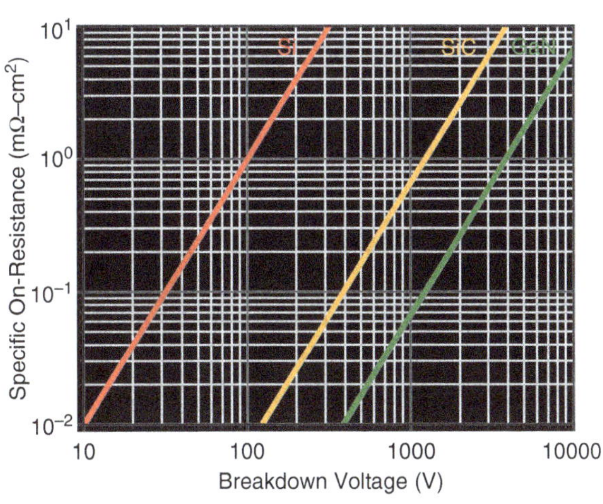

Fig. 4.36 Baliga plot showing the material properties of silicon, silicon carbide, and gallium nitride. (Source: www.odysseysemi.com/technology)

employ GaN wafers. There are future potential markets for RF HEMTs and vertical GaN power devices on GaN wafers. For LEDs, foreign substrates are sufficient so that the added cost of GaN wafers does not justify the marginal device performance improvements.

There are three growth methods for the thick growth of GaN that can be used to make GaN wafers. The most commercially relevant method is hydride vapor phase epitaxy (HVPE). This process uses GaCl and NH_3 gases at high temperature to grow GaN at growth rates of 100 to 150 mm/hr. To fabricate free-standing GaN wafers, HVPE can used to grow small boules, typically less than 1 cm thick, or to delaminate single wafers. The single wafer method can then be divided by the separation process of the GaN wafer from the foreign substrate, which includes interlayer separation and laser liftoff. In addition to HVPE, the other growth methods are ammonothermal and sodium flux. Both of these methods require high pressure and are done at thermodynamic equilibrium, but have seen limited commercial success.

There are less than 20 companies that are producing or developing GaN wafers. Of those, there are only 4 commercially viable manufacturers, 2 in Japan and 2 in China, that can produce GaN wafers in quantities of 1000 or more per year with acceptable quality and prices for manufacturing devices. The commercially viable manufacturers are all using HVPE as the primary growth technique, although HVPE and ammonothermal may be used together to make GaN wafers, where one method makes a seed crystal and another is used to expand it to a boule. The current size of GaN wafers is limited to 50 mm and 100 mm, with some 150 mm wafers having been demonstrated.

4.8.4 Key Applications

High-power transmission and conversion benefit from high voltage. As such, the emergence of higher operating voltages is a consequence of the electrification of everything. Electric vehicles, in order to increase range and reduce charging times, are moving to higher battery voltages, with current battery stacks of 800 V. Because of the headroom required for a robust and reliable circuit, this means that transistors of at least 1200 volts are needed in these applications. These trends, combined with the desire for higher power density make vertical GaN transistors attractive relative to silicon and silicon carbide transistors. In industrial motors, the pursuit of higher power densities are also important and enabled by GaN. One such application is for sinusoidal filtering. GaN is used for high-frequency operation where filters can be small and inexpensive. The motor is powered with sine-waves vs. square-waves, reducing the dv/dt in the motor, and thus reducing eddy currents that generate heat and reduce the overall efficiency of the motors. More generally, vertical GaN will be valuable for any circuit operating in the 1 kV or higher range where efficiency and power density are critical.

4.8.5 *The Challenges of Vertical GaN*

As of the time of this writing, there are no commercially available vertical GaN transistors in the marketplace. Multiple companies have openly been pursuing the development and commercialization of vertical GaN transistors and a significant amount of intellectual property has been created. But the repeatable and reliable process to develop vertical GaN transistors in the 1000 volt range and above has yet to be announced. There is no practical need for vertical GaN transistors below 1000 volt due to the availability of HEMTs. However, we believe it is just a matter of time until vertical GaN transistors are available for the following reasons. First, the material benefits are too significant to ignore. Second, the maturation of bulk GaN wafers continues, which is a critical enabler. Third, companies and researchers remain committed to finding solutions. It's not if, but when.

As mentioned above, area-selected doping is a critical step in building vertical GaN devices. Another technique that had been approached was etch and regrowth. In this approach, sections of the core GaN material are etched away and the desired material is grown in the available section. The challenge here is in the interface that is created. It appears as though this approach has been abandoned.

There is another practical consideration—the material benefits of GaN are significant, but are they practical. For instance, according to the Baliga Plot, a low Rds-on device could be made on a die less than 1 mm × 1 mm. Is this enough area for the conduction of heat away from the device? Would the die have to be enlarged for mechanical or thermal issues? What other system-level challenges will be uncovered once vertical GaN devices are built and testing begins? Time will tell.

4.9 GaN Power Semiconductor Device Dynamic Characterization

Ryo Takeda and Takamasa Arai

In recent years, gallium nitride (GaN) power devices have made significant progress in replacing silicon-based power MOSFETs in various applications such as data centers, small-sized chargers for PCs or smartphones, and potentially electric vehicle on-board chargers, low-voltage electric motors, and more. The fundamental characteristics of GaN enable significant improvements in key areas of power conversion. When comparing GaN to silicon (Si), it is well known that GaN's higher bandgap, higher electron mobility, and larger electric field potential enable important attributes, such as lower losses (i.e., higher efficiency) and faster switching (e.g., >MHz). The high-frequency operation of GaN devices in switch mode power supplies (SMPS) makes peripheral components such as load inductors and capacitors smaller, leading to a significant size reduction in power application products.

However, the fast switching characteristics of GaN cause more interaction with parasitic elements in power electronics circuits than Si or silicon carbide (SiC) power circuits. Ringing, oscillation, or distortion of switching waveforms make it very difficult to accurately characterize the dynamic performance and pure switching behavior of GaN power devices. It is critical to evaluate the dynamic behavior of the device under various conditions by changing variables such as operating temperature and gate resistance to design highly efficient and robust power electronics circuits. Therefore, a dynamic test system for GaN power devices should be able to measure characteristics by reducing ringing or distortion and avoiding oscillation.

GaN power devices are provided in various surface mount device (SMD) packages to take advantage of their fast speed by reducing parasitics related to packaging. The size of some GaN power devices is very small or tiny, in the range of a few square millimeters. Making good contact with such small devices is difficult, and therefore soldering them to the device under test (DUT) board is considered mandatory. Soldering becomes troublesome when multiple devices need to be measured, and unprofessional soldering can degrade the measurement itself.

4.9.1 *Challenges for Dynamic Characterization on GaN Power Devices*

Dynamic characteristics measurement on GaN power devices is very difficult because the fast voltage transition (*dv*/*dt*) and current transition (*di*/*dt*) are extremely fast. Therefore, they interact with parasitic in the test circuit generating excessive voltage ($V = L * di/dt$) or current ($I = C * dv/dt$) resulting in undesired events such as surge current, ringing, oscillation, false-turn on, and others. Figure 4.37 shows example measurement results on GaN field-effect transistors (FET). On the left, the device goes into divergent oscillation right after the turn-off event. On the right, a large ringing and current surge at turn-on event destroys the DUT and the gate driver.

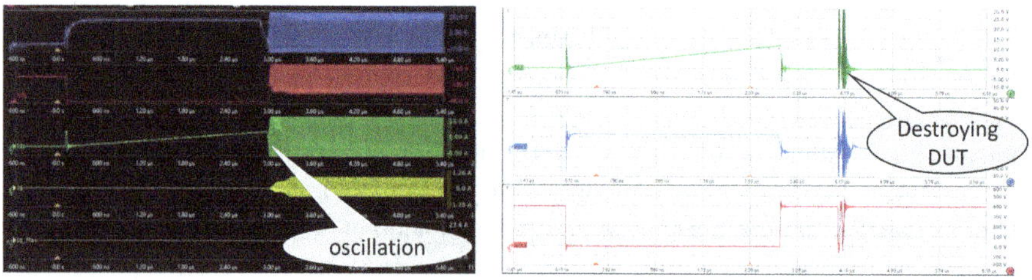

Fig. 4.37 GaN FET switching test results with oscillation and ringing

If the tested waveforms have large oscillations, ringing, or distortion, the true switching waveforms are not visible and the data is not reliable. The DUT can also be damaged if the test comes with surge current or false turn on causing the short circuit of the DUT.

Dynamic characteristics or switching characteristics of GaN power devices are measured with the so-called double pulse test. Using a half-bridge structure of FETs, two pulses are applied to a DUT. Due to the load inductor between the power supply and the DUT, the current gradually increases inversely proportional to the load inductance value when the first pulse is applied. By controlling the width of the first pulse, you can manage the current value for the switching event.

When the first pulse ends, the current cannot flow through the DUT because it is turned off, but it continues flowing in the loop consisting of the load inductor and the reverse conducting channel of high-side GaN FET. When the second pulse is applied, the DUT is turned on again, and the current flows through the DUT to the ground. With this sequence of two pulses, the waveform shown in Fig. 4.38 is obtained. The edge at the end of the first pulse represents off-characteristics, and the edge at the beginning of the second pulse represents on-characteristics. In other words, you can conveniently measure turn-on and turn-off switching characteristics with a single measurement event.

Since there was no commercially available double pulse tester providing accurate and reliable measurements for GaN power devices, it is common for device engineers and researchers to construct a custom double pulse test system. However, those attempting to build such a test system often find it challenging to obtain accurate measurements for GaN power devices.

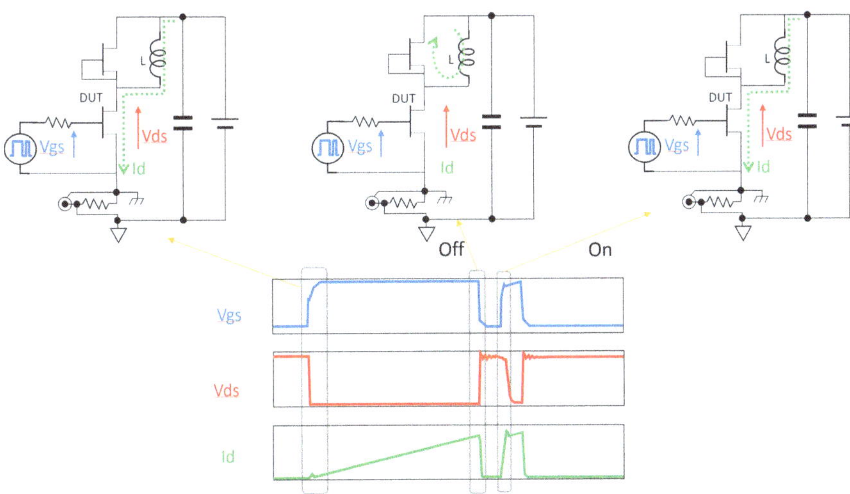

Fig. 4.38 Double pulse test for switching characterization

4.9.2 Concepts and Technologies Enabling Accurate Dynamic Characterization of GaN Power Devices

There are several essential techniques and technologies required for the accurate dynamic characterization of GaN power devices. The paramount concept involves minimizing parasitic inductance in the test circuit, aiming for a lower nH range. The primary principle is to design the test circuit to be as small as possible, facilitating the reduction of power loop inductance and gate loop inductance. While the use of wires for configuring the test circuit is suboptimal, employing a printed circuit board (PCB) with an optimized layout is crucial for precise measurements.

In contrast to real-world applications where GaN devices undergo continuous switching operations, the DUT remains relatively cool during a double pulse test due to the minimal injected power. Consequently, heat dissipation is not a concern, eliminating the need for a heat sink. This absence of thermal issues allows for the overlap of forward and return current traces across the PCB (refer to Fig. 4.39).

The current sensor is of paramount importance. A wide bandwidth, on the order of a couple of hundred megahertz, is imperative to capture transitions lasting only a few nanoseconds. Non-contact current sensors, like the Rogowski coil, fall short in this regard due to their limited bandwidth. Conversely, alternatives with high bandwidth, such as commercially available coaxial shunt resistors, often come with significant insertion inductance ranging from 5nH to 10nH, leading to potential ringing issues. There is a need for a novel current sensor that combines minimal inductance with high bandwidth for optimal performance (Fig. 4.40).

Eliminating skew between probes is crucial, as a shift of just a few nanoseconds between Vds and Id can introduce non-negligible errors in switching loss calculations. Radio frequency (RF) compensation also proves to be effective.

Figure 4.41 illustrates example results obtained from the measurement technologies discussed earlier, applied to a 650 V-rated GaN HEMT (high-electron-mobility transistor).

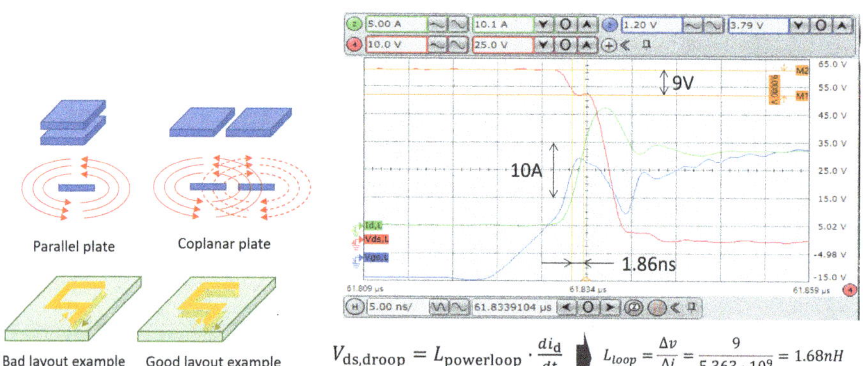

$$V_{\mathrm{ds,droop}} = L_{\mathrm{powerloop}} \cdot \frac{di_{\mathrm{d}}}{dt} \quad L_{loop} = \frac{\Delta v}{\frac{\Delta i}{\Delta t}} = \frac{9}{5.363 \cdot 10^{9}} = 1.68nH$$

Fig. 4.39 Layout example and power loop calculation

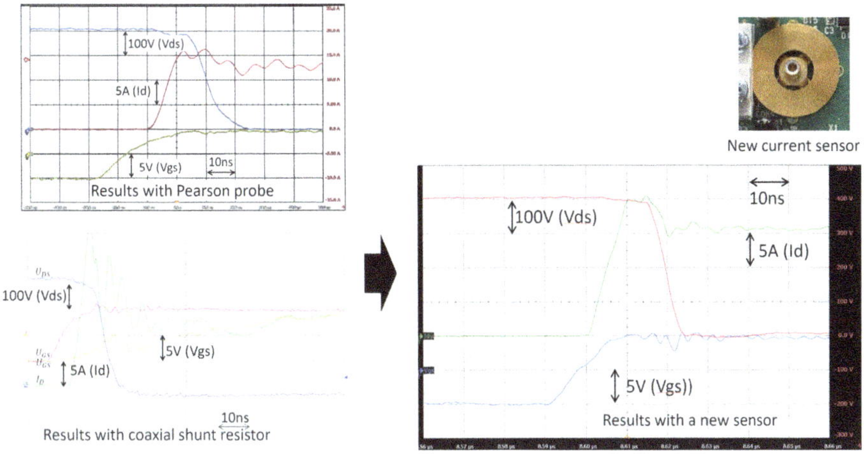

Fig. 4.40 Switching waveforms with different current sensors

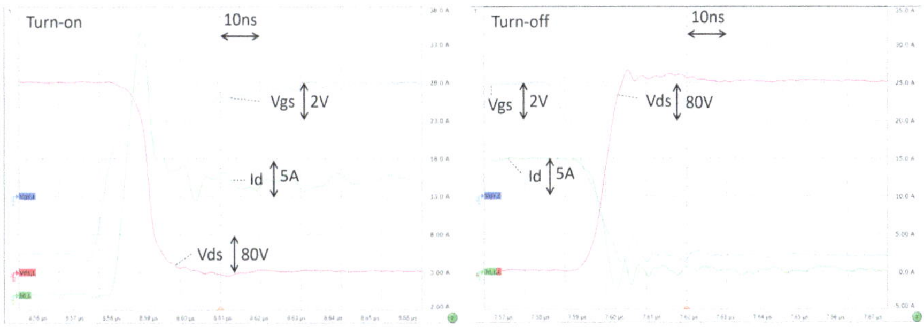

Fig. 4.41 Example test results on GaN HEMT

In addition to the requirements for making accurate measurements, there are a few more considerations to enhance the utility of the test system, such as flexibility and versatility. The ability to change test conditions flexibly, such as using exchangeable gate resistors and a DUT heater, proves beneficial when designing real-world applications with GaN devices.

Beyond standard turn-on, turn-off, and switching loss measurements, dynamic Rds(on) is a crucial characteristic to measure for GaN power devices. In 2019, JEDEC JC-70 released "JEP173 Dynamic ON-Resistance Test Method Guidelines for GaN HEMT-based Power Conversion Devices," emphasizing the necessity of a clamp circuit for accurately measuring dynamic Rds(on). This clamp circuit should be implemented on the test board in close proximity to the DUT. Figure 4.42 displays an example of dynamic Rds(on) measurement results, showing stable measurements 50 ns after the switching event.

There are various SMD types for GaN power devices. The preference is for solderless contact technology due to the time-consuming nature of soldering and

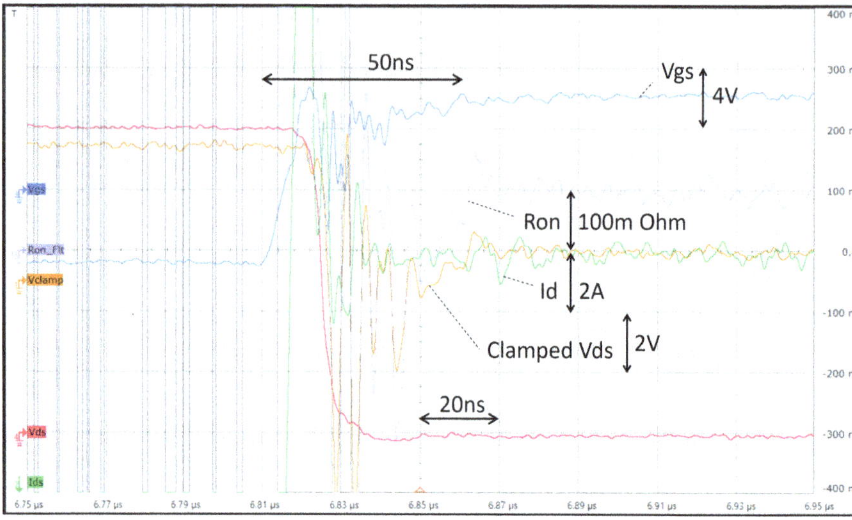

Fig. 4.42 Dynamic Ron measurement example

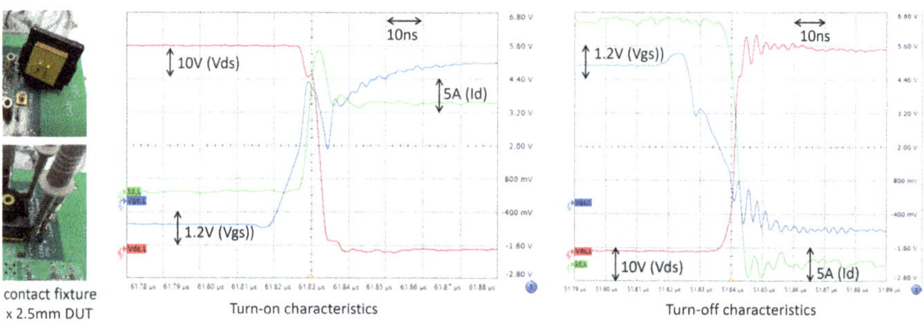

Fig. 4.43 Test results on small-size cascode GaN FET

unsoldering DUTs, which can quickly wear out the test board. Additionally, solderless contact facilitates testing multiple DUTs on the same test board, a crucial aspect for device comparison. Figure 4.43 illustrates an example of measurement results for a very small Cascode GaN FET and its fixture with solderless technology.

4.9.3 Summary

The dynamic characteristics of GaN power semiconductor devices require various measurement technologies and techniques to reveal what has been previously invisible.

For a commercial solution, the Keysight PD1500A series is available, incorporating the technologies and techniques mentioned in this article.

4.10 GaN Integrated Circuits

Stephen Oliver

Early power GaN devices were developed on the same substrate material that LEDs and RF devices were using, primarily sapphire and silicon carbide, respectively. Even with GaN's die-size advantage due to very low specific on-resistance (0.2–0.5 Ohm-mm^2 at 650 V today) compared to state-of-the-art silicon superjunction technology (1–2 Ohm-mm^2), this led to extremely expensive prototypes. Lateral devices were prototyped and GaN-on-SiC showed promise due to improved thermal performance, but devices were still too expensive for power applications. For many years, these devices were limited to depletion mode (dMode or normally on) types and suffered from the current collapse or dynamic $R_{DS(ON)}$ phenomenon and trap-related long-term stress instability that prevented their adoption. Finally, 6″ and 8″ (150 and 200 mm) production-quality GaN-on-Si wafers became available [29] for high-performance enhancement mode (eMode or normally off) lateral GaN devices, using existing foundry processes and high-volume, cost-effective switches and power ICs could be manufactured.

For lateral GaN, a two-dimensional electron gas (2DEG) with AlGaN/GaN heteroepitaxy structure (see Fig. 4.44) gives very high mobility in the channel and drain drift region. A 2DEG is the result of the difference in bandgap voltage between the two layers, which induces a quantum well at the interface. This well collects electrons in a sheet of charged carriers that are free to move laterally with extremely low resistance. The lateral device structure achieves extremely low gate charge (Q_g) due to the very small fraction of the area covered by the gate. The output charge Q_{OSS} determines the amount of energy required to switch the output state of the transistor and is also very low, due to the small size of the transistor, the insulating GaN epi layer, and the short drain drift region enabled by the high critical electric field. In addition, the lateral structure, the isolated substrate, and the ability to electrically isolate one 2DEG region from another by shallow etching or implantation all work together to allow integration of multiple power devices on a chip, as well as drivers and other functions.

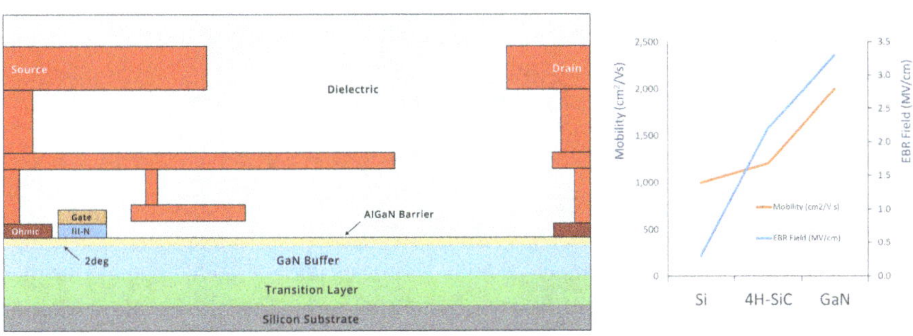

Fig. 4.44 (**a**) Lateral GaN-on-Si device construction, (**b**) material comparison of Si vs SiC, GaN

The dMode devices require a negative V_{GS} to turn off—not a practical solution for off-line applications. This unwanted characteristic was mitigated by the addition of a second, low-voltage "cascode" silicon FET [30, 31] that is used to turn the GaN power device on and off. This essentially converts the dMode device into an eMode device, required to block high bus voltages when a power converter is first turned on. The cascode FET allows for a standard gate drive signal to be used (0 V = OFF). However, the silicon FET frequency characteristics are inferior to GaN and compromise the switching performance of the combined device. Cascoding requires at least a two-chip package that presents a variety of other concerns including complex (multiple and/or stacked die) packaging, high parasitic inductance [32, 33] ceramic interposers for isolation, tendency for oscillation and internal overvoltage stress due to AC and DC mismatch of the GaN device and the silicon device, as well as additional internal passives to mitigate these effects. This results in a reduced overall yield and ultimately a higher cost.

A high level of sensitivity exists at the gate due to a low threshold voltage and limited $V_{GS(MAX)}$ for overdrive above the target on-state gate voltage, so a very clean, low overshoot gate voltage is required. These devices are very susceptible to voltage spikes that can easily occur due to high-frequency and high dV/dt noise from the surrounding switched-mode converter circuit. This results in additional circuitry such as Zener clamps, ferrite beads, and negative gate drive, which increases component count, PCB area and cost, while also limited the switching frequency of the circuit.

4.10.1 Monolithic: The GaN Power IC

The key to improved performance (reduced loss, higher switching frequency) and rapid market adoption (simple, cost-effective designs) is to create what approaches the "ideal switch," i.e., a circuit building block which translates a minimum-energy digital signal into loss-less power delivery. While academic work had progressed at low voltages, a proprietary, lateral 650/700 V eMode GaN technology platform—AllGaN™—has been created which enables the monolithic integration of drive, logic, and FET into a GaNFast™ power IC [35]. This single die can then be packaged into an industry-standard, low-inductance, low-cost QFN package for off-line AC or 400 V DC applications [34]. The integration enables virtually zero loss during turn-off because the turn-off gate-drive loop has essentially zero impedance. In addition, turn-on performance can be controlled and customized for specific application requirements.

4.10.1.1 Application Performance

GaN power ICs combine frequency, density, and efficiency advantages in bridge topologies such as active-clamp flyback, totem-pole PFC, and LLC. If we look at the general loss equation for a primary Si FET under hard-switching, we see several device parameter-related terms (Fig. 4.45). While a step improvement can be made with soft-switching Si [36], the updated equation for soft-switching GaN Power IC

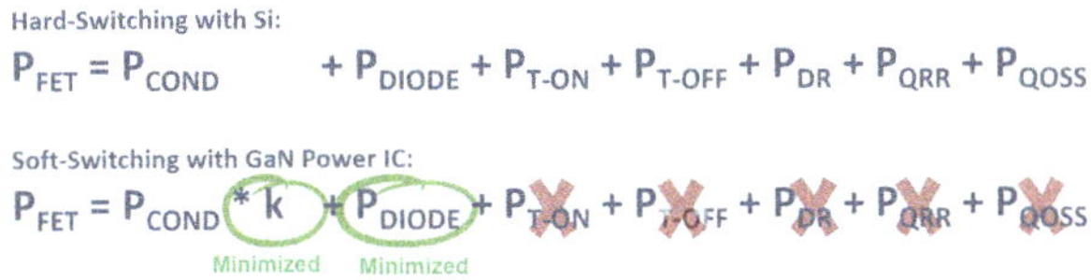

Fig. 4.45 Primary FET losses (hard-switching vs. soft-switching topologies)

(Fig. 4.45b) is a simpler affair, which results in *improved* efficiency at 10x higher frequencies.

Here, soft-switching with GaN has a k term to represent the loss due to increased circulating current, but this is minimized due to GaN's low C_{OSS}. The excess circulating current is used to energize an inductive circuit element to create the voltage swing on the output and charge the C_{OSS}. Because the C_{OSS} is low, the soft switching *dv*/*dt* and achievable operating frequency is higher with GaN. All other elements of the equation are minimal or negligible. Diode losses (P_{DIODE}) are reduced due to synchronous rectification and the high-speed integrated drive enables improved deadtime control. Turn-on loss ($P_{T\text{-}ON}$) is completely eliminated due to soft-switching, and turn-off loss ($P_{T\text{-}OFF}$) disappears due to near-zero loop impedance made possible by the monolithic integration of driver and FET, plus low-inductance SMT packaging. Driver loss (P_{DR}) is nearly negligible for GaN up to 2 MHz at about 0.01% of output power. Navitas eMode GaN has zero reverse-recovery time and no diode-recovery loss (P_{QRR}) since there are no minority carriers (holes) in the conduction path of the device. The output charge loss (P_{QOSS}) is not always zero, even if the C_{OSS} is resonantly charged, because there is an ESR even in a semiconductor junction capacitor. Even worse, there is hysteresis in the charging and discharging of the capacitor in some devices that represents charge stored and lost each cycle. GaN can have more than 10x lower losses when resonantly charging its C_{OSS} compared to silicon superjunction devices that suffer from this effect.

When two power switches are combined into "half-bridge" GaN power ICs, additional features such as level-shifters, boot-strap diode (function), shoot-through protection, etc., can be included with associated efficiency and power-density benefits [37].

4.10.2 GaN Discrete Vs. GaN Power IC

There are several ways to compare the performance of discrete vs. monolithically integrated GaN power devices, including power density, efficiency, component count, PCB area, protection features and EMI signature, etc.

A direct comparison of discrete GaN (in a disparate-technology, multi-chip module) vs. monolithic GaN power IC is shown in Fig. 4.46.

With a GaN power IC, loss-less current sensing and autonomous over-current protection contribute to system efficiency by elimination of discrete R_{SENSE} hot-spots, and an on-chip, closed-loop system (Fig. 4.47) that can detect (measure) and protect (turn-off) within 30 ns—6x faster than any discrete implementation [38].

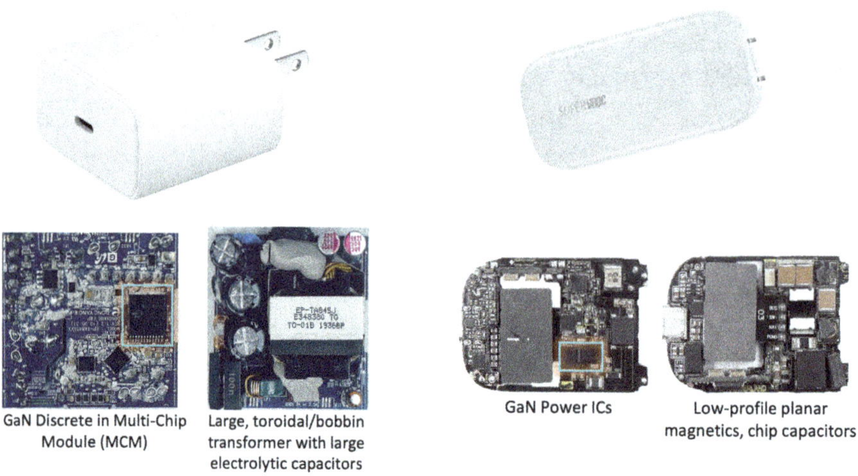

	Discrete GaN	GaN Power IC	
Power (W)	45	50	10% higher
Frequency (kHz)	65	400	6x faster
Dimensions (mm)	52 x 53 x 30	82 x 39 x 11	2.4x smaller
Transformer height (mm)	23	8	2.9x lower
Power Density (W/cc)	0.5	1.5	3x higher

Fig. 4.46 Adoption of new, high-frequency active-clamp topology and GaN power ICs enables shrink of passive components and overall 3x improvement in power density

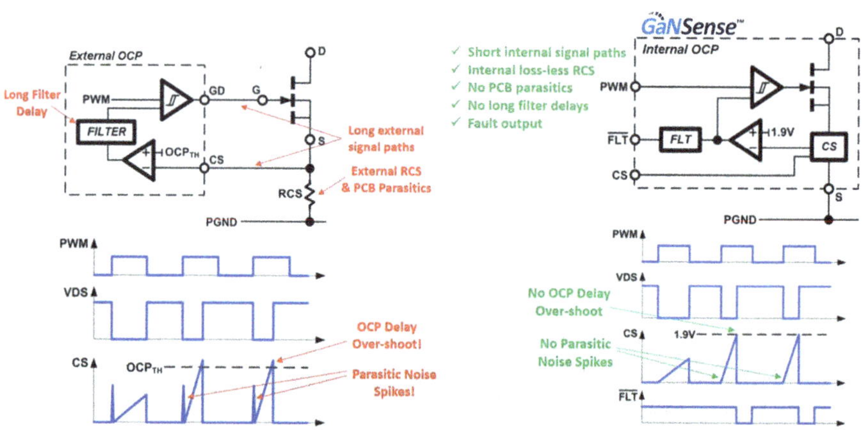

Fig. 4.47 GaNSense OCP vs. discrete implementation (simplified circuits and timing diagrams)

Loss-less current sensing is required in diverse applications from 65 W fly-back converters in laptop power supplies, to three-phase motor inverters. As GaN has no internal body diode, it greatly simplifies motor drive design vs silicon FETs and IGBTs [39], and loss-less current sensing can be integrated into GaNSense™ half-bridge chips, as shown in the high-frequency motor drive shown in Fig. 4.48.

As GaN power ICs enter high-power, high-reliability applications, such as solar micro-inverters, EV on-board chargers and AI data center power supplies, reliability is critical. Application-specific test requirements and protection features such as "double-pulse" and short-circuit protection (Fig. 4.49) are key features of high-power GaNSafe™ power ICs [41].

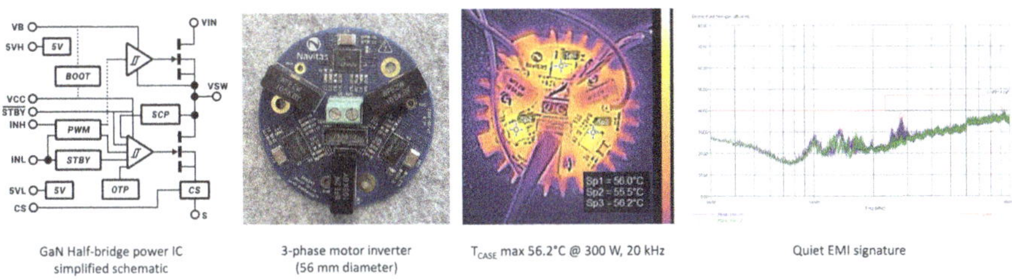

Fig. 4.48 GaN half-bridge power IC (NV6245M) schematic, and motor inverter example with cool-running performance [40]

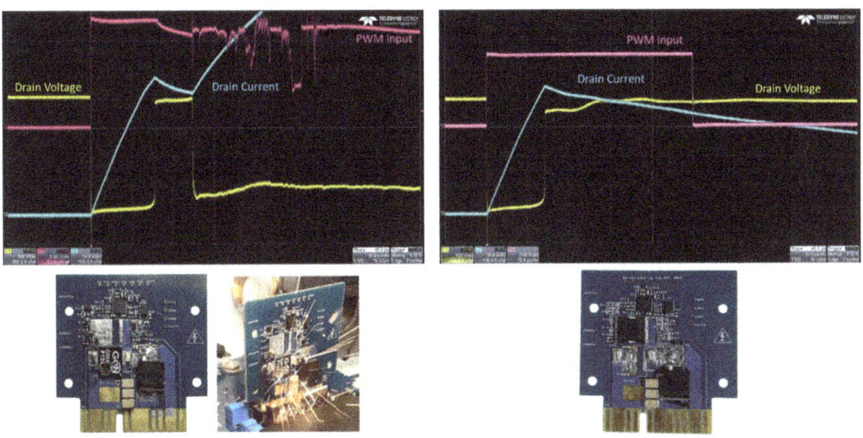

Fig. 4.49 Short-circuit test: GaN discrete (left) explodes and fails short, whereas the GaNSafe™ Power IC (NV6514, right) output safely ramps down

4.10.3 Summary: Monolithic GaN Power Integration Delivers Speed, Performance, and Reliability

Gallium nitride is the key to high-frequency performance, vastly exceeding the capabilities of legacy silicon power chips. Since the first production of GaN charger in 2018, the fast-charger market became a beachhead for GaN, now with 10 of the top 10 mobile OEMs in mass production with GaNFast power ICs. GaN power ICs, with monolithic integration of GaN FET(s) with GaN driver(s), voltage regulators, level-shifters, protection, etc., are the key to overcome GaN's "weak gate" Achilles' heel, and to break through the material's "glass ceiling" into higher-power, higher reliability systems.

References

1. K.J. Chen et al., GaN-on-Si power technology: devices and applications. *IEEE Trans. Electron Devices* **64**, 77
2. B.J. Baliga, Power semiconductor device figure of merit for high-frequency applications. IEEE Electron Device Letters **10**(10), 455 (1989)
3. A. Johnson, Physical limitations on frequency and power parameters of transistors. RCA Review **26**, 163–177 (1965)
4. F. Udrea, G. Deboy, T. Fujihira, Superjunction power devices, history, development, and future prospects. IEEE Trans. Electron Devices **64**, 720–734. ISSN 0018-9383, (2017)
5. H. Kang, F. Udrea, True material limit of power devices—applied to 2-D Superjunction MOSFET. IEEE Trans. Electron Devices **65**, 1432–1439. ISSN 0018-9383, (2018)
6. G. Longobardi, Vertical GaN Power Devices, in *Modern Power Electronic Devices: Physics, Applications, and Reliability*, ed. by F. Iannuzzo, (IET Energy Engineering Series, 2020)
7. D. Pagano et al., Suppression of substrate coupling in GaN high electron mobility transistors (HEMTs) by hole injection from the p-GaN gate. Appl. Phys. Lett. **115**, 203502 (2019). https://doi.org/10.1063/1.5121637]
8. A. Lidow, M. de Rooij, J. Strydom, D. Reusch, J. Glaser, *GaN Transistors for Efficient Power Conversion*, 3rd edn. (Wiley, 2020) [Online]. Available: https://epc-co.com/epc/Products/Publications/GaNTransistorsForEfficientPowerConversion.aspx
9. D. Reusch, J. Strydom, *Understanding the Effect of PCB Layout on Circuit Performance in a High Frequency Gallium Nitride Based Point of Load Converter* (OT Twenty-Eighth Annual IEEE Applied Power Electronics Conference and Exposition (APEC), Long Beach, CA, 16–21 March, 2013), pp. 649–655
10. B. Becker, How a driver source pin improves switching efficiency. EE Times Magazine (2021) [Online]. Available: https://eepower.com/industry-articles/how-a-driver-source-pin-improves-switching-efficiency/#
11. Efficient Power Conversion Corporation, EPC8010 datasheet (2023), [Online]. Available: https://epc-co.com/epc/Portals/0/epc/documents/datasheets/EPC8010_datasheet.pdf
12. J.S. Glaser, A. Helou, *PCB Layout for Chip-Scale Package GaN FETs Optimizes Both Electrical and Thermal Performance*, IEEE Applied Power Electronics Conference and Exposition (APEC), 20–24 March 2022
13. Y. Zhang, Y.F. Udrea, H. Wang, Multidimensional device architectures for efficient power electronics. Nat Electron **5**, 723–734 (2022)
14. S. Preimel, *600 V CoolMOS™ G7 and 650 V CoolSiC™ G6 come in a new top-side cooling package—The DDPAK*, Application Note, AN_1802_PL52_1803 (2018)

15. Infineon Technologies, Recommendations for board assembly of Infineon packages with dual row gullwing leads, Rev. 5.0 (2020)
16. Infineon Technologies, TO-leaded top-side cooled (TOLT) package automotive power MOSFET, Application Note v1.1, Z8F80044621 (2021)
17. Infineon Technologies, “TOLL vs. TOLT,” Application Note, Z8F80127016 (2022)
18. Infineon Technologies, Innovative top-side cooled package solution for high-voltage applications, Application Note v1.0, AN_2101_PL52_2103_112902 (2021)
19. S. Preimel, 600 V CoolMOS CFD7 comes in a new top-side cooling package—The QDPAK. Application Note, AN_1802_PL52_1803_120439 (2018)
20. B. Zojer, CoolMOS™ gate drive and switching dynamics, Application Note v1.0, AN_1909_PL52_1911_173913 (2020)
21. Infineon Technologies, *Thermal Design for Packaged GaNPX® Devices*, Application Note GN002 (2020)
22. Infineon Technologies AG: Half-bridge evaluation board designed with CoolGaN™ GIT HEMTs and EiceDRIVER™ 2EDB8259Y, more at www.infineon.com/GaN
23. D. Veereddy et al., *Reverse Bias Lifetime Reliability Assessments of HV GaN Power Device* (IEEE International Reliability Physics Symposium (IRPS), 2017)
24. M. Glavanovics et al., *Application Related Reliability Test Concept for GaN HEMT Power Devices*, Workshop on Compound Semiconductor Devices and Integrated Circuits (WOCSDICE) (2017)
25. T. McDonald, *State of Demonstrated HV GaN Reliability and Further Requirements* (IEEE Applied Power Electronics Conference and Exposition (APEC)) 2015 Industry Session
26. T. McDonald et al., Reliability and qualification of high-voltage CoolGaN™ GIT HEMTs, *Infineon Whitepaper*, www.infineon.com/GaN (2022)
27. JC-70.1 GaN Power Electronic Conversion Semiconductor Standards, www.jedec.org/committees/jc-701
28. AIP Advances 7, 095302, Effect of Si doping on the thermal conductivity of bulk GaN at elevated temperatures—Theory and experiment (2017). https://pubs.aip.org/aip/adv/article/7/9/095302/941663/Effect-of-Si-doping-on-the-thermal-conductivity-of
29. GaN as a Displacement Technology for Silicon in Power Management, Lidow, EPC, IEEE ECCE (2011)
30. Active Driving of Normally on, Normally off Cascoded Configuration Devices through Asymmetrical CMOS, US Patent US 7,408,399 B2; Salato, Soldano (2008)
31. Commercialization of High 600V GaN-on-Silicon Power HEMTs and Diodes, Parikh et al, Transphorm Inc., IEEE EnergyTech, May 2013
32. Evaluation of High-Voltage Cascode GaN HEMT in Different Package, Liu et al, VPT (2014)
33. Evaluation and Applications of 600 V/650 V Enhancement-Mode GaN Devices, Huang et al, CPES (2016)
34. GaN Power IC Technology: Past, Present, and Future, Kinzer, ISPSD (2017). https://navitassemi.com/download/gan-power-ic-technology-past-present-future/?wpdmdl=37065&ind=1563179445605
35. GaN Matures for Industry with Monolithic Power ICs, *Power Electronics Europe*, May 2016. http://www.power-mag.com/pdf/issuearchive/82.pdf
36. Next-generation GaN Isolators/Level-Shifters for High Frequency, Oliver and Giandalia, APEC (2017). https://navitassemi.com/download/next-generation-gan-isolators-level-shifters%ef%bf%bdfor-high-frequency/?wpdmdl=36536&ind=1561120399092
37. New GaNFast™ Power ICs with GaNSense™ Technology Loss-Less Current Sensing & Autonomous Protection, Navitas application note AN015, Ribarich, https://navitassemi.com/wp-content/uploads/2021/12/AN015-GaNSense-FINAL-12-20-2021.pdf
38. Motor Drive Based on III-Nitride Devices. US patent, Battello and Oliver, US 8,390,241. https://ppubs.uspto.gov/dirsearch-public/print/downloadPdf/8390241
39. Reducing System Cost with GaN FETs in Motor Drive Applications, Hesener, APEC (2024). https://navitassemi.com/download/reducing-system-cost-with-gan-fets-in-motor-drive-applications/?wpdmdl=987501731&ind=987501730

40. Navitas GaNSafe: World's most protected GaN power. https://navitassemi.com/navitas-gansafe-worlds-most-protected-gan-power-semiconductor-opens-multi-billion-dollar-data-center-solar-and-ev-opportunities/
41. Navitas' GaNFast™ Enables World's Thinnest Travel Adapter, March 2018. https://navitas-semi.com/navitas-ganfast-enables-worlds-thinnest-travel-adapter/

Dr. Florin Udrea is the Chief Technology Officer and a co-founder of Cambridge GaN Devices, Ltd. The company is a spin-out from Cambridge University with around 60 people (in 2024) aiming to bring to market innovative GaN technologies. Florin Udrea is also a professor in semiconductor engineering and head of the High Voltage Microelectronics and Sensors Laboratory at University of Cambridge. He is currently leading a research group in power semiconductor devices and solid-state sensors that has won an international reputation during the last 30 years. He has published over 600 papers in journals and international conferences. He holds over 300 patents in power semiconductor devices and sensors. Prof. Florin Udrea founded five companies, Cambridge Semiconductor (Camsemi) in power ICs—sold to Power Integrations, Cambridge CMOS Sensors (CCS) in the field of smart sensors—sold to ams, Cambridge Microelectronics in Power Devices, Cambridge GaN Device and Flusso in flow and temperature sensors. For his "'outstanding personal contribution to British Engineering,"' he has been awarded the Silver Medal from the Royal Academy of Engineering. In 2015, Prof. Florin Udrea was elected a Fellow of Royal Academy of Engineering. In 2018, Prof. Udrea has been awarded several major prizes, including the Mullard medal from the Royal Society. In 2020, he received his first Ohmi award (ISPSD) as a co-author of a paper on SiC FinFET technology. In 2021, Prof. Udrea was named "the Academic Entrepreneur of the Year" by Business Weekly. In 2023. he received his second Ohmi award for his work on Silicon Carbide FinFET and he was inducted in the ISPSD IEEE Hall of Fame (2023).

Kengo Ohmori is the technical marketing manager responsible for Rohm's Wide Band Gap high-power device products for the Americas market. He received the bachelor's degree and M.S. degree in physics from Kwansei Gakuin University. Before his current role, he was leading GaN product marketing/design team at ROHM HQ in Japan. His role at HQ was searching and understanding the requirement from the market and design/develop products based on that (Almost everything of GaN strategy and development). He has broad semiconductor experiences covering device engineering, product marketing/planning, creating road map, competitor analysis, customer promotion, and business development. He holds thirteen patents including four U.S. patents in power devices and modules.

Dr. Ming Su is the technical marketing manager responsible for Rohm's SiC and high-power device products for the Americas market. He received the bachelor's degree in electronic engineering from Tsinghua University, and the M.S. and Ph.D. degrees in electrical engineering from Rutgers University. During his doctoral research, he developed normally off SiC JFET transistors for power switching applications and power IC. Prior to joining Rohm, Dr. Su worked as a research engineer at Ford Motor Company, responsible for discovering and evaluating emerging power semiconductor technologies in support of the company's vehicle electrification strategies. He has contributed over 35 scientific journal articles and conference presentations in the area of silicon and wide bandgap semiconductors, and holds two U.S. patents in power devices and modules.

Dr. Michael de Rooij is currently Vice President of Applications Engineering at Efficient Power Conversion Corporation (EPC) headquartered in El Segundo, CA, U.S.A. He received the B. Eng., M. Eng. and, D. Eng. degrees from The University of Johannesburg fka Rand Afrikaans University, South Africa, in 1992, 1994, and 1998, respectively, all in electronics and electrical engineering with major in power electronics. He worked as a post-doctoral researcher at the Center for Power Electronics (CPES), Virginia Tech in 2001. Dr. de Rooij has authored and co-authored over 50 papers at various conferences and journals and has been granted over 58 US and international patents. He is author and co-author of 7 books. His experience covers highly resonant loosely coupled wireless power, high-density DC to DC converters, GaN IC development, RF power amplifiers, solid-state high-frequency power converters and devices, utility applications of power electronics, uninterruptible power supplies, integration of power electronic converters, power electronic packaging, induction heating, photovoltaic converters, Magnetic Resonance Imaging (MRI) Systems, and gate drivers with protection features. Dr. de Rooij is senior member of the IEEE.

Sam Abdel-Rahman received his PhD and M.S. degrees in Electrical Engineering from University of Central Florida in 2007 and 2005, respectively, and his B.S. degree in Electronics Engineering from Princess Sumaya University for Technology (Jordan) in 2003. He joined Infineon in 2011, currently he is a System Architect Engineer and Technical Lead responsible for developing the application roadmap of Servers/Data center SMPS and Renewable applications.

Walter Balzarotti holds a M.S. in Electrical/Power Electronics at Politecnico di Milano. He has wide experience in the power management semiconductors industry with skills in Drivers Devices for silicon and wide bandgap switches for industrial and automotive applications.

Doug Bailey joined Power Integrations in November 2004 as vice president of marketing. Prior to joining PI, Mr. Bailey served as vice president of marketing at ChipX, a structured ASIC company. His earlier experience includes Business Management and Marketing Consultant for Sapiential Prime, Inc.., Director of Sales and Business Unit Manager for 8 × 8, Inc., application engineering management for IIT, Inc., and design engineering roles with LSI Logic, Inmos, Ltd. and Marconi.

Dr. Oliver Häberlen received his M.S. and Ph.D. degree in semiconductor physics from the University of Munich and Technical University of Munich, respectively. In 1995, he joined Infineon Technologies Austria AG in Villach. He is Infineon Fellow for Power Semiconductor Devices & Technology and responsible for advanced technology concepts evaluating future silicon and wide bandgap power device concepts for improved energy conversion solutions. His research areas include low and medium voltage trench power MOSFETs, high-voltage superjunction devices, GaN power devices, and power device reliability. He is a senior member of IEEE and has served as technical program committee member for many IEDM and ISPSD conferences and served as general chair for ISPSD 2020. He is author and co-author of over 100 granted international patents in the field of power semiconductors.

Tim McDonald is currently Senior Consulting Advisor for Infineon's CoolGaN™ program. Tim serves as Chair of the JEDEC JC-70 committee on standards for reliability qualification, test methods, and datasheet parametrics for wide bandgap (GaN and SiC) power conversion devices and is also sub-committee Vice Chair for JC-70.1 (GaN standards). Tim also serves on the Board of Directors for the Power Sources Manufacturers Association (PSMA). Tim has over 40 years of diversified experience (mostly at Infineon Technologies and International Rectifier) in power conversion/management and has held senior-level positions in device engineering management, product and market development, product engineering, device characterization, test platform development, and operations. Tim holds a Bachelor of Science degree in Physics from the University of California at Los Angeles (UCLA).

Mark Davidson has spent over 20 years in the analog and power semiconductor industry. He currently works with QPT, pioneering advancements in motor drive using GaN technology. Prior, Mark served as Chief Executive Officer of Odyssey Semiconductor, leading their pursuit of high-voltage vertical GaN power transistors. Prior to these roles helping to accelerate the adoption of WBG semiconductors, he held various positions at Intel, Altera, Texas Instruments and National Semiconductor. In all cases, Mark focuses on maximizing the business results of innovative technology. He received a B.S. in Electrical Engineering from Pennsylvania State University.

Troy Baker is an expert in the growth of GaN by HVPE, having worked in the field for over 20 years. He studied at the University of California, Santa Barbara under Prof. Shuji Nakamura, and did his PhD thesis on HVPE growth of semipolar GaN. In 2015, he founded Eta Research, based in China. Eta Research developed 4 inch free-standing GaN wafers in both semi-insulating and n-type. The company is one of the few commercial suppliers of GaN wafers. Currently, he is developing vertical GaN power devices at a new startup.

Mr. Ryo Takeda is a Solution Architect at the Automotive and Energy Solutions of Keysight Technologies. He joined Hewlett-Packard in 1989 and initially served as an Application Development Engineer and Manager for semiconductor parametric test solutions. Transitioning into product planning, he has guided power device analyzer product definition. Currently, his focus is on projects that integrate power device measurement, modeling, and circuit simulation, as well as the development of power device dynamic testers. He holds both a BSEE and MSEE from Keio University.

Mr. Takamasa Arai joined Keysight Technologies Japan (formerly Agilent Technologies Japan) in 2007, initially serving as an Application Engineer supporting impedance measurement and semiconductor parametric tests. Since 2021, he has worked as a Solution Application Engineer at Keysight Technologies, providing global technical support for power semiconductor dynamic testing, power device modeling, and power electronics circuit simulation solutions. He received his B.S. and M.S. degree in physics from Tokyo Institute of Technology, Tokyo, Japan, in 2005 and 2007, respectively.

Stephen Oliver is VP Corporate Marketing & Investor Relations at Navitas Semiconductor, with over 30 years of experience in the power semiconductor and power supply industries worldwide. Prior to Navitas, Stephen served as VP of the VI Chip product line at Vicor, and earlier at Motorola, Philips (NXP), and International Rectifier (Infineon). He holds a B.Eng. (Hons) in Electrical & Electronic Engineering from Manchester University (UK) and an MBA in Global Marketing & Strategy from the UCLA Anderson School of Management. Stephen hold several patents in power semis, is a Chartered Engineer and has been VP, President, and Chairman of the Power Source Manufacturers' Association (PSMA).

Chapter 5
Applications of GaN Technology

Michael de Rooij, Alejandro Pozo, Stephen Colino, John Glaser, Marco Palma, Martin Wattenberg, Max Zafrani, Alex Lidow, Robert Strittmatter, Tony Marini, Scott Durkin, Jim Honea, Alex Q. Huang, Emad Nazerian, Peng Han, and Maurizio Di Paolo Emilio

In this chapter, various authors from different organizations and institutions discuss the diverse applications of gallium nitride (GaN) technology across different sectors.

Michael de Rooij and Alejandro Pozo, representing Efficient Power Conversion, emphasize the utilization of GaN technology in two prominent converter types: the buck converter and the LLC converter. They underscore GaN transistors' exceptional performance and versatility in meeting evolving power demands.

Stephen Colino and John Glaser, also from Efficient Power Conversion, delve into the crucial role of GaN technology in time-of-flight lidar systems in Sect. 5.2. They explain how GaN devices facilitate both direct and indirect time-of-flight approaches, laying the groundwork for understanding efficient lidar operation.

Marco Palma, Director of Motor Driver at Efficient Power Conversion, explores the significance of GaN technology in revolutionizing motor drive systems in Sect. 5.3. Palma highlights GaN transistors' superiority in enhancing performance and efficiency, particularly in two-level inverters, covering various aspects such as benefits, applications, layout considerations, and testing results.

Martin Wattenberg, a System Application Engineer at Infineon Technologies, discusses the pivotal role of GaN HEMTs in motor drive systems in "Three-phase

M. de Rooij (✉) · A. Pozo · S. Colino · J. Glaser · M. Palma · M. Zafrani · A. Lidow
R. Strittmatter · T. Marini
EPC, El Segundo, CA, USA

M. Wattenberg
Infineon Technologies Austria, Villach, Austria

S. Durkin · J. Honea
Nexperia, Goleta, CA, USA

A. Q. Huang · E. Nazerian · P. Han
University of Texas, Austin, TX, USA

M. Di Paolo Emilio
Power Electronics News, Pescara, Italy

© The Author(s), under exclusive license to Springer Nature Switzerland AG 2024

M. Di Paolo Emilio (ed.), *GaN Technology*,
https://doi.org/10.1007/978-3-031-63238-9_5

inverters for motor drives." He explores the benefits of GaN technology, focusing on enhanced efficiency, reduced switching losses, and optimized design practices.

Section 5.5, authored by Max Zafrani, Alex Lidow, Tony Marini, and Robert Strittmatter of Efficient Power Conversion, delves into the application of GaN technology in space applications. The authors explore GaN devices' resilience and performance in harsh radiation conditions, emphasizing their superiority over silicon MOSFETs and their potential for revolutionizing power systems in space environments.

Scott Durkin and Jim Honea from Nexperia focus on adapting a continuous-conduction mode boost PFC controller to the totem-pole PFC topology in Sect. 5.6. They discuss the intricacies of modifying the boost converter for power factor correction, highlighting the significance of dynamic variations in duty cycle and voltage rectification.

In Sect. 5.7, Alex Q. Huang, Emad Nazerian, and Peng Han from The University of Texas at Austin introduce the commercial potential of GaN in electric vehicle (EV) power systems. They compare GaN's advantages over silicon carbide (SiC) and discuss its application challenges, setting the stage for exploring GaN's role in high-voltage and high-power EV converters. Last section will analyze general aspects of foundry services.

5.1 DC-DC Conversion

Michael de Rooij

5.1.1 Introduction

In virtually every application of electricity in this millennium, there is a DC-DC converter in the path of electrons. In many applications, the demands for extremely high power density, light weight, thin form factors, low cost, and high reliability have caused the bounds of power transistors to be pushed beyond the capability of the aging silicon power MOSFET.

This chapter looks at two of the most popular types of DC-DC converters using GaN devices, a hard-switching buck converter and a resonant LLC converter. In both examples, state-of-the-art performance is achieved.

5.1.2 Buck Converters

Automotive 48 V/12 V converters are essential in modern hybrid electric vehicles, as the energy is exchanged between the 48 V and 12 V buses. This two-voltage system accommodates legacy 12 V systems and enables important weight and energy savings at the system level by operating the more power-demanding loads from the 48 V bus, such as vacuum and water pumps, electric superchargers, and steering system. Among all the requirements for the 48 V/12 V converter, efficiency, power density, size, and cost are on the top of the list. This example addresses these design criteria by employing a GaN integrated circuit, the EPC23101.

GaN FETs used in 48 V applications are rated between 80 V and 100 V, and usually have a four times better figure of merit ($A_{Die} \cdot R_{DSon}$) compared to equivalent MOSFETs [1]. For the same gate voltage of 5 V, GaN FETs have at least five times lower gate charge than equivalent MOSFETs. Other important advantages of GaN FETs include lower C_{OSS}, faster voltage transition, and zero reverse recovery (Q_{RR}), and are physically smaller.

Examples of monolithic integration of GaN power FET and gate drivers have been established [2]. One such example, the EPC23101, integrates only the upper GaN FET of a half-bridge with a complete half-bridge gate driver, as shown in Fig. 5.1. The on-resistance of the integrated power FET is 3.3 mΩ, with a maximum voltage rating of 100 V.

To form a half-bridge, an external low-side GaN FET is required. The EPC2302 is the good choice for this example application. The EPC2302 is a 100 V-rated 1.8 mΩ device that is available in a similar QFN package as the EPC23101. This asymmetrical half-bridge yields an optimal balance between switching and conduction losses for 48 V/12 V converter [3]. A photo of this chipset (EPC23101 and EPC2302) and a functional block diagram is shown in Fig. 5.2.

5.1.3 Converter Design

A 2-phase synchronous buck converter topology demonstrates well the capability of GaN integrated circuit technology. This converter can also operate as a boost in reverse. A simplified schematic diagram is shown in Fig. 5.3 that includes the power stage and support functions such as current sensors, temperature sensor, digital controller, and 5 V and 3.3 V auxiliary power. Each phase is designed to carry up to 70 A current on the 12 V bus. This equates to 1 kW at 14.3 V. The total maximum output power of this design is 2 kW, or 140 A into the 12 V bus. Furthermore, the design is scalable, so two boards can be paralleled for a 4 kW system.

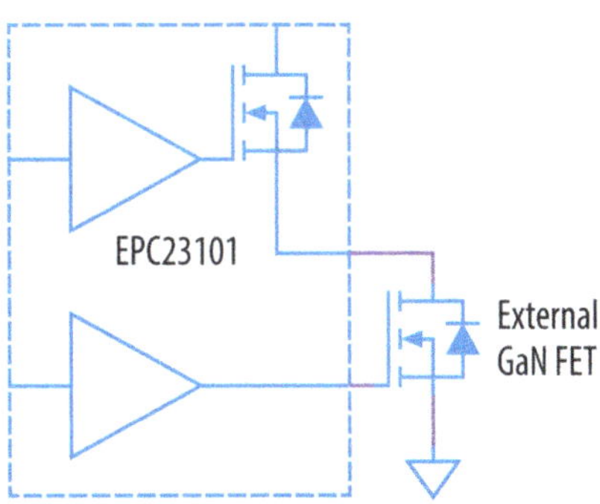

Fig. 5.1 EPC23101 concept diagram

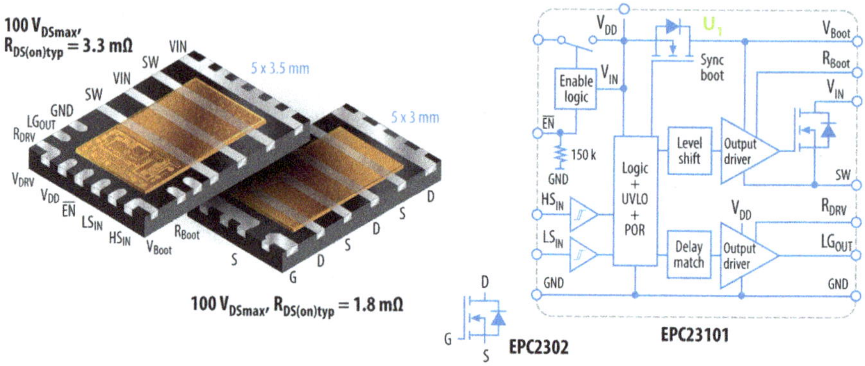

Fig. 5.2 Photo of EPC23101 and EPC2302 chipset with functional block diagram

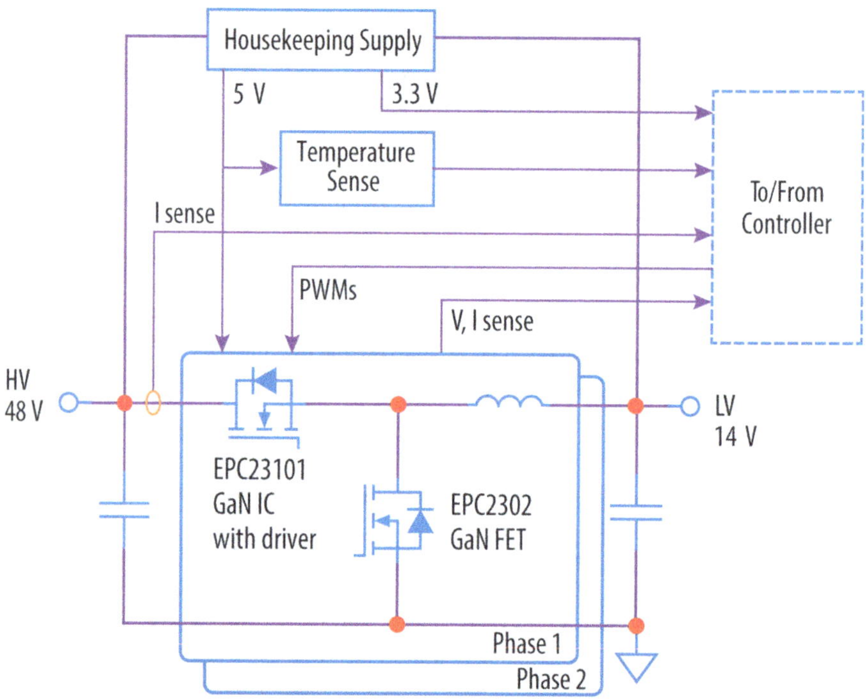

Fig. 5.3 Simplified schematic diagram of the 2-phase buck-reverse-boost converter

5.1.4 Performance Evaluation

Figure 5.4 shows a photo of the converter (EPC9170). With the heatsink installed and 2000 LFM airflow, the converter was operated at 48 V input, 14.3 V output, and 500 kHz switching frequency. The measured efficiency is shown in Fig. 5.5, compared with the converter built with discrete EPC2206 GaN FETs and external gate

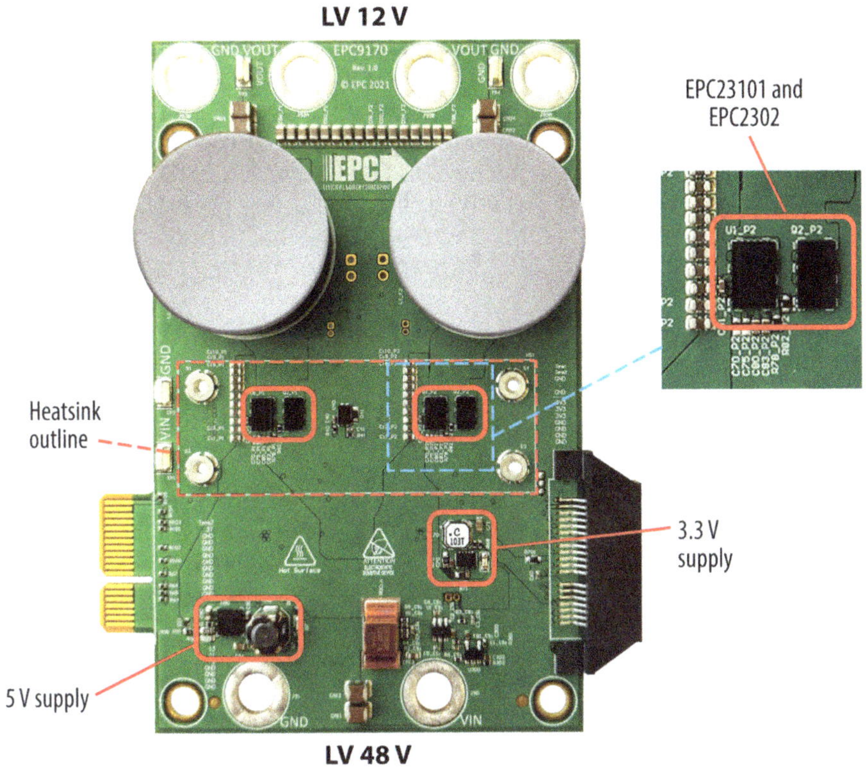

Fig. 5.4 Photo of the EPC9170 with EPC23101 and EPC2302 ePower stage chipset

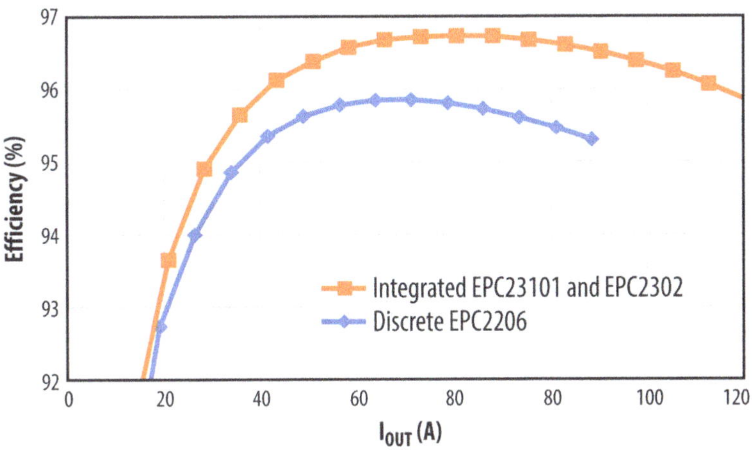

Fig. 5.5 Measured efficiency of EPC9170 (integrated EPC23101 and EPC2302) compared with EPC9137 (EPC2206 discrete transistors)

driver [4]. By using the integrated solution, this converter achieved higher peak efficiency, at 96.7%, and higher full-load power. At full load of 1 kW/phase, the efficiency is 95.8%.

Previously, discrete GaN FETs demonstrated superior performance in a 48 V/12 V hard-switched buck converter applications over equivalent MOSFETs. With the integrated power stage chipset of EPC23101 and EPC2302, converter performance was further improved while reducing size and cost through reduced component count.

5.1.5 LLC Converters

The rapid growth in demand for digital services, Internet traffic, the surge in artificial intelligence systems, and the continuous increase in computational power are driving significant technological improvements in power management solutions for datacenters and server racks. Energy distribution at 48 V reduces losses significantly but requires very efficient and high-power dense converters to be located as close as possible to the loads. LLC resonant converters have emerged as the preferred topology for an intermediate 48 V to 12 V conversion due to their high efficiency, high power density, and good dynamic response. The outstanding performance resulting from the combination of this topology with GaN transistors has been demonstrated in the past [5–8]. The example given in this chapter surpasses the performance of the converters in these references.

The EPC9159KIT [9] achieves more than 5 kW/in^3 of power density when assembled into a module. Peak efficiency of more than 97.5% and full-load efficiency over 95.5% when delivering 1 kW into a 12 V load are realized using a novel partial power technique [10, 11].

5.1.6 Converter Overview

The LLC resonant converter featured in EPC9159KIT, shown in Fig. 5.6, is based on a primary-side full bridge, with a 3:1 transformer and center-tapped full-wave synchronous rectifier. Four EPC2619 [12, 18] GaN transistors are used in the primary and six EPC2067 [13] GaN transistors in the secondary. The resonant tank is tuned to approximately 1.8 MHz using the transformer leakage inductance and two additional external inductors.

In addition, the EPC9159KIT, shown in Fig. 5.7, takes advantage of the inherent electrical isolation of the LLC topology to implement a partial power technique [10, 11] and enable higher efficiency and higher current capability. When configured in partial power mode, the converter has an input-to-output voltage ratio of 4:1 and can process up to 1 kW into a 12 V load from a 48 V supply. In this mode, the return of the primary full bridge (0VHV) floats above the output voltage (V_{OUT}) and the transformer only processes 75% of the output power.

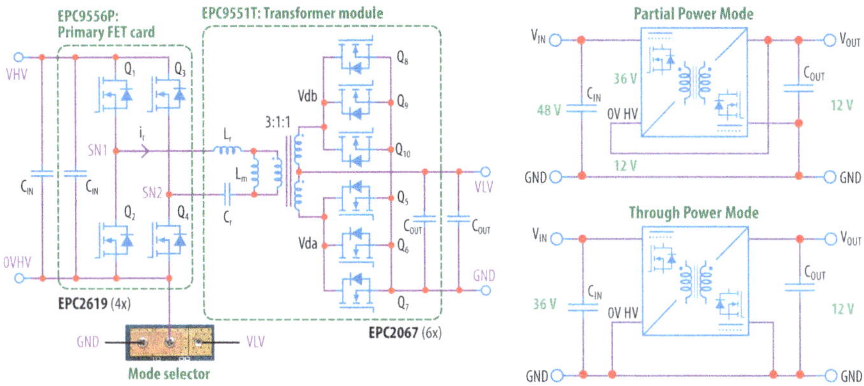

Fig. 5.6 (left) Topology of the LLC converter, (right-top) connection diagram for partial power mode, (right-bottom) connection diagram for through power mode

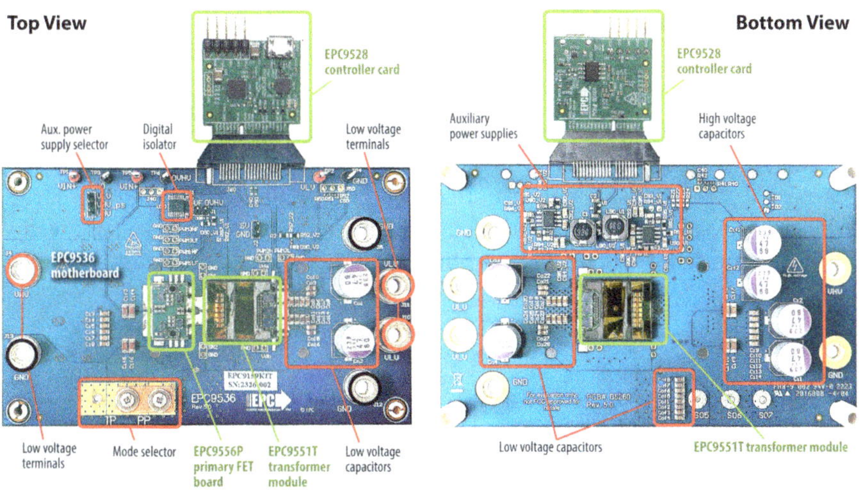

Fig. 5.7 Top and bottom views of the EPC9159KIT with the main components highlighted

5.1.7 Power Transistors, Gate Drivers, and Controller

The primary-side full-bridge features four 100 V-rated 3.3 mΩ EPC2619 [12] used in conjunction with two uP1966E [14], a half-bridge gate driver IC. For the secondary side, a total of six 40 V-rated 1.55 mΩ EPC2067 [13] are used as synchronous rectifiers. These six transistors are divided into two branches, with an LMG1020 [15] low-side gate driver for each branch. All power transistors and gate drivers are in WLCSP (wafer-level chip scale package) format for minimum size.

GaN transistors, and specifically the EPC2619, offer the lowest $R_{DS(on)}$ * C_{OSS} figure of merit available in the market for a >80 V-rated FET. Low $R_{DS(on)}$ is needed to minimize conduction losses, and a small C_{OSS} helps in reducing the transition times required to maintain ZVS in the primary while keeping the magnetizing current as low as possible. A die size of only 2.5 × 1.5 mm² is also a key feature to realize the desired power density.

A total of six independent PWM signals are generated using a Microchip dsPIC33CK256MP503 controller to compensate propagation delay mismatches in gate drivers and achieve near-perfect synchronization between the primary transistors and the synchronous rectifiers. This is done with a time resolution of 250 ps in a very compact size with a 5 × 5 mm² QFN package. A 4-channel digital isolator is also included to level-shift the four PWM signals for the primary when operating in partial power mode.

5.1.8 *Transformer Module: Transformer Core, Resonant Tank, and PCB*

The core of the EPC9159KIT is the EPC9551T transformer module shown in detail in Fig. 5.8. In it, the transformer windings and resonant tank are built into two 12-layer PCBs with a 17.5 mm × 23 mm footprint. This provides a total of 20 layers in 2 oz. of Cu thickness for the transformer windings, with the remaining four layers dedicated to routing and connections to components. Both primary and secondary windings form concentric turns around the transformer core's center post. Examples

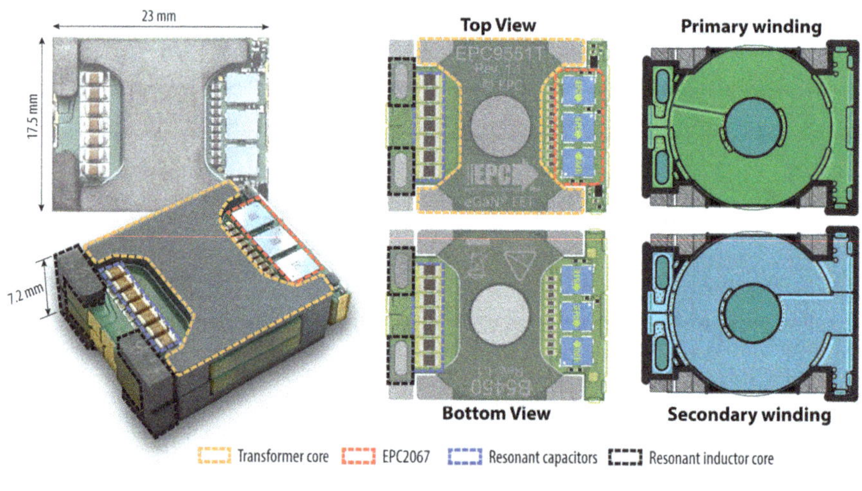

Fig. 5.8 (left) Photo of the EPC9551T transformer module, (center) top and bottom views of the module, (right) example layers for one of the primary turns of the transformer, and example layer for one of the secondary turns

of both are also provided in Fig. 5.8. The layers used for the three primary turns and the single secondary turn of the transformer are interleaved to minimize proximity losses by alternating the magnetomotive force. And the rectifier FETs are integrated in the transformer windings to minimize leakage inductance, which was found to have a significant impact on efficiency in previous work [8]. The center post goes through both PCBs with two plates at the ends that provide a magnetic path for the flux to return through four small satellite posts. The geometry of the end plates was optimized for uniform and adequate flux density while freeing enough board space for the resonant capacitors and secondary synchronous rectifier FETs. The transformer core is split into two identical halves, top and bottom, and manufactured using ML91S material from Proterial. Two layers of 3.5 mil polyimide tape separate each half to set the magnetizing inductance to approximately 1.8 μH. This, combined with the C_{oss} of EPC2619, results in a transition time of ~23 ns when the voltage across the primary is 36 V.

The resonant tank is also integrated in the transformer module, and it is tuned at approximately 1.8 MHz. The resonant capacitance consists of fourteen 22 nF ceramic capacitors in parallel. These capacitors were selected to use U2J dielectric for minimum equivalent series resistance (ESR), high stability over temperature and voltage, and high capacitance compared to C0G [16]. The resonant inductance of ~26 nH is comprised the transformer leakage inductance of ~1.6 nH and two external inductors of ~12.2 nH each. These inductors are built into the PCB using a total of 22 layers and a custom magnetic core in ML95S material from Proterial.

5.1.9 Testing Results

Figure 5.9 shows waveforms of the converter operating in partial power mode and through power mode at various currents. The switching frequency is set to slightly above 1.6 MHz for operation at resonance, considering the transition times of ~23 ns and resonant frequency of ~1.8 MHz. At no load, the resonant current is just the magnetizing current, and it has a triangular shape. As the load increases, the resonant current becomes closer to a sinusoid. It is important to note that the value

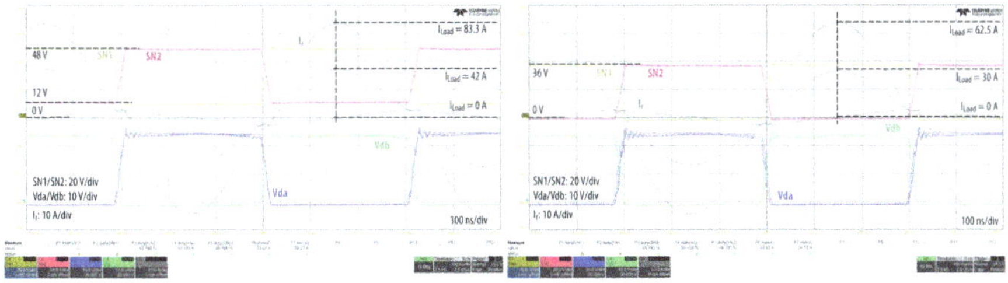

Fig. 5.9 (left) Measured waveforms in partial power mode, (right) measured waveforms in through power mode

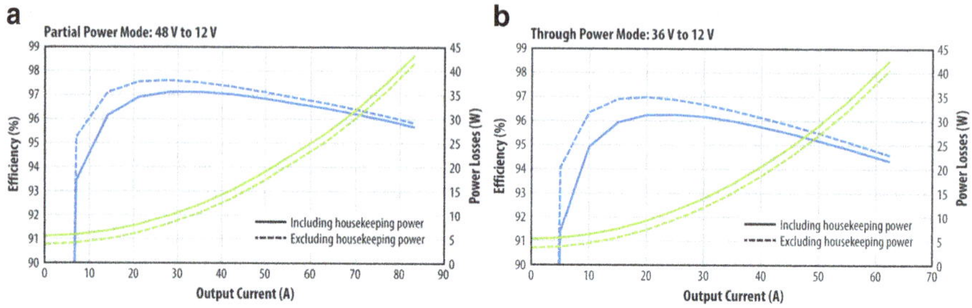

Fig. 5.10 (left) Measured efficiency and power loss in partial power mode, (right) measured efficiency and power loss in through power mode

of the resonant current at the beginning and end of the transition times is the same regardless of the load current, which indicates operation at resonance. In addition, the shape of the resonant current and the secondary drain voltages are independent of the operating mode. It is not the case for the primary switch nodes (SN1 and SN2), where in partial power mode with a 48 V supply, they have a trapezoidal shape between 48 V and 12 V (V_{OUT}), and in through power mode a trapezoidal shape between 36 V and 0 V. The waveforms in Fig. 5.9 are labeled according to the node designators shown in Fig. 5.1.

Moreover, Fig. 5.10 provides efficiency and power loss curves for the EPC9159KIT operating in both modes up to the maximum load. In partial power mode, excluding housekeeping power supplies, peak efficiency surpasses 97.5% at 25 A load and is 95.8% at full load (1 kW). Similarly, in through power mode, peak efficiency reaches 97% at 20 A and is 94.5% at a full load (750 W). As mentioned previously, the power processed by the converter is the same in both modes relative to their full load. Consequently, the power losses relative to the full load are also the same, which can be observed in Fig. 5.10. Details of the thermal kit, including the thermal interface materials (TIMs) and heatsinks, are provided in Ref. [9].

This converter demonstrates how to achieve 48 V to 12 V power conversion with a power density beyond 5 kW/in^3 using GaN transistors. GaN FETs offer the lowest relevant figure of merit for LLC converters, enabling outstanding performance. EPC2619, a 100 V-rated GaN FET in a WLCSP package of 2.5 mm × 1.5 mm in size, offers more than four times lower $R_{DS\ (on)}$ per unit area than an equivalent 100 V-rated Si MOSFET. Similarly, EPC2067, a 40 V-rated GaN FET in a WLCSP package of 2.85 mm × 3.25 mm in size, offers similar $R_{DS(on)}$ per unit area of an equivalent 40 V-rated Si MOSFET with less than half the gate charge.

5.1.10 Summary

In this chapter, two DC output power conversion topologies using GaN devices were examined for their ability to meet the ever-increasing demands from emerging applications for the extremely high power density, light weight, thin form factors, low cost, and high reliability. These constantly increasing requirements have pushed the bounds of power transistors beyond the capability of the aging silicon power MOSFET.

The buck converter was first analyzed, exploring the bounds of efficiency and size using state-of-the-art GaN integrated circuits. A second example discussed the LLC converter that is quickly becoming the preferred topology for high-density applications from 30 to 60 V_{IN} where power levels exceed about 300 W.

In each case, GaN devices enabled the state-of-the-art performance.

5.2 GaN for Time-of-Flight Lidar

Stephen Colino and John Glaser

5.2.1 Introduction

This chapter looks at two types of time-of-flight (ToF) light detection and ranging (lidar) approaches to machine vision where GaN devices have been an enabling building block for both direct time-of-flight (DToF) and indirect time-of-flight (IToF) lidar.

5.2.2 Lidar Fundamentals

A lidar controller/detector triggers the emission of light from lasers. The light is steered through optics, reflected by surrounding objects, and detected to produce 3D point cloud, as shown in Fig. 5.11 [1]. 3-D point cloud photo courtesy of Kim Ingram, University of California, Agriculture and Natural Resources.

5.2.3 Lasers for Time-of-Flight Lidar

Lasers have a set of fundamental characteristics that determine laser driver requirements. Figure 5.12 shows representative curves for lasers at two temperatures. Note that laser performance decreases as temperature increases with elevated temperature shown in red.

A laser must deliver sufficient optical power to meet the sensor's range and field-of-view needs. Current and temperature determine a selected laser's forward voltage drop. The impact of this voltage will be discussed later in the chapter.

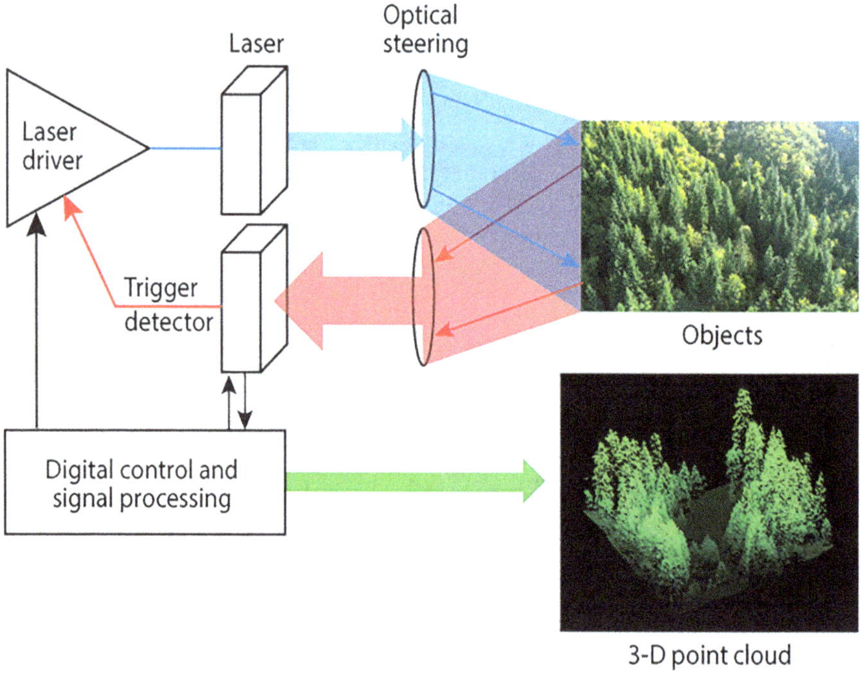

Fig. 5.11 Overview of a lidar 3D imaging system

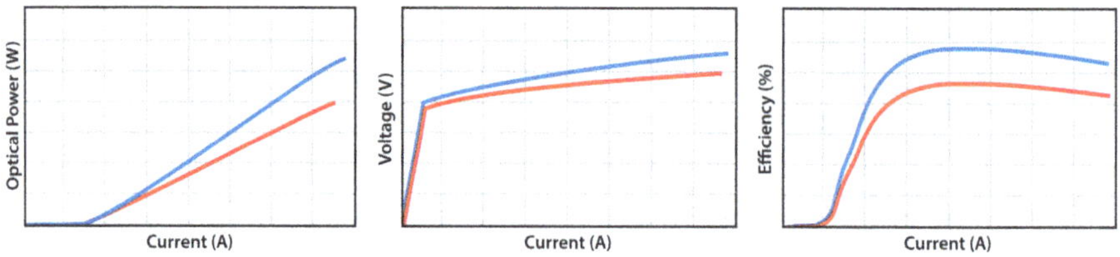

Fig. 5.12 Laser characteristic curves for room temperature (blue) and high temperature (red). (**a**) Optical power vs. current. (**b**) Voltage vs. current. (**c**) Efficiency vs. current

Figure 5.13 shows a simplified model of a laser containing anode and cathode terminals. In addition to the curves of Fig. 5.12, which represent the resistance and voltage drop of the junction, there is parallel capacitance and series inductance. Capacitance is a part of the physical structure of the junction and is determined by the type of laser, the semiconductor materials used, the design power level, and the packaging. It is usually insignificant, but it can start to have an influence in the case of large VCSEL arrays. Capacitance diverts current away from the junction. Series inductance comes from the wire bonds, which impedes a rise in current, decreases emitter speed, and reduces resolution. Wire bonds, which become a part of the laser power loop inductance, should be short and paralleled to minimize inductance.

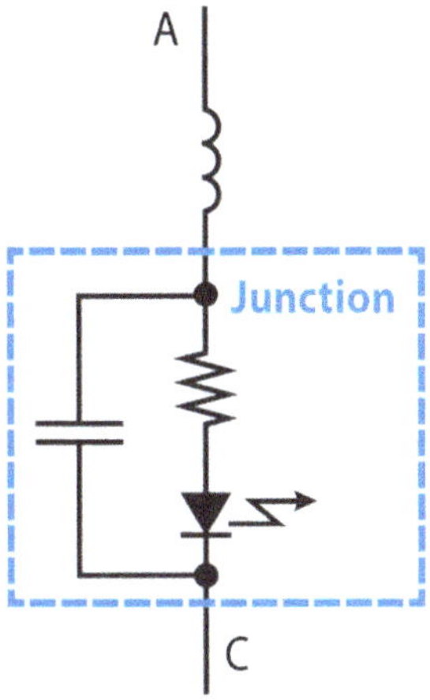

Fig. 5.13 Generic laser model

Laser models from laser manufacturers are essential to predicting emitter performance for reduced design cycles and have been improving, though there is a need for continued improvement in quality of the models.

5.2.4 Laser Drivers

A laser driver must deliver a large amount of current to the laser very quickly. It typically consists of an energy source (usually a capacitor bank), switch (GaN FET or IC), and power loop (stray) inductance as shown in Fig. 5.14 [17]. While there are several variations of the circuit, they all work in the same manner.

With the laser driver preventing current flow, the capacitor voltage is seen across the switch. Turn-on begins with a signal on the input commanding the laser driver to fire, causing its voltage collapse. Subtracting laser voltage, the remaining voltage is applied across the stray inductance according to Eq. 5.1. Laser current increases based on the voltage across the stray inductance as shown in Eq. 5.2.

$$V_{L(\text{str})} = V_{c(LD)} - V_L - V_{LD} \tag{5.1}$$

$$\frac{di_L}{dt} = \frac{V_{L(\text{str})}}{L_{\text{stray}}} \tag{5.2}$$

Considering laser turn-on of 1 ns, 100 A from 80 V laser drive voltage, the maximum allowable power loop stray inductance is 800 pH. Laser turn-on time is a major cause of lidar resolution degradation because it increases timing error. A first-order estimate of this error is about 0.15 m/ns.

Laser turn-on time has three variables: laser driver propagation delay, stray inductance, and laser drive voltage. Propagation delay determines how the laser

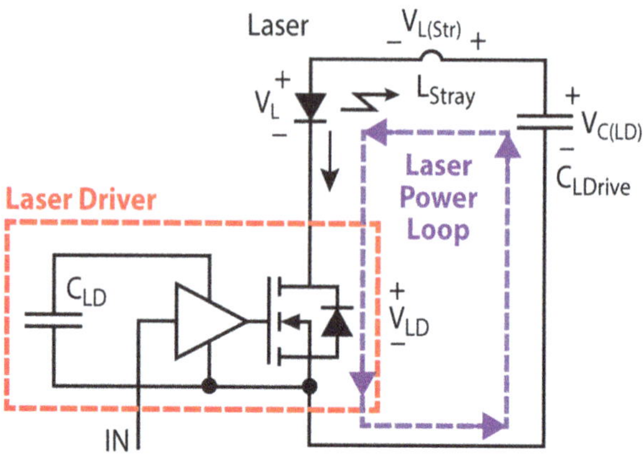

Fig. 5.14 Fundamental Pulsed Laser Drive Circuit

drive voltage is applied to the stray inductance, which determines laser di/dt. Laser driver voltage collapse causes current to flow, which initiates light emission. As laser driver voltage collapses, di/dt is limited by the stray power loop inductance. This stray inductance comes from laser wire bonds, switch interconnections, capacitor, and stray PCB inductance. Techniques for minimizing PCB inductance will be described later in the chapter.

The laser driver must deliver sub-nanosecond laser speed. The laser driver consists of two elements: a switch comprised a field effect transistor (FET) and a gate driver. In order to understand laser driver capabilities and limitations, it is necessary to look at its components, surrounding circuitry, and parasitic elements in detail, as shown in Fig. 5.15.

When the gate driver is triggered, it closes the turn-on gate loop, allowing charge to be delivered from the gate drive capacitor into the FET gate-to-source capacitance (C_{GS}) and gate-to-drain capacitance (C_{GD}), thus opening the FET channel. This transfer of charge is impeded by the gate driver and gate resistances and multiple wire bonds, solder bumps, and other stray inductances. Once the channel begins to open, it starts to conduct current, which discharges the FET's drain-to-source capacitance, and collapses its drain-to-source voltage (V_{DS}). As V_{DS} collapses, voltage is applied to the laser power loop stray inductance, which begins its current ramp and laser optical emission. During this time, charge must continue to flow into the gate to bring V_{GS} up to a voltage where the FET can support the peak laser current. Minimizing R_G, C_{GS}, and C_{GD} for a given maximum V_{DS} and peak current capability is essential to increase laser driver performance.

As laser current increases, voltage is induced across inductance common to the power and gate drive loops (common-source inductance or L_{CS}). This voltage is subtracted from the gate drive voltage, which limits the FET's current-carrying capability while impeding charging of its gate capacitance. For high-resolution lidar, power loop di/dt is typically in the 50A/ns to 100 A/ns range. Values of L_{CS} in

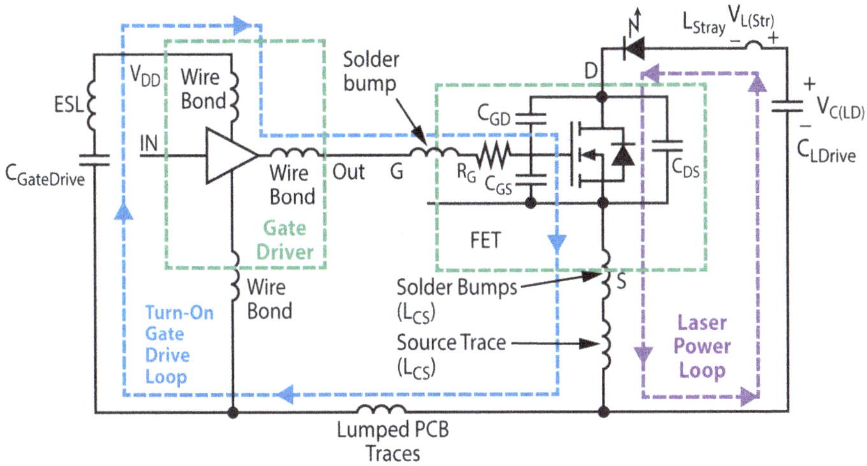

Fig. 5.15 Detailed laser power and turn-on gate drive loops of a discrete laser driver solution

the range of 10 pH to 20 pH (pH, not nH!) are required to achieve the necessary di/dt values in practice.

Design of a laser drive circuit for lidar starts with proper selection of the power element. It must have sufficient peak current capability to deliver the required optical power at 125 °C. While the laser will be the hottest element in the system, the low inductance needed for high speed puts the laser driver very close to the laser on the PCB.

The data in Table 5.1 show that GaN technologies are superior to the latest MOSFET technology. While this is a good semiconductor technology comparison, the packaging of the technologies must be different. MOSFETs are vertical devices; therefore, one of the power terminals must be looped from the top of the chip to the PCB, thus adding substantial inductance. GaN devices are lateral with the active area and all interconnects on one side of the chip allowing wafer-level packaging (WLP) using solder bumps, allowing the active area to be placed very close to the PCB. Additionally, there are multiple source and drain bumps offering ultra-low picohenry inductance to the board. QFN packages typically have the source on top connected to the terminal using a clip, which puts the source inductance in the nanohenry range. While inverted and multiple source terminal packaging are becoming available, fundamental figures of merit for MOSFETs are still far inferior to those of GaN FETs.

The gate driver must have high peak current capability with low resistance and low inductance. The absence of wire bonds gives a WLP gate driver lower inductance and better performance.

Recent breakthroughs in GaN integration have allowed the gate driver to be integrated with the power FET on a single chip for a monolithic laser driver integrated

Table 5.1 FET parametric comparison

Description	EPC2619 [12] (V_{GS} = 5 V)	EPC2204 [19] (V_{GS} = 5 V)	ISC080N10NM6 [20] (V_{GS} = 7 V)
Manufacturer	EPC	EPC	Infineon
Technology	eGaN® FET Gen 6[1]	eGaN® FET Gen 5[1]	OptiMOS™ 6[2]
$V_{DS(max)}$ (V)	100	100	100
$I_{D(PK)}$ (A, 25 °C)	164	125	110 [3]
$I_{D(PK)}$ (A, 125 °C)	145	125	100 [3]
Q_G (nC)	8.5	4.7	14
Q_G Normalized[4]	0.052	0.037	0.127
R_G (Ω)	0.4	0.4	1.2
Q_{OSS} (nC, V_{DS} = 50 V)	27	25	35
Q_{OSS} Normalized[4]	0.164	0.2	0.318
Area (mm²)	3.75	3.75	30
Package Type	WLP	WLP	QFN

Note 1 eGaN is a registered trademark of Efficient Power Conversion Corporation
Note 2 OptiMOS is a trademark of Infineon Technologies AG
Note 3 From datasheet, diagram 7 transfer characteristics curve
Note 4 Normalized charge is $I_{D(PK)}$ @ 25 °C × Q

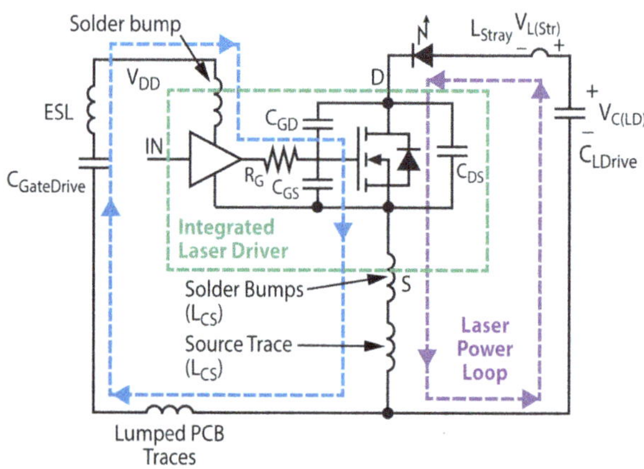

Fig. 5.16 Detailed laser power and turn-on gate drive loops of an integrated laser driver solution

circuit (IC). The advantages of an integrated laser driver over a discrete solution are reduced gate drive loop inductance, matching of the gate driver to the power FET, and smaller size. Laser driver ICs are designed for layout that minimizes common-source inductance. A schematic showing a laser drive circuit with stray inductances is shown in Fig. 5.16 and can be compared with the discrete solution in Fig. 5.15. The size advantage is of particular value in systems with multiple laser channels from reduced PCB routing [21].

5.2.5 *Laser Driver Layout*

Obtaining the low stray inductances needed for lidar requires particular attention to PCB layout. Both power loop and gate drive loop must be considered, minimizing overlap (L_{CS}). This requires the use of short, wide traces, and inductance cancelation where possible. A practical layout is shown in Fig. 5.17 with detail of power and gate drive loops and current flow shown in Fig. 5.18.

Figure 5.18a shows a laser driver schematic with laser power and gate drive loops. Figure 5.18b shows the top layer of the PCB with current paths. Note that the current of the loops goes in opposite directions immediately upon exiting the power FET, leaving the FET source bumps as the only common path, which brings the common-source inductance to an absolute minimum. Figure 5.18c shows the return path to the drain of the FET on the second layer directly beneath the first layer for maximum inductance cancelation.

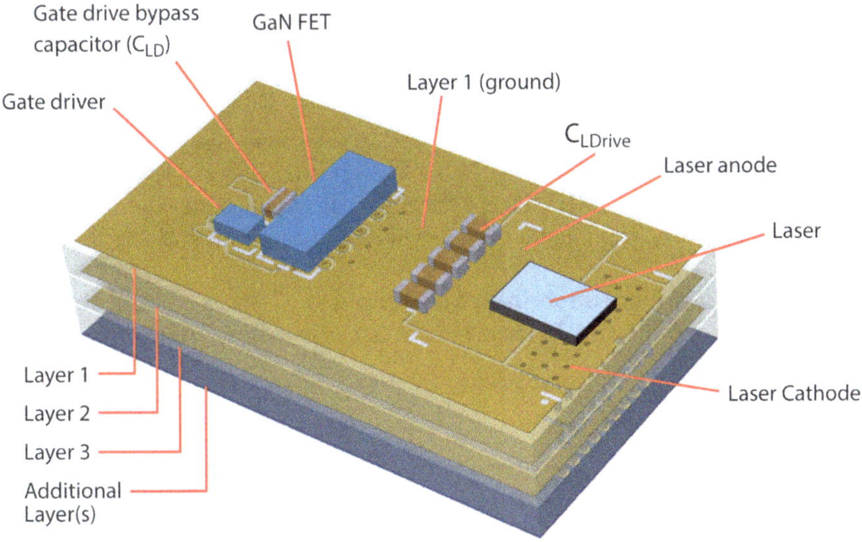

Fig. 5.17 3D image of an example laser driver PCB

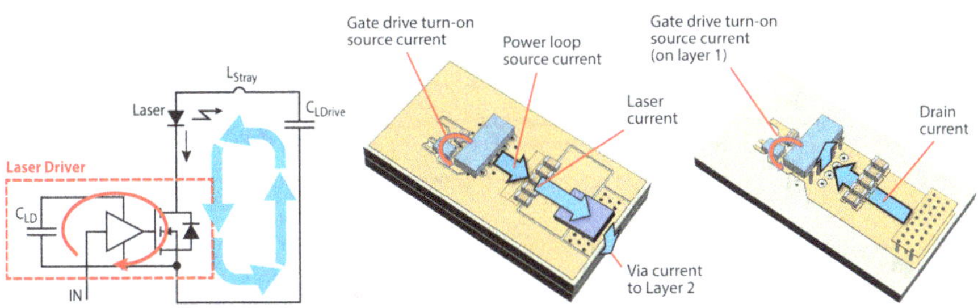

Fig. 5.18 Power and gate drive loops of (**a**) schematic, (**b**) layer 1, and (**c**) layer 2

5.2.6 Direct Time of Flight

DToF is a technique where a timer is started at laser emission and terminated upon the photons being sensed by the detector. Distance is simply the time divided by the speed of light divided by two. There are two critical parameters of the light pulse—pulse shape and pulse peak optical power [22]. Pulse peak power determines the range considering the optical field of view. Pulse shape impacts distance resolution. There are several sources of resolution error in this type of measurement.

The timer start can be the input into the laser driver, another node on the system, such as the power FET gate or drain, or the emitted light reflected off a nearby object, such as a lens. Reflection from the lens would be best as error due to the light path being extremely short and consistent. Using a node in the electrical path adds propagation delay error. This is okay if the delay is predictable. Sources of error are part-to-part variation and temperature. To compensate for these errors, the system may have to use a calibration sequence or temperature feedback.

Pulse shape is a source of error. At time zero, optical emission begins and then optical power rises, peaks, and falls. Simple detectors do not know the point in the pulse when returning photons were emitted. Considering two objects in close proximity, a narrow pulse would return two distinct pulses allowing identification as individual objects. Wide pulses could overlap, returning one indistinguishable pulse. This is illustrated in Fig. 5.19. Pulse widths for DToF are typically 1 to 10 ns with frequency of up to 1 MHz. Long-range lidar can require current pulses through the laser of up to several hundred amps, while short-range lidar can require 10 A or less.

To achieve such high current in such a short period of time, a resonant discharge laser driver topology is used. Referring to Fig. 5.4, resonance is formed by the series

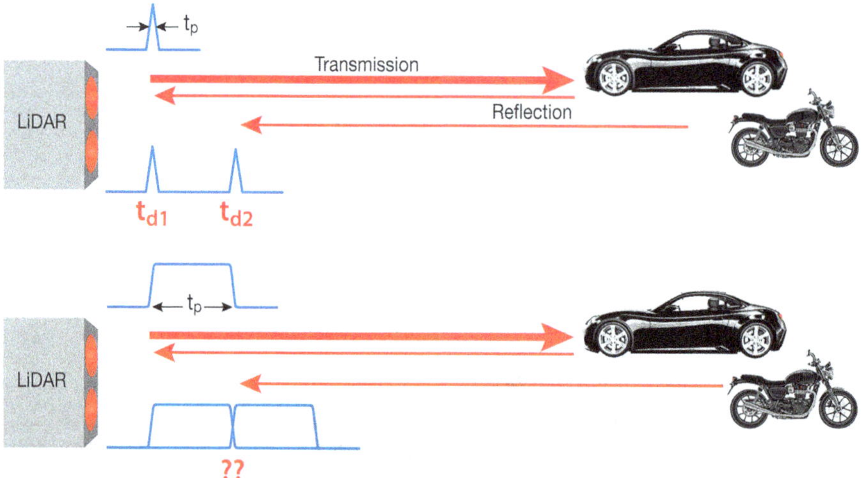

Fig. 5.19 Effect of pulse width on distance resolution for DToF lidar

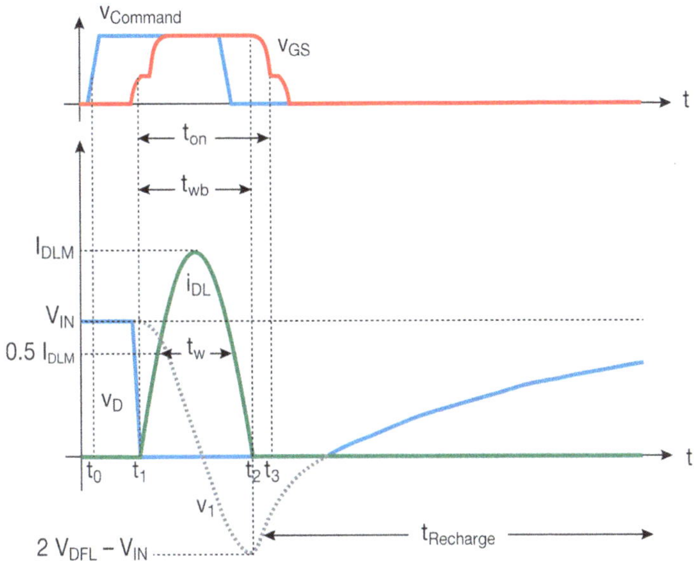

Fig. 5.20 Theoretical waveforms of a resonant discharge laser driver

combination of C_{LDrive} and L_{Stray}. Theoretical waveforms are shown in Fig. 5.20 with pulse width typically considered at full width, half max (FWHM) and are depicted as t_W in the image. Resonant frequency and peak current are estimated by Eqs. 5.3 and 5.4.

As several of the parameters are dynamic and difficult to quantify, iteration of design may be required, but it is always based on minimizing laser power loop inductance, typically in the hundreds of pH range. Capacitors must be of high quality and paralleled where possible to minimize their inductance. NPO or silicon capacitors should be used because of their voltage, temperature, and aging stability. To refine the system, variation in capacitance and voltage can be used. Reducing capacitance decreases the pulse width, peak current, and energy consumption. Increasing voltage increases peak current and energy consumption.

$$f_{\text{res}} = \frac{1}{2\pi\sqrt{L_{\text{stray}} C_{\text{LDrive}}}} \tag{5.3}$$

$$I_{DLM} = \frac{V_{C(LD)} - V_L}{\sqrt{\frac{L_{\text{stray}}}{C_{\text{LDrive}}}}} \tag{5.4}$$

Efficient Power Conversion manufactures GaN power FETs and laser driver ICs, along with demonstration boards using the techniques described above. The EPC9150, shown in Fig. 5.21a, is a demonstration board using a discrete EPC2034C

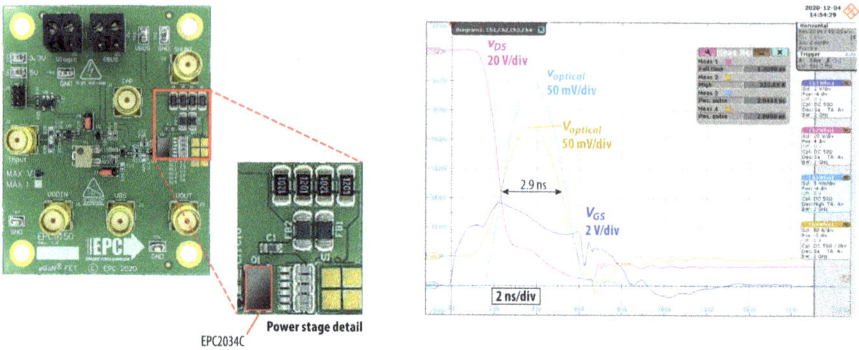

Fig. 5.21 (**a**) EPC9150 laser driver demo board (**b**) associated waveforms

and an LMG1020 driver. Figure 5.21b shows associated waveforms with a 160 V laser driver voltage and capacitance of 2.35 nF.

Peak current reaches 221 A, which produces approximately 500 W of peak laser optical power with an FWHM pulse width of 2.9 ns.

5.2.7 *Indirect Time of Flight*

IToF uses a high-frequency burst of pulses and determines distance by measuring phase shift against the return signal compared with the emitted signal. Bursts are typically in the 100 MHz range. Higher frequency increases the signal-to-noise ratio of the system but introduces the challenge of aliasing. While modern detectors can resolve aliasing challenges, power consumption limits IToF to short ranges.

IToF introduces unique challenges to the laser driver. Because of the inability to recharge the capacitance between pulses at high frequency, a resonant topology cannot be implemented. The laser driver must deliver a high-speed, active turn-off. In this case, the gate driver shorts the gate to source, thus pulling charge out of the gate capacitance and turning the FET off. With the FET blocking current, the power loop current will flow into FET's C_{DS}, causing drain voltage to rise. Laser current will continue to increase until the drain voltage plus laser voltage exceeds the laser drive voltage, which applies a negative voltage on the power loop stray inductance causing the current to decrease. While the laser power loop remains the same as turn-on, the gate drive loop is significantly different giving a large advantage to the use of integrated laser driver ICs.

Figure 5.22 shows the power loop and gate drive turn-off loops of (a), discrete gate driver solution and (b), integrated laser driver solution. In the integrated laser driver solution, the turn-off gate drive loop is entirely within the IC, thus eliminating all wire bond, solder bump, and board trace inductances, which brings common-source inductance to near zero.

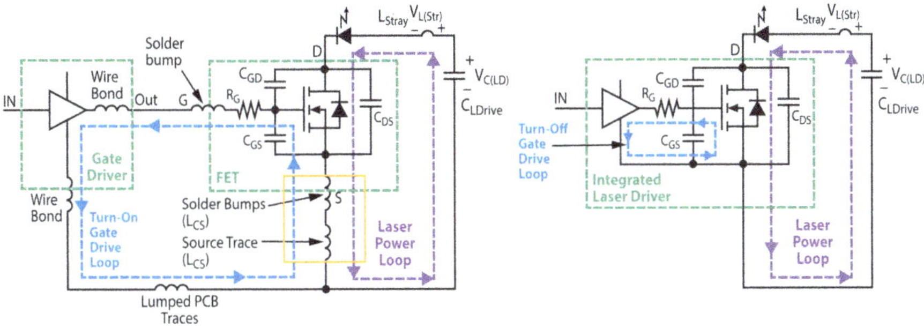

Fig. 5.22 Schematics of laser driver circuits with parasitic inductances for (**a**) discrete laser driver and (**b**) integrated laser driver

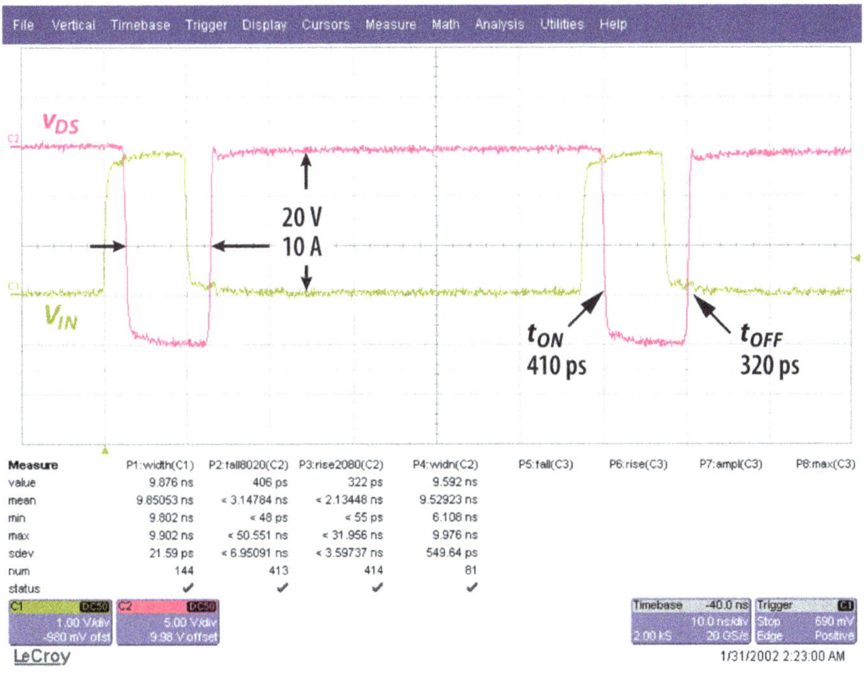

Fig. 5.23 Oscillogram from EPC9154 demonstration board with 2-ohm resistive load

Efficient Power Conversion's EPC21601 is a 15 A, 40 V integrated GaN laser driver capable of over 100 MHz. Its 1 mm × 1.5 mm wafer-level package is compact; yet, its solder bumps offer low inductance. Typical variation in pulse width distortion from 0 °C to 100 °C is 100 ps, which is critical to achieving high signal-to-noise ratio [8]. The EPC9154 is a demonstration board using the EPC21601 [23]. Waveforms from the EPC9154 demonstration board show 10 A switching of a low inductance, 2 Ω load from 20 V with drain fall, and rise times of 410 ps and 320 ps in Fig. 5.23 [24].

5.2.8 Conclusions

The extraordinary capabilities of GaN technology are an enabling factor in the development of affordable lidar. Pulsed lidar is among the most challenging switching power circuits with extreme demands of speed and current on both components and PCB design. The benefits of laser drivers of GaN FETs over silicon MOSFETs are clear, with even further benefit coming from integrated GaN laser drivers. Minimizing inductance with techniques of careful separation of gate and power loops and inductance cancelation is necessary to achieve high performance while minimizing cost.

5.3 Motor Drive Applications

Marco Palma

5.3.1 Introduction

Today, the permanent magnet motor, also known as DC brushless motor (BLDC), is widely used and offers higher torque capability per cubic inch and higher dynamics when compared to other motors. So far, silicon-based power devices have dominated in inverter electronics, but today, their performance is nearing its theoretical limits [1, 25]. There is an increasing need for higher power density.

Gallium nitride (GaN) transistors and ICs have the best attributes to satisfy BLDC inverter needs. The superior switching capability of GaN helps to remove dead time and increase PWM frequency to obtain unmatched sinusoidal voltage and current waveforms for smoother, silent operation with higher system efficiency. Power density increases by substituting electrolytic capacitors with smaller, less expensive, and more reliable ceramic capacitors in the input filter. GaN transistors and ICs provide the capability to design motor drives that operate smoother while reducing size and weight. These advantages are critical for many motor drive applications.

EPC has developed a series of reference designs to ease the design of motor drive control applications and help reduce time to market.

5.3.2 Three-Phase, Two-Level Inverters

Many applications are becoming battery-operated with the advent of high-capacity and high-current cells for batteries, such as lithium-ion cells. Some applications are being converted from combustion engines to electric engines, and others that were initially connected to the power grid are now becoming wireless. Application examples include power tools, forklifts, automatic guided vehicles (AGV) in warehouses or agricultural fields, urban mobility electric vehicles, humanoid robots, drones for cargo transportation, and, in the near future, drones to transport people.

The battery voltages for these applications range from 36 V to 96 V. There is a trade-off between the system weight and the duration of the machine operation with a single battery charge. The longer the operation time with one single charge, the higher the battery weight. This compromise is challenging, especially with flying machines, where the weight must be optimized to the last gram.

Today, all these applications are based on a two-level inverter with the power and size dependent on the application. The motor's phase current may range from 25 A_{RMS} for pedal-assisted bikes to 200 A_{RMS} for a more demanding system designed to lift heavy weights.

The power section of a two-level inverter is shown in Fig. 5.24, where the EPC9173 reference design block diagram is shown as an example. All EPC reference designs have on-board auxiliary power supplies, current sensing, over-current protection, voltage sensing, and temperature sensing circuits. Some reference designs (as the EPC9173 shown in Fig. 5.24) provide two methods to sense the current, either in the leg or phase shunts. However, in motor drives, to exploit the full capabilities of the GaN circuits, it is recommended to sense the phase current

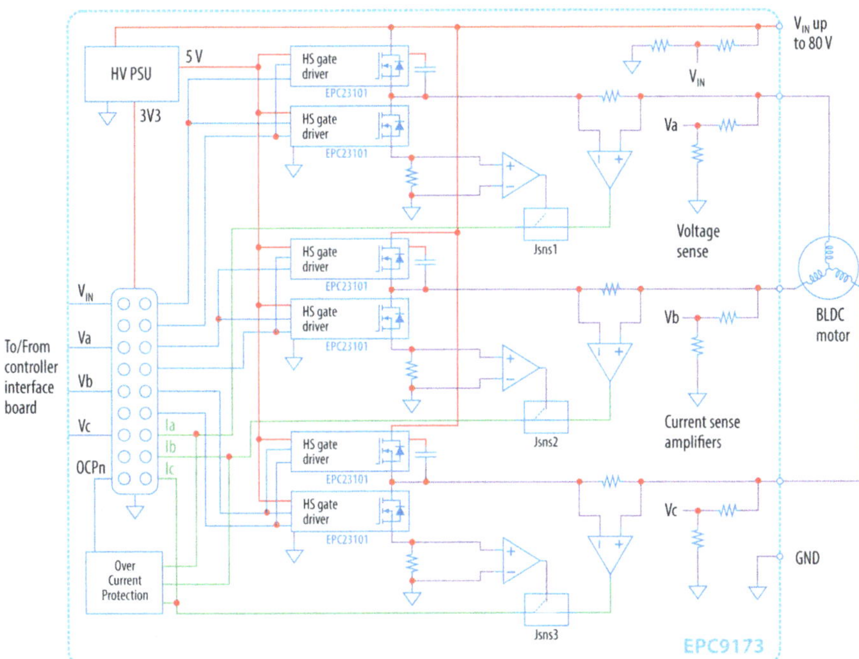

Fig. 5.24 Block diagram of EPC9173 board in BLDC drive example application [26]

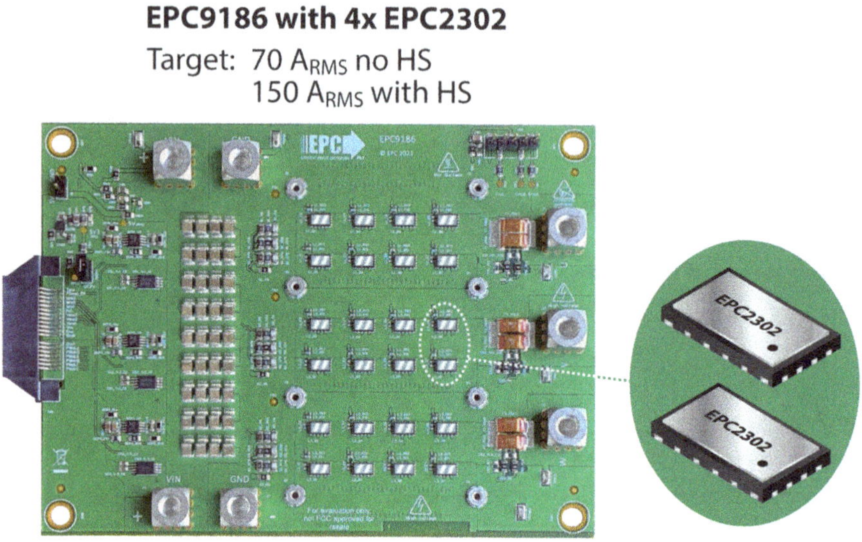

Fig. 5.25 EPC9186 is a three-phase inverter with four EPC2302 in parallel per switch [27]

directly. When the applications require a high current of hundreds of Amperes, several GaN transistors are put in parallel to increase the current capability. The EPC9186, shown in Fig. 5.25, is an example of an EPC reference design that shows how to parallel GaN devices.

5.3.3 GaN Benefits to Motor Drives

GaN integrated circuits and GaN FETs bring several advantages to motor drive applications. The most straightforward benefit is reduction of inverter size due to an intrinsically smaller dimension of a GaN FET and ICs versus an equivalent MOSFET. However, to get the most out of the new technology, it is better to operate a motor at a higher PWM frequency and reduce the dead time [28].

High switching losses and commutation behavior constrain conventional silicon MOSFET inverters, which usually cannot be conveniently operated above a 40 kHz PWM frequency and have a dead time higher than 200 ns. GaN-based inverters are not limited in this sense. They can increase the efficiency of the motor because of lower ohmic dissipation due to reduced current ripple (Fig. 5.26, [29]) and fewer torque harmonics that are responsible for motor vibrations [28].

Moreover, increasing the switching frequency helps to reduce the input filter and removes the need for electrolytic capacitors. A comparison between two inverters, one running at 20 kHz, 500 ns dead time, and the other, based on GaN, running at

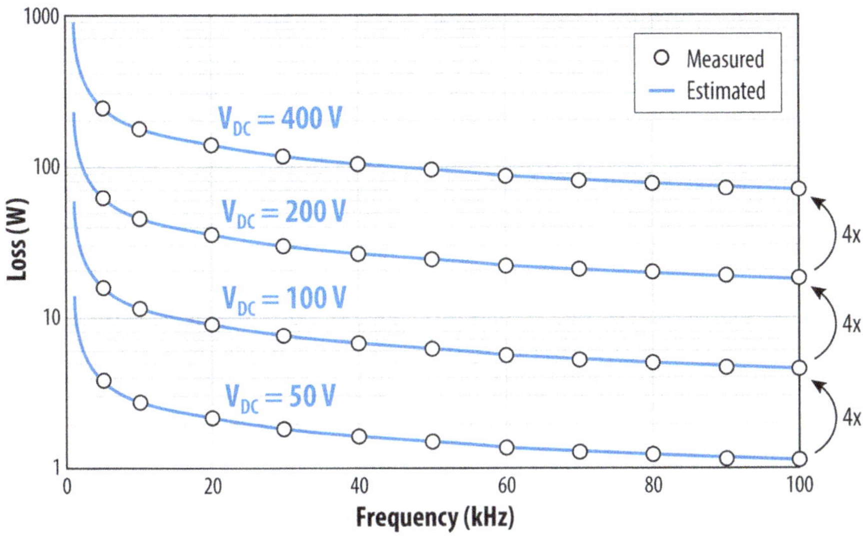

Fig. 5.26 Comparison between measured and estimated PWM-induced losses in a motor with 50% duty cycle square wave voltage excitation [29]

Table 5.2 Comparison between 20 kHz, 500 ns inverter and 100 kHz, 14 ns GaN-based inverter with reduced input filter [28]

Setup	Inverter 20 kHz, 500 ns dead time, 400 RPM, 5 A_{RMS}	Inverter 100 kHz, 14 ns dead time, 400 RPM, 5 A_{RMS}
Input inductance	2.7 μH	None
Input capacitor	660 μF electrolytic	44 μF ceramic
P_{in}	121.3 W	113.3 W
P_{out}	119.6 W	111.3 W
$\eta_{inverter}$	98.50%	98.20%
Speed	42.25 rad/s	41.94 rad/s
Torque	1.876 Nm	1.940 Nm
P_{mech}	79.3 W	**81.36 W**
η_{motor}	66.30%	73.10%
η total efficiency	65.30%	**71.80%**

100 kHz, 14 ns dead time, is shown in Table 5.2. The GaN inverter has no input inductor and uses only two ceramic capacitors. Both inverters operated in the same setup in the same conditions, and the motor was more efficient because many energy-wasting harmonics were removed in the GaN-based inverter case.

5.3.4 Applications with Low L/R Time Constant

All applications that require high electrical frequency and fast dynamics, such as drone propellers and e-bike in-pedal motors, use very low inductance (less than 10 μH range) motors. With the advent of more efficient magnetic circuit designs realized by better materials and higher strength permanent magnets, the number of turns on an electromagnetic phase can be reduced and still produce the same back-emf. Permanent magnet brushless motors generate back-emf voltage, e, proportional to speed, ω ($e = Ke \bullet \omega$), and the maximum speed a given motor can run is directly related to the DC bus voltage and the back-emf voltage constant, Ke. To increase the speed, it is necessary to lower Ke, by reducing the phase coil turns count, which reduces the inductance by the square of the difference. Limiting the current ripple to less than 10% of the phase current is a good design practice and can only be achieved by increasing the PWM frequency.

The current rise with time is related to the ratio of voltage to inductance, and as inductance decreases, the current rises quicker, as does the PWM-induced current ripple. As a result, the higher ripple increases the generated heat in the motor and creates additional EMI noise, which is undesirable. These motors generally have a small time constant, $\tau = L/R$, and benefit from a 100 kHz PWM frequency. It is a good design practice to measure the motor impedance across frequencies (i.e., from 20 kHz to 120 kHz) and select the proper PWM frequency to the point where the motor impedance decrease becomes negligible or where the increased switching losses in the inverter jeopardize the marginal benefit to the motor of PWM increase. MOS-based inverters have no choice: Operating above 40 kHz is not recommended due to the high switching losses. GaN inverters operate perfectly at 100 kHz, or even higher.

5.3.5 Input Current and Voltage Ripple

As per Eq. 5.5, the input voltage ripple increment Δv_{in} in an inverter is proportional to the output phase current and input capacitance.

$$\Delta v_{\text{in}} \propto \frac{1}{f_{PWM}} \frac{I_{\text{phase}}}{C_{\text{in}}} \tag{5.5}$$

The desired Δv_{in} ripple depends on the EMI constraint given by the emissions generated by the cables from the DC source to the inverter. If the PWM frequency is in the range of 20 kHz, the required input capacitance C_{in} can only be practically obtained by using electrolytic capacitors that are bulky and less reliable than ceramic capacitors. Moreover, the electrolytic capacitors are limited by the RMS current that can flow through them, thus requiring more capacitors in parallel and resulting in a total input capacitance that is higher by more than one order of magnitude than needed for a design. When the frequency is increased to 100 kHz, designers can use ceramic capacitors such as X7R, keeping in mind, as a design rule, the effective capacitance drops to half of the specified value when the applied voltage is half of the rated voltage.

5.3.6 Power Transistors and GaN ICs

In the case of GaN, the drift region can be 10 times smaller than in silicon for the same breakdown voltage. Given that the critical field of the crystal is 10 times higher, it follows that the electrical terminals can be 10 times closer together. This leads to a clear differentiation factor between GaN and silicon—medium voltage gallium nitride devices can be built on a planar technology, while this is cost-prohibitive for silicon devices. This further integration in GaN technology improves the power density in miniature motors such as the ones used in humanoid robot joints. With GaN, designers can choose between a conventional approach of discrete components and a more integrated and compact design built on power-integrated circuits. The ICs such as EPC2152 [30], EPC23102 [31], and EPC23101 [32] are the leading edge of the GaN technology.

5.3.7 Layout Rules for Motor Drive Inverters

GaN transistors switch faster than equivalent Si MOSFETs. But with great power comes great responsibility; the layout must be carefully designed. The drain-source power loop circuits and the gate-source loop circuits are sensitive to parasitic inductances. This is important in power converters that must switch above 200 kHz. Motor drive GaN inverters switch up to 100 kHz, and the switching dv/dt is set to less than 10 V/ns to be compatible with motor winding insulation high-frequency breakdown and EMI compatibility requirements. While most GaN-specific layout considerations still hold, other layout design aspects can be relaxed without compromising the outcome.

Reducing the parasitic inductances and following a symmetric approach is essential without compromising the modularity. Three types of parasitic inductance adversely affect the inverter operation:

(a) Common-source parasitic inductance (CSI)
(b) Power loop parasitic inductance (L_{Loop})
(c) Gate loop parasitic inductance (L_{Gate})

For these reasons, the *internal-vertical* layout is recommended that, with the GND plane return on the first inner layer below the GaN devices, has been demonstrated to yield the lowest inductance. One consequence of these rules is that it is common for high-frequency GaN converters to parallel the half-bridges instead of paralleling the transistors.

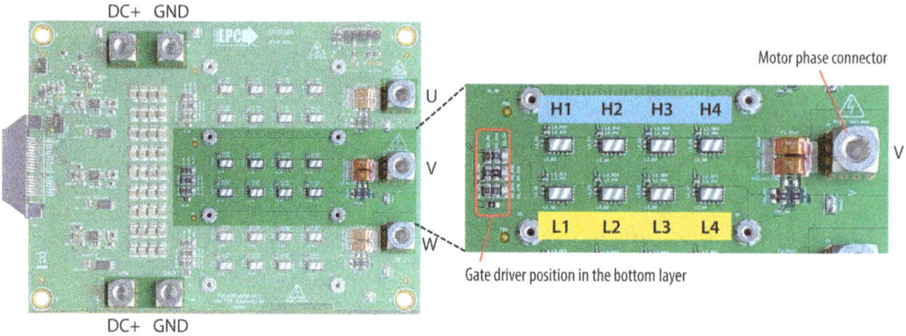

Fig. 5.27 EPC9186 100 V, 200 A_{RMS} per phase motor drive board. Switching cell details on the right

However, this guideline of paralleling the half-bridges instead of the individual transistors can be relaxed in motor drives, and the EPC9186 reference design is an example of how to lay out a 200 A_{RMS} inverter with minimal rules. The EPC9186 switching cell has four EPC2302 transistors connected in parallel, placed with a simplified layout. The gate driver is on the left side of the switching cell; the low-side and high-side GaN FETs are placed in two rows toward the phase output connector.

Figure 5.27 shows the EPC9186 board and the switching cell details, where L1 and H1 are the low-side and high-side transistors nearest the gate driver, and L4 and H4 are the transistors farthest from the gate driver. The motor phase output connector is shown in the same figure.

5.3.8 *Testing Results*

Experimental tests showed that the simplified switching cell power section layout results in clean gate signals in all transistors (Fig. 5.28), independent of the distance from the gate driver. Figure 5.29 shows the EPC9186 steady-state phase current capability as a function of the accepted temperature rise. The steady-state current depends on the thermal condition, and the EPC9186 was tested using a heatsink cooled by 400 LFM airflow.

5.3.9 *Summary*

DC and battery-powered motor applications are moving from conventional Si MOSFETs and low PWM frequency inverters to GaN-based, high-frequency PWM and small dead time inverters. The advantages are a higher system efficiency and removal of the electrolytic capacitors and input inductor.

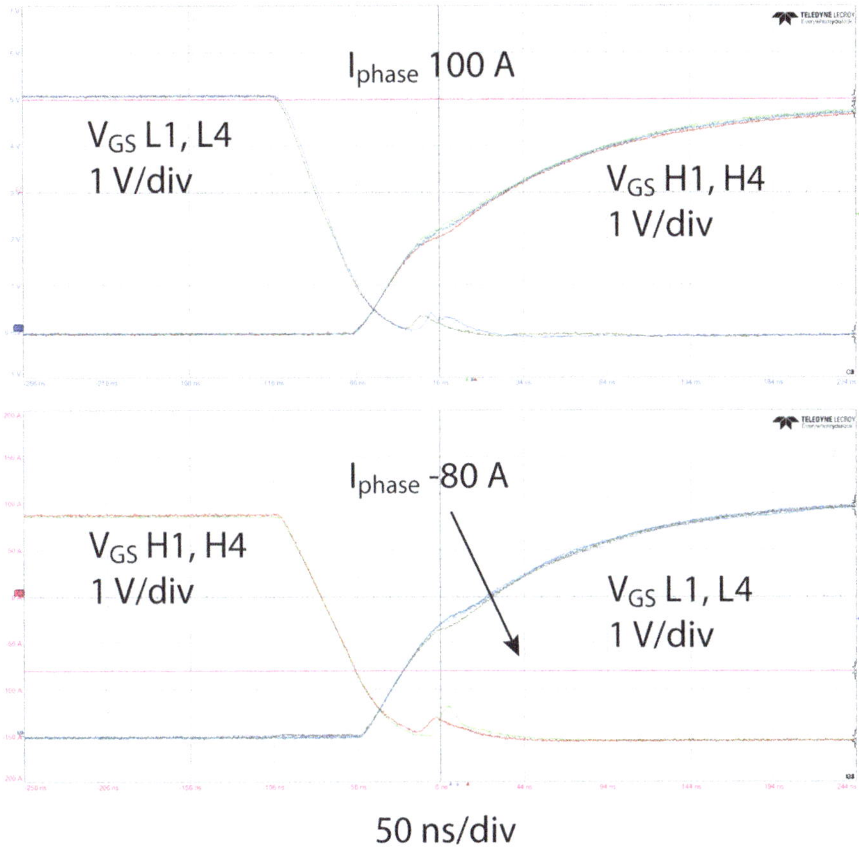

Fig. 5.28 Gate signals H1-L1 vs. H4-L4 at +100 A and −80 A motor phase current. Time 50 ns/div

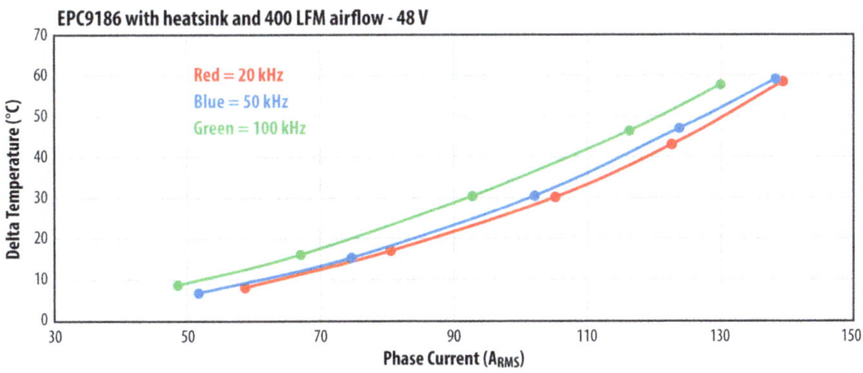

Fig. 5.29 EPC9186 current vs. delta temperature and frequency

GaN FET devices are smaller than silicon MOSFETs and have lower thermal resistance to the case, allowing double-side cooling. They allow more devices to be connected in parallel that can conduct more current in the same inverter enclosure volume, while showing superior thermal management. This paralleling of devices benefits battery-powered industrial vehicles such as power tools, forklifts, automatic guided vehicles (AGV) in warehouses or agricultural fields, urban mobility electric vehicles, humanoid robots, and drones demanding higher currents in smaller volumes.

5.4 Three-Phase Inverters for Motor Drives

Martin Wattenberg

Today's world is more automated than ever, from industrial processes and machinery to the growing field of self-driving vehicles and consumer applications such as robotics and drones. At the heart of it all is the electric motor, thanks to its high power density, energy efficiency, and flexibility in design and control. Crucially, when powered by green energy, these drives play a key role in the transition to a greener and more sustainable economy. The DC/AC inverter at the heart of any modern drive system has an impact on the energy consumption, lifetime, and cost of ownership of the drive.

Gallium nitride (GaN) high-electron-mobility transistors (HEMTs) improve the efficiency and power density of the inverter by enabling faster switching transitions and therefore higher switching frequencies. Operating the inverter at higher switching frequencies provides system-level benefits: (1) lower current ripple in the windings, (2) smaller DC bus and filter size, and (3) faster control response and tighter control of phase current and voltage. In this chapter, we explore best practices for designing a GaN-based motor drive and its impact on system efficiency.

5.4.1 *Benefits of GaN HEMTs*

Compared to conventional silicon-based power transistors used in motor drives, GaN HEMTs provide several improved key properties, due to their wide bandgap nature. Some of these benefits include the following:

- Lower Specific On-Resistance: Due to the wider bandgap of the GaN structure, GaN HEMTs have the potential to significantly outperform silicon (Si) MOSFET with respect to specific on-resistance ($R_{DS(on)} \times A$). Thus, transistors can be made smaller, reducing the device's parasitic capacitance. This further reduces losses and enhances performance. Additionally, the GaN transistors occupy less space on the board.

- Reduced Switching Losses: Switching losses occur every time a transistor turns on or off, and they can significantly impact the overall efficiency of motor drive. GaN's reduced switching losses mean that less energy is wasted in the form of heat leading to higher overall system efficiency as well as smaller and lighter inverter.
- No Reverse Recovery: The HEMT structure does not contain the PN junction found in MOSFETs and therefore does not have any reverse recovery charge. This makes GaN very attractive for hard-switching motor drive applications where the recovery charge increases switching loss, ringing, and often leads to increased EMI.

In the context of motor drives, GaN transistors are particularly well suited to field-oriented control (FOC). Compared to simple block commutation, FOC can maintain a constant torque to the electrical machine and avoids distortions in the current waveform. However, it typically requires higher switching frequencies with increased switching losses compared to block commutation [33]. This is where the significantly lower switching losses of GaN provide the greatest benefit.

It should be noted that GaN HEMTs perform less favorably in block commutation. Here, long free-wheeling periods combined with higher forward voltage can potentially increase switching losses compared to Si MOSFETs (Fig. 5.30).

5.4.2 *Using GaN for Improved Motor Drives*

To implement these benefits in the real world, a recent study provided a comprehensive analysis of a GaN-based motor drive system in FOC applications for permanent magnet synchronous motors (PMSM) [34].

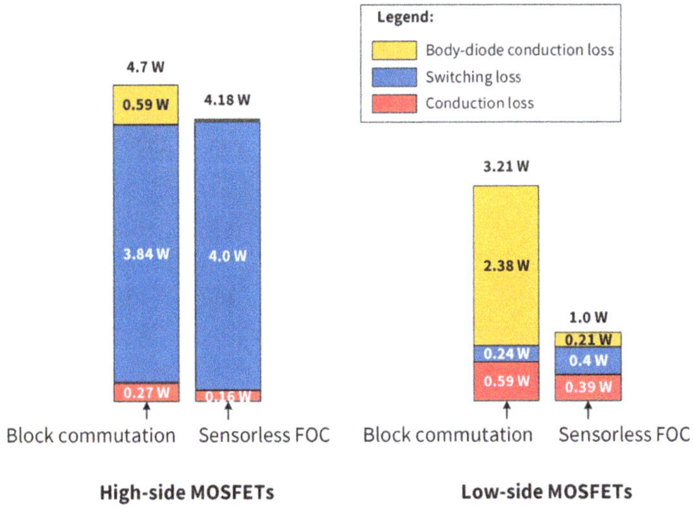

Fig. 5.30 Loss distribution of a conventional Si motor drive for different modulation schemes

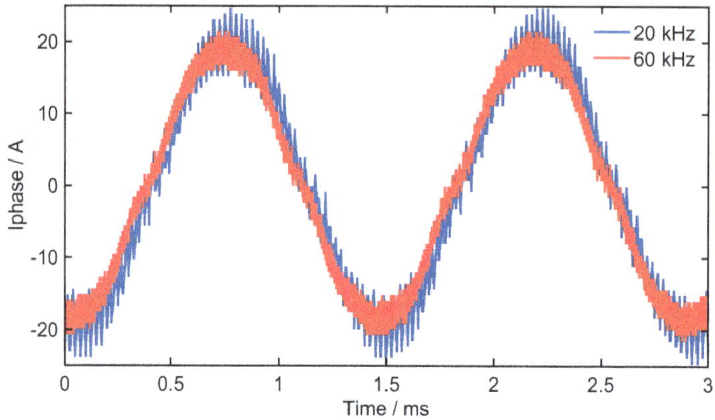

Fig. 5.31 Phase A current waveform at 3000 RPM and 300 W, switching at 20 and 60 kHz

The research compared the impact of two different switching frequencies (20 and 60 kHz) at a motor speed of 3000 RPM and 60 and 100 kHz at 6000 RPM. The motor used was an off-the-shelf drone motor with a phase-to-phase inductance of 40 μH. The findings showcased a significant reduction in current ripple (Fig. 5.31) within the motor windings when operating at higher frequencies. This reduction in ripple was directly correlated with a decrease in winding and core losses, which in turn led to a substantial reduction in motor temperature by approximately 10 °C. Such a decrease in temperature is crucial for enhancing the motor's longevity and performance, especially in demanding operational environments.

However, the study also highlighted a critical trade-off associated with higher switching frequencies: an increase in switching losses that scales linearly with frequency, increasing both power dissipation and temperature of the GaN transistors. In traditional designs based on Si MOSFETs with relatively high switching losses, the best overall performance is often found in the range of 10 to 40 kHz. With the significantly lower switching losses of GaN HEMTs, this optimum is pushed further up in frequency, increasing motor efficiency beyond what is traditionally possible. It is therefore essential to look beyond the performance of the inverter to a holistic design approach when using GaN.

5.4.3 Increased Current Handling with Paralleling GaN Devices

In the context of motor drives, paralleling refers to the technique of connecting multiple power transistors in parallel within an inverter to share the current load. This approach enhances the system's ability to handle higher currents and power

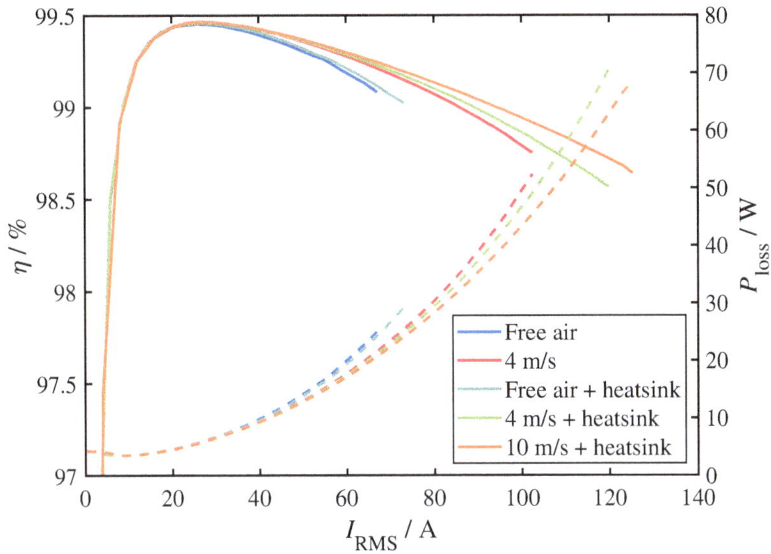

Fig. 5.32 Measured efficiency of a design with 8 devices in parallel under different cooling conditions

levels without exceeding the limits of individual devices. Paralleling improves the overall thermal management and reduces the stress on each transistor, leading to more efficient operation and a longer device lifespan.

To realize the benefits of paralleling, another study presented an ultra-compact three-phase inverter design incorporating eight GaN devices per half-bridge [35]. This resulted in a combined output power of 7.3 kW at 48 V (Fig. 5.32). The inverter's design and layout were optimized to enable parallel switching of the GaN devices using a highly symmetrical layout, which included a "butterfly" structure for balanced gates and power loops.

This approach allowed the system to maintain high switching speeds and low switching losses, which is critical for efficient operation. The high switching frequency, in turn, permitted the reduction of the size of the DC-link capacitor, significantly contributing to the inverter's overall compactness. The peak efficiency of the system was measured to be over 99.3 percent, demonstrating the inverter's ability to deliver high output power with minimal airflow. This efficiency was achieved even under free convection, with the device temperature limited to 100 °C, where the inverter reached an output power of 4 kW, maintaining the efficiency above 99 percent across the full range of operation.

Using active cooling, the inverter was found to be capable of delivering the full 7.3 kW while keeping device temperatures around 60 °C.

5.4.4 Optimizing PCB Layout for GaN Inverters

As demonstrated in the paralleling study, the PCB layout has a significant impact on the overall performance of GaN-based inverters.

A well-designed PCB layout is instrumental in optimizing the electrical performance and thermal management of an inverter, directly impacting its efficiency, reliability, and form factor. In GaN inverters, where high-frequency operation is common, the layout becomes even more crucial due to the potential for increased electromagnetic interference (EMI) and parasitic effects that can degrade performance.

For this reason, the highly symmetrical "butterfly layout" was proposed (Fig. 5.33). This layout ensures that the gate and power loops are balanced, which is essential for maintaining a stable and uniform current flow through all GaN devices. By adopting this symmetrical layout, the electrical stress is shared among many devices, significantly extending the overall current-handling capability.

The layout technique also focused on minimizing parasitic inductance and maximizing thermal performance within a compact form factor. Large copper areas with opposite polarity are interleaved in the PCB to cancel out the parasitic inductance of the traces as much as possible.

5.4.5 High-Voltage Applications and Future Directions

While much of this chapter has focused on the use of GaN transistors in low-voltage FOC applications, these devices have the same benefits in high-voltage applications such as 650 V motor drives. However, the transition to such higher voltage applications introduces new challenges and considerations.

Fig. 5.33 Butterfly layout offers a balanced gate and power loops

For example, scaling GaN technology to higher voltages necessitates careful management of slew rates and dv/dt, the rate of change of voltage over time. These parameters become increasingly critical in high-voltage systems as they can impact the efficiency and reliability of the drive system, as high dv/dt can lead to overshoots and ringing, potentially stressing the devices as well as motor winding insulation. Therefore, designing for higher voltage applications requires a nuanced understanding of GaN device behavior under these conditions, ensuring that the benefits of GaN technology (higher efficiency, reduced size, etc.) are retained without compromising system performance or reliability.

5.4.6 Conclusion

As we have seen, GaN's superior material properties—high-electron-mobility and reduced switching losses—translate into tangible benefits for motor drives, offering more precise control, enhanced thermal management, and greater overall system efficiency. Through good design practices such as optimal PCB layouts and device paralleling, motor drive engineers can harness GaN's full potential.

From low-voltage motor drives to high-voltage power systems, GaN's influence permeates a broad spectrum of applications, promising a future where power electronics are more efficient, more compact, and more capable than ever before.

5.5 GaN in Space

Max Zafrani, Alex Lidow, Robert Strittmatter and Tony Marini

5.5.1 Introduction

Gallium nitride (GaN) power device technology enables a new generation of power converters, sensors, and motor drives for operation in harsh radiation conditions of space. These power devices exhibit superior radiation tolerances compared with silicon MOSFETs. GaN devices operate at higher frequencies and higher efficiencies, and have greater power densities than ever achievable before.

This chapter discusses the capabilities of GaN devices that have been specifically designed for critical applications in high-reliability or commercial satellite space environments. Failure mechanisms in GaN, and how they impact radiation performance, will be explored. The electrical performance of radiation-hardened (rad-hard) GaN transistors is compared with the most popular rad-hard MOSFETs in the market.

GaN-based integrated circuits offer an extraordinary opportunity for improving performance, size, and radiation tolerance compared with silicon alternatives. Much like in commercial applications, combining multiple power devices on a single chip with drivers, level-shifting circuits, and protection circuits can enable performance otherwise unobtainable through a discrete design.

5.5.2 *Radiation Effects in GaN*

There are three primary types of radiation experienced by semiconductors used in space applications. Regardless of whether devices are being employed in satellites orbiting around Earth or incorporated in exploration satellites visiting the most distant parts of our solar system, all semiconductors experience some form of high-energy radiation bombardment. These types of radiation are gamma radiation, neutron radiation, and heavy ion bombardment.

An energetic particle can cause damage to a semiconductor in three fundamental ways; it can cause traps in non-conducting layers, it can cause physical damage to the crystal or insulator, also called displacement damage, or the particle can generate a cloud of electron-hole pairs that will cause the device to momentarily conduct, and possibly burn out in the process [36].

5.5.2.1 Gamma Radiation: Trapping in Silicon MOSFETs

Gamma radiation consists of high-energy photons that interact with electrons. Figure 5.34a is a cross section of a typical silicon MOSFET. It is a vertical device with the source and gate on the top surface and the drain on the bottom surface. The gate electrode is separated from the channel region by a thin silicon dioxide layer.

In a silicon-based MOSFET, the gamma radiation knocks an electron out of the silicon dioxide layer leaving behind a positively charged "trap" in the gate oxide [36, 37]. The positive charge reduces the threshold voltage of the device until the

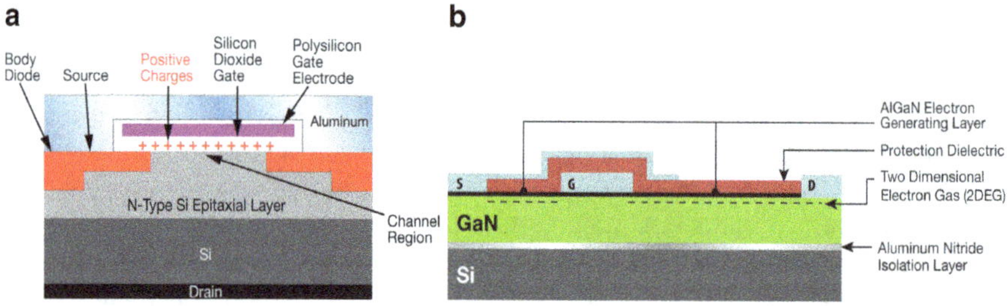

Fig. 5.34 (**a**) Cross section of a typical silicon MOSFET, (**b**) cross section of a typical enhancement-mode GaN (eGaN®) device

transistor goes from normally off—or enhancement mode—to normally on, which is a depletion-mode state. At this point, the system will need a negative voltage to turn the MOSFET off. Typical ratings for rad-hard devices range from 100 kRads to 300 kRads. In some cases, devices can be fabricated for a 1 MRad rating, but these tend to be very expensive, difficult to produce, and generally perform poorly during single-event testing as it is a design trade-off between total ionizing dose (TID) and single-event effects (SEE).

5.5.2.2 Gamma Radiation: Enhancement-Mode GaN Transistors

Enhancement-mode GaN devices are built very differently from silicon MOSFETs. As shown in Fig. 5.34b, all three terminals, gate, source, and drain, are located on the top surface. As in a silicon MOSFET, conduction between source and drain is modulated by biasing the gate electrode from zero volt to a positive voltage—usually 5 V. In enhancement-mode GaN devices, the gate is separated from the underlying channel by an aluminum gallium nitride layer. Unlike silicon MOSFETs, the gate does not accumulate charge under gamma radiation because photo-generated mobile carriers are able to leave the gate to restore equilibrium.

As an example of the performance of rad-hard GaN devices, 100 V GaN-on-Si transistors were subjected to 500 kRad of gamma radiation. Throughout the testing, leakage currents from drain to source and gate to source, as well as the threshold voltage and on-resistance of the devices at various checkpoints along the way were measured, to confirm that there are no significant changes in device performance. Since the initial testing, these same GaN devices have been subjected to 50 MRads, confirming that enhancement-mode GaN devices built as shown in Fig. 5.34b will not be the first component in any space system to fail due to gamma radiation. Testing results are shown in Fig. 5.35. Additional testing and analysis can be found in reference [38].

5.5.2.3 Neutron Radiation

The primary failure mechanism for devices under neutron bombardment is displacement damage [39]. High-energy neutrons will scatter off atoms in the crystal lattice and leave behind lattice defects. Figure 5.36 shows the impact of neutron radiation at doses up to 4×10^{15} per square cm. As with gamma radiation, the impact of neutrons on the GaN crystal and the entire device structure is minimal.

The reason for GaN's superior performance under neutron radiation is that GaN has a much higher displacement threshold energy compared with silicon. The displacement energy of a crystal is proportional to the bond strength of the crystalline elements. The bond energy between gallium and nitrogen is significantly higher than the bond energy between silicon atoms in a silicon power MOSFET as shown in Fig. 5.37. In addition, the electronic trap states that form when crystal damage does occur are far more benign to device operation than they are in silicon [40].

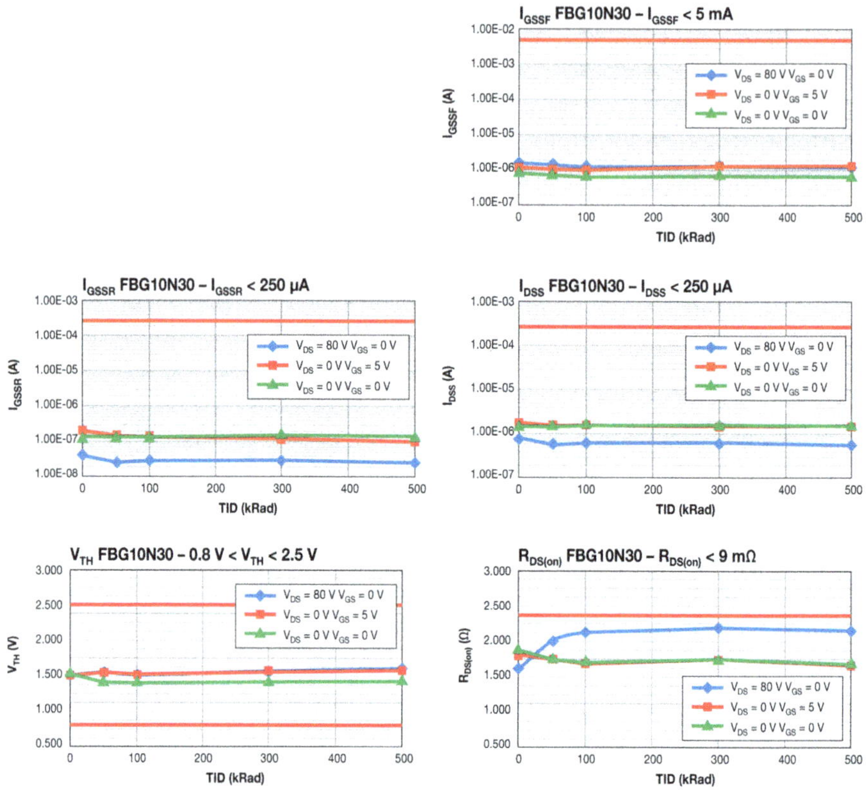

Fig. 5.35 Results of gamma radiation testing of eGaN devices to 500 kRads

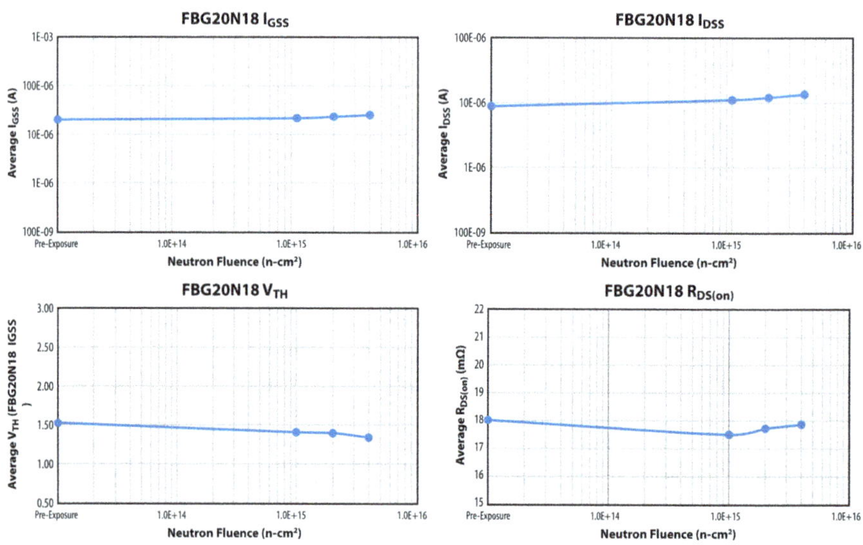

Fig. 5.36 Impact of neutron radiation on eGaN devices at doses up to 4×10^{15} per square cm

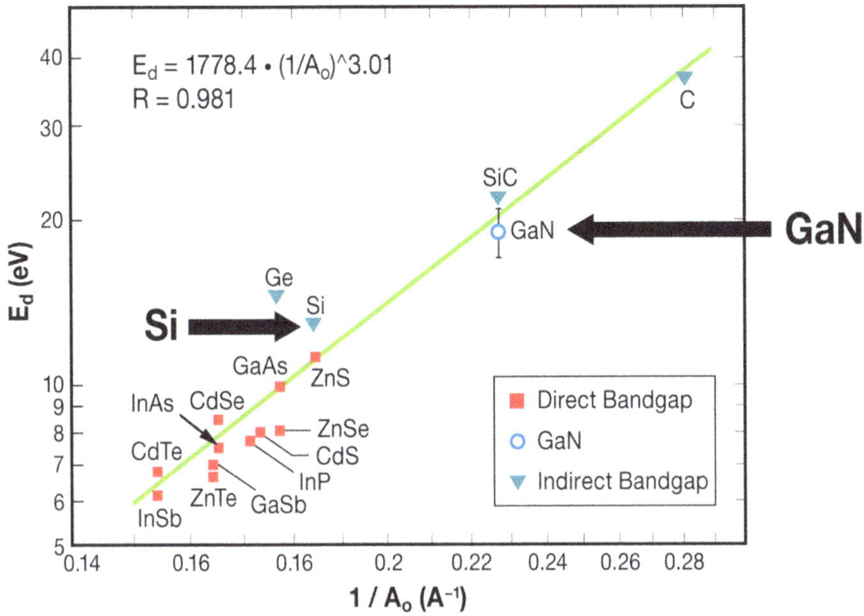

Fig. 5.37 Graph of displacement threshold energy versus the inverse of the lattice constant for different materials taken from Ref. [40]

5.5.2.4 Single-Event Effects (SEE): Si MOSFETs

SEE are caused by heavy ions generated by the impact of galactic cosmic rays, solar particles, or energetic neutrons and protons. This terrestrial condition can be simulated by using a cyclotron to create beams of different ions. Two of the most common ions used to evaluate radiation tolerance of electronics components are Xenon, with a linear energy transfer (LET) of about 50 MeV·cm^2/mg, and gold, with an LET of about 85 MeV·cm^2/mg in silicon.

In a silicon MOSFET, there are two primary failure mechanisms caused by these heavy ions, single-event gate rupture (SEGR) and single-event burnout (SEB). SEGR is caused by the energetic atom creating such a high transient electric field across the gate oxide that the gate oxide ruptures, as illustrated in the upper cross section in Fig. 5.38. However, SEB is caused when the energetic particle transverses the drift region of the device where there are relatively high electric fields. The energetic particle loses its energy while generating a large number of hole-electron pairs.

These hole-electron pairs crossing the drift region cause the device to momentarily short circuit between drain and source, as shown in the bottom cross section in Fig. 5.38. This short circuit can either destroy the device, which is a single-event burnout, or the device can survive, appearing as a momentary short circuit that can cause damage to other components in the system. This latter case is called single-event burnout (SEB).

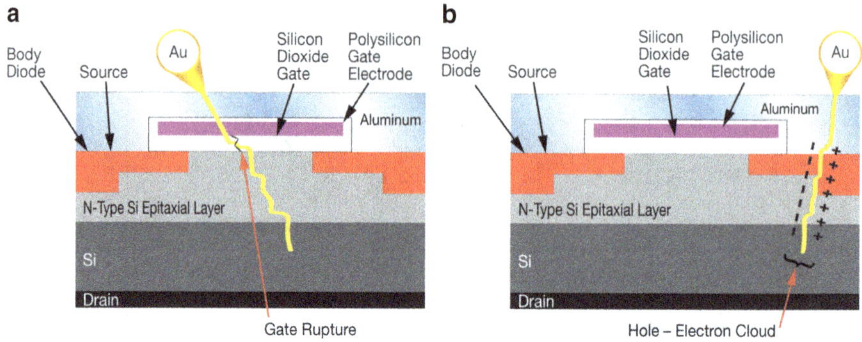

Fig. 5.38 (top) In Si MOSFETs, heavy ions impacting gate regions can cause ruptures in the thin gate oxide region. This phenomenon is called single-event gate rupture (SEGR). (bottom) Heavy ions penetrating drain and source regions can cause clouds of hole-electron pairs to be generated that appear as a momentary short circuit. This phenomenon is called single-event burnout (SEB)

5.5.2.5 Single Event: GaN Devices

Since enhancement-mode GaN devices built, as in Fig. 5.34b, do not have a gate oxide, they are not prone to single-event gate rupture. Also, since GaN devices do not have the ability to conduct large numbers of holes very efficiently, nor do they have a body diode, they are not prone to single-event burnout.

The primary failure mechanism for enhancement-mode GaN devices under heavy ion bombardment is caused by energetic particles crossing the drift region of the device where there are relatively high electric fields. The conditions are about the maximum conditions possible, with an 85 LET beam of gold atoms pummeling the device biased at the maximum data sheet limit. As shown in Fig. 5.39, in testing, the gate leakage does not go up during bombardment. The drain-source leakage, however, does start to rise as the displacement damage from the heavy ions increases. For more research on GaN device SEE performance, see references [41–46].

5.5.2.6 SEE Safe Operating Areas

Many specially produced enhancement-mode GaN transistors have been tested for SEE under varied conditions—40 V and 100 V product did not fail under any conditions up to full-rated voltage and 87 LET. Figure 5.40 shows the results for several commercially available FBG20N18, 200 V products and FBG30N04, 300 V products from EPC Space. For the 200 V products, the first failures occurred at 85 LET and 190 V, as shown in the red circle on the left. The FBG30N04 300 V product failed at 85 LET and 310 V, as shown in the red circle on the right.

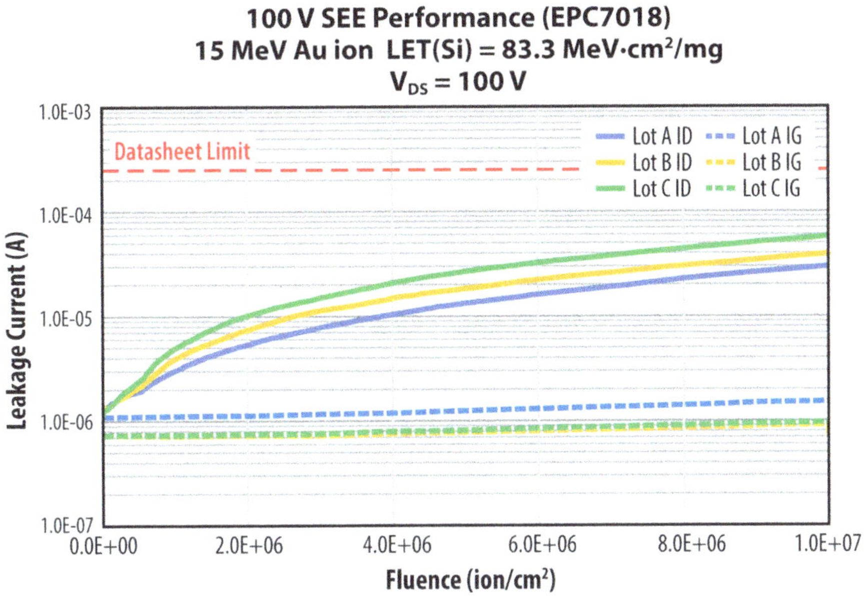

Fig. 5.39 IDSS and IGSS progression for EPC7018G GaN transistor during single-event testing up to a fluence of 1 × 107 ions/cm2 AU at 83.3 LET and biased at 100 VDS

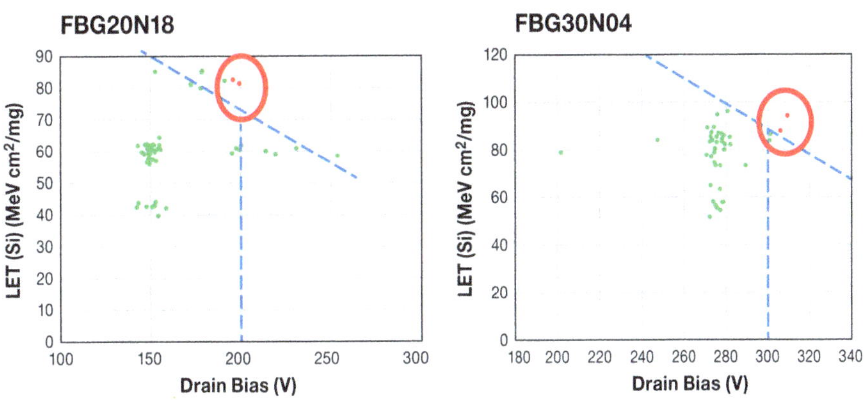

Fig. 5.40 Results from several FBG20N18 200 V products (left) and FBG30N04 300 V products (right)

5.5.2.7 Electrical Performance Comparison

In addition to the superior rad-hard advantages of gallium nitride over silicon, GaN has superior electrical performance. As an example, the electrical performance comparisons of 60 V, 100 V, and 200 V rad-hard GaN transistors against rad-hard power MOSFETs are shown in Table 5.3.

Table 5.3 (**a**) Electrical performance comparisons rad-hard GaN transistors from EPC Space against power MOSFETs from Infineon (60 V). (**b**) Electrical performance comparisons rad-hard GaN transistors from EPC Space against power MOSFETs from Infineon (100 V). (**c**) Electrical performance comparisons rad-hard GaN transistors from EPC Space against power MOSFETs from Infineon (200 V)

(a) 60 V	Parameter	EPC7014	IRHLUB730Z4/JANS F2N7616UB	Units
	V_D	60	60	V
	I_D	2.4	0.8	A
	$I_{D\ pulsed}$	4	3.2	A
	$R_{DS(on)}$	340	680	mΩ
	$Q_{G\ max}$	0.2	3.6	nC
	Q_{GS}	0.03	1.8	nC
	TID Radiation Level	**>10M**	300 k	Rad (Si)
	SEE at LET = 84	60	60	V
	Size	100	9	mm^2

(b) 100 V	Parameter	EPC7018	IRHNA67160	Units
	V_D	100	100	V
	I_D	90	56	A
	I_D pulsed	345	224	A
	$R_{DS(on)}$	**6**	10	mΩ
	$Q_{G\ TYP}$	**11.7**	170	nC
	Q_{GD}	**2.1**	80	nC
	Q_{OSS}	**63**	138	nC
	Q_{RR}	**0**	5500	nC
	Hard-switching FOM ($Q_G \cdot R_{DS(on)}$)	**70**	1700	nC mΩ
	Soft-switching FOM ($Q_{OSS} \cdot R_{DS(on)}$)	**380**	1380	°C/W
	TID Radiation Level	**>10 M**	300 k	Rad (Si)
	SEE at LET = 84	100	100	V
	Size	**44**	236	mm^2

(c) 200V	Parameter	EPC7020G	IRH NA67260	Units	Parameter	FBG20N18	IRHNJ67230	Units
	I_D	76	56	A	I_D	18	16	A
	I_{DM}	200	224	A	I_{DM}	72	64	A
	BV_{DSS}	200	200	V	BV_{DSS}	200	200	V
	$R_{DS(on)}$	**7**	28	mΩ	$R_{DS(on)}$	**26**	130	mΩ
	Q_G	**13.5**	240	nC	Q_G	**6**	50	nC
	Q_{GD}	**2.5**	60	nC	Q_{GD}	**1.95**	20	nC
	Q_{RR}	**0**	11.7	µC	Q_{RR}	**0**	3.5	µC
	$R_{\theta JC}$	1.5	0.5	°C/W	$R_{\theta JC}$	2.12	1.67	°C/W
	TID Radiation Level	**>10M**	300 k	Rad (Si)	TID Radiation Level	**>10 M**	300 k	Rad (Si)
	SEE at LET = 84	190	170	V	SEE at LET= 84	170	170	V
	Size	**14**	236	mm^2	Size	**23**	78.5	mm^2

The 60 V EPC7014 and the 100 V FBG10N30 packaged part have half the on-resistance compared to the silicon MOSFET, yet each is one-tenth the size and has about one-twentieth the gate and gate-drain charges that determine switching speed than their comparable MOSFET devices. In addition, the radiation resistance is significantly higher for these two GaN devices.

At 200 V, the difference in electrical performance of the GaN transistors is even greater. Note that the GaN device listed on the left side of the 200 V section of Table 5.3 has similar on-resistance to its MOSFET counterpart, yet is one-tenth the size and has about 30 times better switching performance while demonstrating superior radiation resistance.

GaN power transistors and ICs are the best choice for power conversion applications in spaceborne systems. Enhancement-mode GaN devices have proven to be more rugged than rad-hard MOSFETs when exposed to various forms of radiation. In addition, the electrical performance of GaN devices is many times superior to the aging silicon power MOSFET.

5.5.3 Summary

Space is a challenging environment for semiconductor components. Constraints such as size and weight are coupled with the need for a high tolerance of various forms of radiation. Solutions in silicon have been available for many years but require compromise in size and weight due to reduced electrical performance and the requirements for various amounts of shielding from harmful radiation.

GaN transistors, hybrid circuits, and integrated circuits create more than an order of magnitude improvement in electrical performance as well as radiation resistance—properly designed and fabricated they may be thought of as essentially inert to radiation damage. This opens up new opportunities to improve existing systems and potentially to change power architectures that further improve efficiency, size, and weight—GaN transistors represent a true gateway technology.

In the final chapter, the journey of GaN power devices from their first commercial launches in 2010 to the date of these writings will be discussed. This journey is still at the early stages.

5.6 Designing a GaN Totem-Pole PFC with a Simple Boost PFC Controller

Scott Durkin and Jim Honea

The bridgeless totem pole is increasingly recognized as the topology of choice for single-phase power factor correction (PFC) due to its elegant simplicity. Perhaps not so well recognized is that the totem-pole PFC is fundamentally just a type of boost

PFC which can be implemented using commercially available boost PFC controllers. The following explains the process of adapting a continuous-conduction mode boost PFC controller to the totem-pole PFC topology.

The basic function of any boost converter is to increase the magnitude of an input supply voltage. The simple DC/DC topology is normally presented using positive voltages, as shown in Fig. 5.41, where *Vin* and *Vout* are positive. But the same functionality can be achieved with negative voltages, as in Fig. 5.42, where *Vin* and *Vout* are still positive valued, but the node voltages are negative with respect to ground. In both cases, energy is transferred from the input to the inductor during the period when switch *S* is turned on (ton = *DT*, where *D* is duty cycle and *T* is the switching period), and released to the output when *S* is turned off and rectifier *D* clamps the inductor-induced voltage at the switching node, Vs, to the capacitor voltage, *Vout*, where *Vout* is of the same sign as *Vin*, but greater in magnitude: *Vout* = *Vin*/(1-*D*).

To use this basic boost converter for power factor correction, given a sinusoidally varying *Vin*, two modifications are required. First, the duty cycle, *D*, needs to be dynamically varied in a way to cause the current drawn from *Vin* to follow the sinusoidal waveform, and also in accordance with the continuously changing boost ratio (*Vout/Vin*). Second, *Vin* needs to be rectified so that the boost converter can always operate in boost mode, whether with positive voltage as in Fig. 5.41, or negative voltage as in Fig. 5.42. The conventional positive-voltage boost converter, including bridge rectifier, is shown in Fig. 5.43. For a universal AC input (85Vac < *Vin* < 264Vac), *Vout* must be at least 375 V if the converter is to always be in boost mode. More typically *Vout* = 380-400Vdc [47, 48].

If a device existed which at different times could serve either the function of the diode *D* or the switch *S* in Figs. 5.41 and 5.42, without adding excessive switching

Fig. 5.41 DC/DC boost converter with positive node voltages

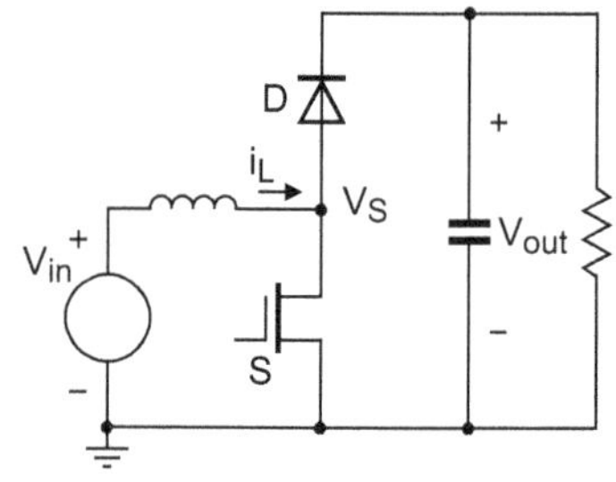

Fig. 5.42 DC/DC boost converter with negative node voltages

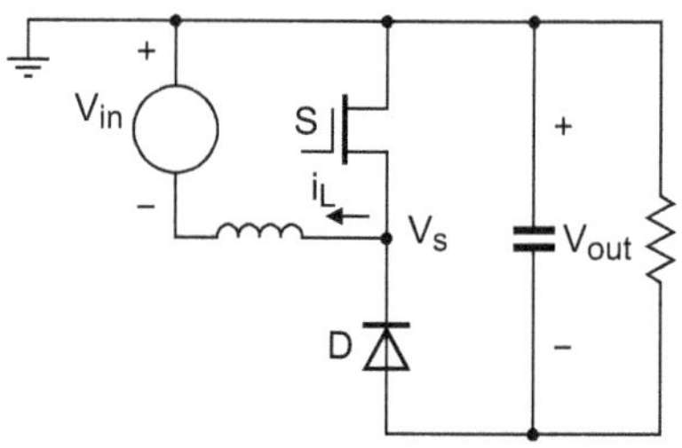

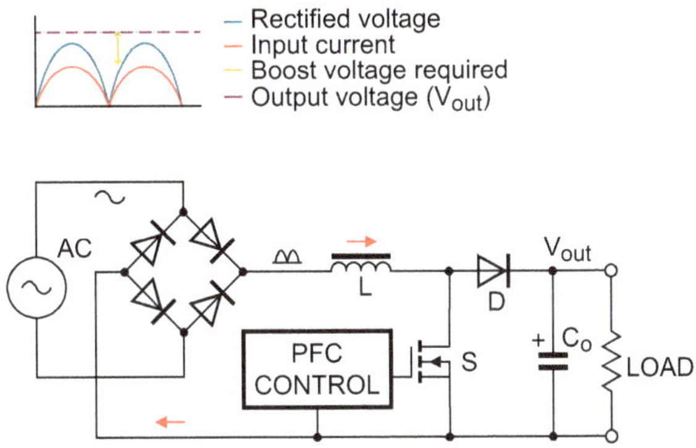

Fig. 5.43 Positive-voltage boost PFC converter, including bridge rectifier

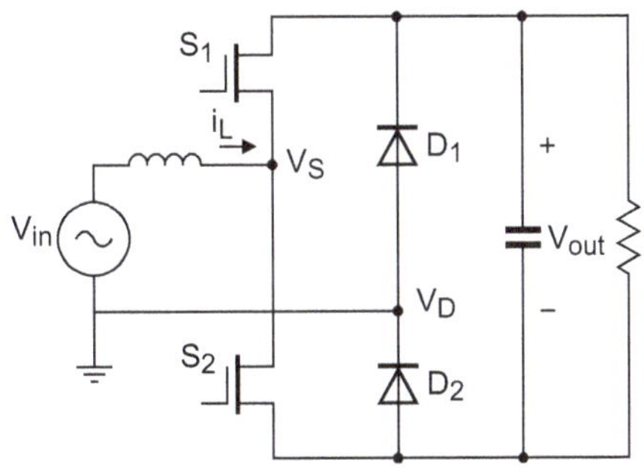

Fig. 5.44 Bridgeless totem-pole PFC converter

loss, then the diode bridge of Fig. 5.43 would not be necessary and the converter could boost *Vin* to *Vout* for either polarity of *Vin* with respect to the AC neutral line (designated as ground in the following figures). The only extra requirement placed on the controller for this topology would be to know which device to drive as the switch and which to drive as the rectifier, based on the polarity of *Vin*. Figure 5.44 shows such a converter.

S1 and *S2* cannot be 600 V/650 V superjunction silicon MOSFETs because the recovery charge associated with the body diodes is too large to permit the MOSFETs to be used as synchronous rectifiers.* High-voltage GaN FETs, on the other hand, have extremely low recovery charge and work perfectly well as either switches or rectifiers.

During the positive half-cycle of the grid, when *Vin* > 0, *S2* acts as the switch and *S1* as the rectifier, corresponding to Fig. 5.41. During the negative half-cycle, the roles reverse; *S1* is the switch and *S2* is the rectifier, corresponding to Fig. 5.42. Diodes *D1* and *D2* effectively move the ground return connection from the positive to negative terminal of the output capacitor, as indicated in Figs. 5.41 and 5.42. During the positive half-cycle of *Vin*, current returns through *D2*, while *D1* is reverse-biased. During the negative half-cycle, current returns through *D1*, and *D2* is reversed-biased.

A further improvement to the circuit is to use MOSFETs as synchronous rectifiers in place of *D1* and *D2*, as shown in Fig. 5.45. Since the current transitions between *SD1* and *SD2* occur at the grid frequency (50/60 Hz), these can be any technology FET, selected for low Rds(on).

The key point is that the operation of the converters in Figs. 5.44 and 5.45 is always that of a boost PFC. Any boost PFC controller can be used, provided the routing of the switching signals can be changed dynamically based on polarity of the input voltage. As will be explained in the following, a little logic and modification of the sensing circuit accomplishes this.

Since the purpose of any power factor correction circuit is to regulate the input current, sensing the input current is a necessary function. Figure 5.46 illustrates how this is typically done in the simple boost PFC. A sense resistor is placed in the return path and the voltage across it is used as the feedback signal. Since the bridge rectifier guarantees that this current is always positive, as indicated by the arrow, the controller can use this signal without concern for polarity.

In contrast, with the bridgeless totem-pole PFC of Figs. 5.44 and 5.45 the return current to Vin does change polarity. A single-sense resistor could still be used, provided an amplifier was configured to alternately invert or not invert the signal, depending on phase of the AC input. A more elegant approach is shown in Fig. 5.47, where a second-sense resistor is added. Since only *SD1* or *SD2* conducts at a given time, current only flows in *Rs1* or *Rs2* at a given time. So there is no penalty of double-power dissipation. A little thought confirms that the sense signal, *Vsense*,

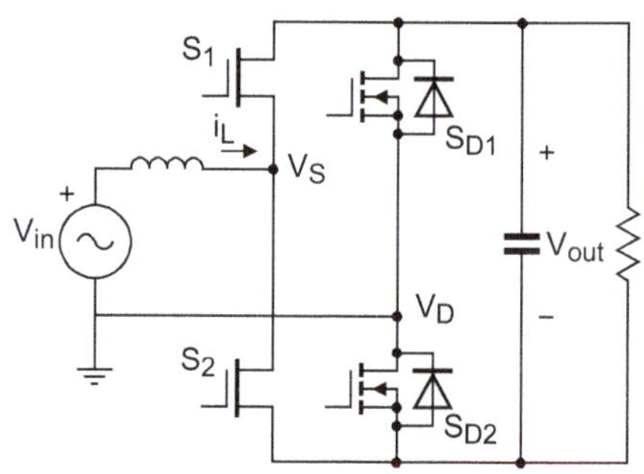

Fig. 5.45 Bridgeless totem-pole PFC with synchronous rectifiers in the slow leg

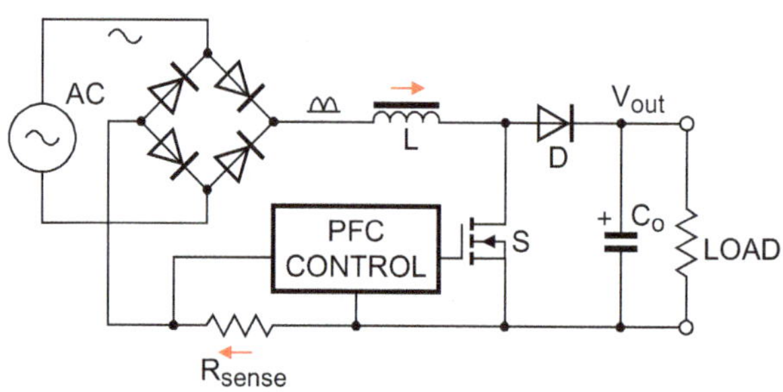

Fig. 5.46 Current sensing in a simple boost PFC

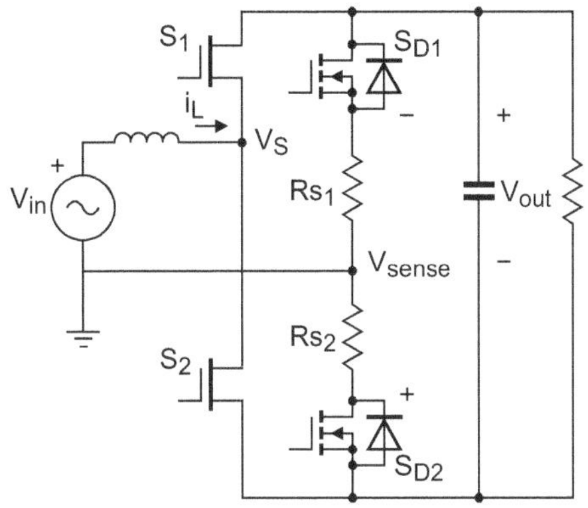

Figure 5.47 A simple approach to current sensing in a bridgeless totem-pole PFC, $R_{S1} = R_{S2} = R_S$

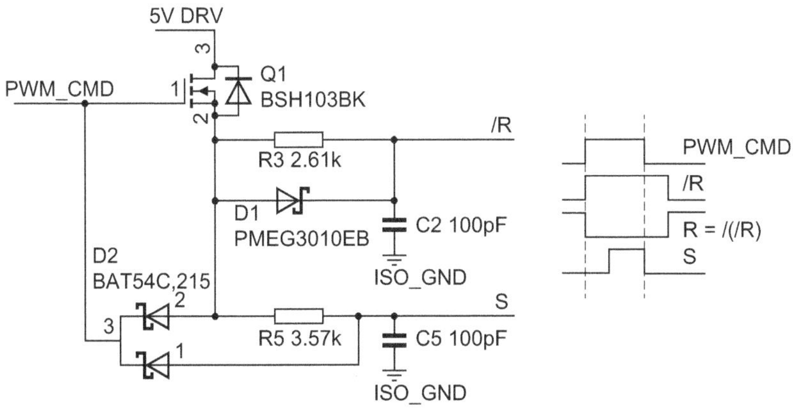

Fig. 5.48 Generating nonoverlapping command signals from a single PWM command

must always be positive and equal to IL*Rs, regardless of the input phase (note the polarity designation—drain of SD2 with respect to source of SD1).

The PWM output command of a simple boost PFC controller is intended to drive the gate of the low-side FET, which is always the active switch in the simple boost circuit (S in Fig. 5.46). For the totem-pole PFC, this command needs to alternately route to *S1* or *S2*, based on the phase of the AC input. In addition, a nonoverlapping command signal must be generated for the opposite FET, if it is to act as a synchronous rectifier. These logic signals are designated S and R for switch and rectifier, indicating the associated function. The circuit of Fig. 5.48 generates the nonoverlapping command signals. Note that in this circuit, both outputs are in phase with the command signal (PWM_CMD). The command for the rectifier needs to be inverted, and for that reason, the signal here is designated */R*. As will be seen, it is not actually necessary to create signal *R* from */R* in the logic as finally implemented.

A voltage comparator circuit is used to generate a logic signal to provide the status of the AC phase, which we will call simply *P*, for polarity ($P = 1$ for $Vin > 0$, $P = 0$ for $Vin < 0$). The gate command for *S1* and *S2* will then be

$$S1\_\text{cmd} = P \cdot R + /P \cdot S$$

$$S2\_\text{cmd} = P \cdot S + /P \cdot R$$

The gate drive commands for *SD1* and *SD2* are even simpler, being simply the status of *P*:

$$SD1\_\text{cmd} = /P$$

$$SD2\_\text{cmd} = P$$

The approach described so far is quite simple. One complication needs to be added, however, and that relates to the zero-crossings of the AC input. Just before *Vin* transitions from positive to negative, for example, the duty cycle of the *S*2_cmd is nearly 100%, for maximum boost, and rectifier *SD2* is holding the negative node of *Vout* at ground potential (AC neutral). Just after the transition, it is the command to *S1*, *S*1_cmd, which is nearly 100%, and *SD1* holds the positive terminal of *Vout* at ground potential. It is not desirable for the mid-node of the slow leg, *VD* in Fig. 5.45, to make an instantaneous transition from 0 to *Vout (+),* with respect to *Vout(−).* If all switching functions cease at the time of the zero-crossing, then the inductor current itself could conceivably cause the desired soft transition. To facilitate this functionality at the zero-crossings, an additional logic signal is required to flag that a zero-crossing is in progress. The *S* and *R* command logic signals should be gated by this signal, as should also the SD1_cmd and SD2_cmd signals. We will assume that this zero-crossing disable is an active low signal, and so we will designate it as /ZC.

It can be tricky to guarantee that the inductor current will create the desired transition, however. An alternative approach for handling the zero-crossing is to make use of the soft-start feature of the controller to effect a gradual transition of the duty cycles for *S*1 and *S*2. The details of this approach depend on the soft-start functionality of the specific controller used.

One final logic signal will be used, and that will be to inhibit turning on the synchronous rectifier when the inductor current is very low (IL < ~300 mA, for example). This will keep the current from reversing in the synchronous rectifier, mimicking the behavior of a diode. Effectively then, the CCM converter operates in discontinuous mode rather than triangular mode for low currents. A voltage comparator can be used to compare the *V_SENSE* (=|IL|*Rs) signal of Fig. 5.47 to a reference voltage. Since the synchronous rectifier must turn off and on within a portion of the switching cycle, a fast-acting comparator (propagation delay of 100-150 ns) is required. We will designate the output signal as LC, for low current, assuming the comparator output is high when *V_SENSE* < *Vref.* To use this signal to inhibit the synchronous rectification, the rectifier command, *R*, should be ANDed with /LC.

Including the signals for zero-crossing and low current, the logic for the FET gate drive commands become

$$S1\_\text{cmd} = P \cdot R / \text{LC} \cdot /\text{ZC} + /P \cdot S \cdot /\text{ZC}$$

$$S2\_\text{cmd} = P \cdot S \cdot /\text{ZC} + /P \cdot R \cdot /\text{LC} \cdot /\text{ZC}$$

$$SD1\_\text{cmd} = /P \cdot /\text{ZC}$$

$$SD2\_\text{cmd} = P \cdot /\text{ZC}$$

A possible implementation of this logic is shown in Fig. 5.49, where DeMorgan's theorem has been used to convert (*R** /LC) to /(/*R* + LC). Note that all four gate drive commands are ANDed with /ZC. If the gate drivers have an unused enable input which can be employed to force the driver outputs low, then /ZC could be connected to those inputs, and the AND gates would not be needed.

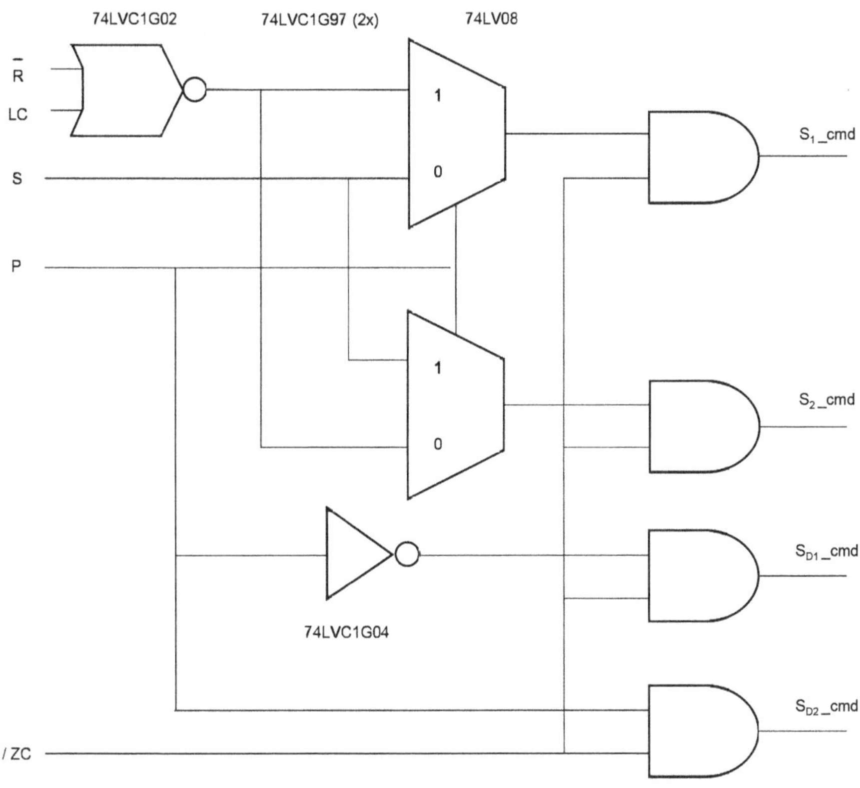

Fig. 5.49 Logic schematic. P polarity, S switch command, /R rectifier command (active low), /ZC zero-crossing (active low), LC low current

Some of the specific circuit implementations presented here have been respectfully adapted from reference [49].

* This limitation is avoided if the ripple current in the inductor is large enough that iL goes to zero or changes sign. In these cases, the MOSFET body diode is not conducting at the moment of switching. Such a large ripple is generally undesirable, however, especially at higher power levels.

5.7 High-Power EV Applications of GaN

Alex Q. Huang, Emad Nazerian, and Peng Han

5.7.1 *Introduction*

The commercialization of wide bandgap (WBG) devices, specifically gallium nitride (GaN) and silicon carbide (SiC), has ushered in a new era for the design of highly efficient and compact power converters. GaN has emerged as the preferred choice for lower voltage and power applications, demonstrating significant success in fields such as data center power supply units (PSUs) and cell phone chargers.

GaN's superiority in these applications stems from two key advantages over SiC. Firstly, the presence of a two-dimensional electron gas (DEG) results in substantially higher channel mobility compared to SiC MOSFETs, enabling the development of low-voltage GaN power switches, with voltage levels as low as 30 V where channel resistance dominates. Secondly, the use of GaN-on-SiC facilitates a lateral power device construction, leading to a reduction in drain-to-source capacitance (Cds). Coupled with GaN's high critical electric field characteristic, GaN switches boast a significantly better figure of merit (FOM) than SiC, as illustrated in Table 5.4 for a 600 V GaN switch.

The comparison utilizes various FOMs, where smaller values indicate superior performance. FOM1 represents gate loop speed, FOM2 denotes drain loop speed, and FOM3 reflects the reverse recovery performance of the body diode. Clearly, from an electric performance standpoint, GaN power switches emerge as the preferable choice. However, practical considerations such as thermal performance, cost, and reliability must also be weighed when selecting a device for real-world applications.

Despite its merits, GaN faces a significant challenge in the realm of high-voltage applications. While lower voltage GaN designs are relatively straightforward, the same cannot be said for their higher voltage counterparts. Presently, 650 V GaN devices are available from various vendors, but offerings for 1200 V GaN are limited. The lateral GaN device construction poses challenges due to a pronounced electric field crowding effect, complicating the design of devices with voltages

Table 5.4 Si, SiC, and GaN device comparison at 600 V. Several device figures of merit (FOMs) are used for this comparison

600V FETs	Ron (mohm)	Ciss (nF)	FOM 1 (Ron*Ciss)	Coss (nF)@400V	FOM2 (Ron*Coss)	Qrr(uC)	FOM3 (Ron*Qrr)
Si SJ	37	7.24	267	0.38	14	36	1332
SiC MOS	120	1.2	144	0.09	10.8	0.053	6.3
GaN HFET	25	.52	13	0.13	3.25	0	0

Si SJ: Infineon IPW65R037C6. SiC MOSFET: Rohm SCT2120AF GaN HFET: GaNSystem GS66516T

exceeding 1200 V. This limitation becomes a notable drawback for GaN when compared to SiC, which boasts commercialized MOSFETs ranging from 650 V to 3.3 kV, thanks to its more robust vertical junction structure.

Furthermore, GaN power switches are predominantly available in discrete package form, with limited availability of GaN modules. This scarcity constrains their application in high-current scenarios. In this article, we address these challenges by presenting two design examples showcasing the use of GaN in high-voltage and high-power applications, shedding light on potential solutions and advancements in this promising field.

5.7.2 *Power Electronics Applications in EV*

One of the most important high-voltage and high-power applications is those found in the electric vehicles (EVs). Historical design practices predominantly employed silicon power MOSFETs and silicon IGBTs (insulated gate bipolar transistors). However, there has been a notable shift in recent years, driven by a strong interest in enhancing power conversion efficiency. This transition involves replacing traditional silicon power devices with silicon carbide (SiC) power MOSFETs. As part of this evolution, gallium nitride (GaN) power devices are also emerging as a natural consideration, particularly if they can be engineered for significantly high-voltage and power applications.

The examples presented in this article specifically pertain to the challenges and solutions relevant to EVs. Traditionally reliant on silicon-based components, the industry is actively exploring SiC power MOSFETs to improve overall power conversion efficiency. GaN power devices, with their distinctive advantages, are becoming an integral part of this consideration, provided they can be effectively designed for the demanding requirements of high voltage and high power in the EV domain.

This shift toward SiC and GaN in EV power systems reflects a broader trend in the pursuit of more efficient and advanced technologies. As the industry continues

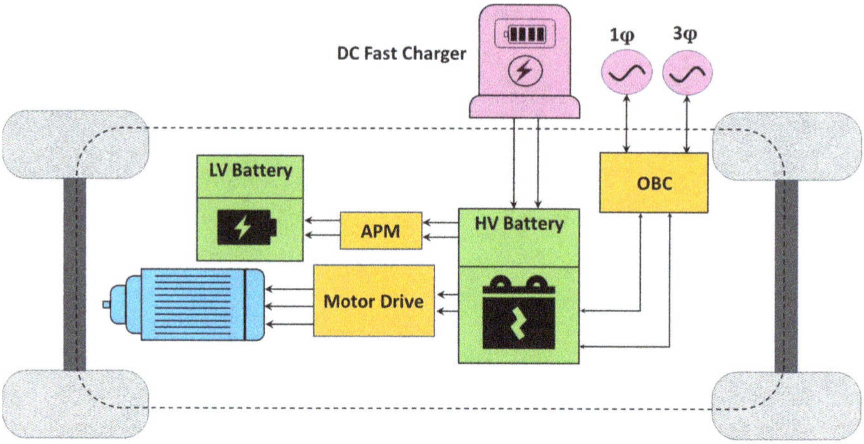

Fig. 5.50 Key components of the EV powertrain

to innovate and adapt, these developments signify a crucial step forward in the evolution of electric vehicle technology, promising enhanced performance and efficiency in the realm of power electronics.

As depicted in Fig. 5.50 below, the auxiliary power module unit (APM), onboard battery charger (OBC), and traction inverter are the three major power converters utilized in an EV powertrain. While the traction inverter has the highest power in this list and the APM has the lowest power level, all of these power converters share three common characteristics: They operate at the same DC bus voltage (HV Battery), hence requiring high-voltage design that are compact and highly efficient.

5.7.3 High-Current Design Example: 400 V/135 kW EV Traction Inverter

Presently, electric vehicle (EV) motor drives predominantly rely on silicon (Si) IGBT technology, which has undergone 40 years of development. The appeal of Si IGBT lies in its cost-effectiveness, well-established reliability, and robust supply chain. In contrast, wide bandgap (WBG) power devices like silicon carbide (SiC) and gallium nitride (GaN) offer superior performance compared to Si IGBT, albeit with a less proven track record in terms of reliability.

In recent times, some automotive manufacturers have begun embracing SiC power MOSFETs as the next-generation devices to replace IGBTs in traction inverter applications. This shift is attributed to demonstrated advantages in terms of higher efficiency and reliability. However, the adoption of GaN-based EV inverters has been relatively limited due to concerns regarding reliability and other factors as discussed above.

GaN power devices with voltage ratings spanning from 30 V to 650 V are currently available, with ongoing efforts by certain vendors to develop 1200 V products. While 650 V devices are commercially accessible, the supply chain for 650 V GaN devices remains restricted, mostly offered in discrete packaging with relatively low-current ratings. As of now, only 28 models boast current ratings exceeding 30A. Consequently, the development of high-power GaN converters necessitates two key initiatives:

Approach 1: Introduction of High-Power GaN Power Modules: Device manufacturers need to take a cue from the SiC space and initiate the production of high-power GaN power modules. This step would expand the availability and accessibility of high-power GaN solutions in the market.

Approach 2: Repackaging Discrete GaN Power Devices: Another crucial effort involves repackaging discrete GaN power devices into high-power units at the converter or board level. This approach addresses the current limitation of discrete packaging, enabling the creation of more robust and powerful GaN converters for EV applications.

In summary, while SiC has gained traction in replacing IGBTs in certain EV applications, the potential of GaN remains largely untapped. Overcoming supply chain limitations and enhancing the form factors of GaN devices are critical steps toward realizing the full potential of GaN in high-power EV converters.

A 400 V GaN traction inverter was developed by our team based on the second approach mentioned above. The design example has a targeted specification summarized below:

- Topology: two-level six-switch inverter
- DC bus voltage: 400Vdc
- Nominal power: 135 kW
- Nominal phase current: 295A
- Peak current: 430A
- Device technology: GaN
- Peak efficiency: >99%
- Cooling system: double-side liquid cool, T_{liquid} = 70 °C

The motivation of using GaN in this application is shown in Fig. 5.51 which compares the losses of GaN and IGBT. The calculated loss at full load (135 kW) is about 300 W lower than that of IGBT.

Figure 5.52 shows the developed 650 V, 2.1 $m\Omega$, PCB-based GaN traction inverter half-bridge power module. The key specifications of the proposed module are summarized in Table 5.5. The design is based on Approach 2: repackaging of discrete GaN devices to achieve high-current ratings. So, the major barrier is how to properly parallel multiple low-current devices inside the power module while maintaining good electrical and thermal performance. In this example, 12 GaN HEMT devices in parallel are used in each switch position, so a total of 24 GaN devices are used in the half. The module uses a 6-layer PCB for power loop connections. Isolated thermal interfaces to the external heatsink/cold plate are achieved by using

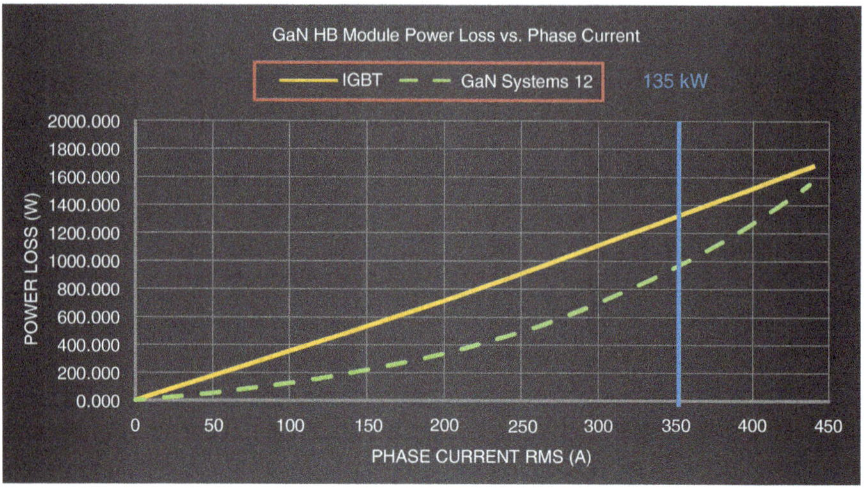

Fig. 5.51 Predicted power loss reduction in the GaN traction inverter

Fig. 5.52 GaN high-power half-bridge module designed for 135 kW EV traction inverter

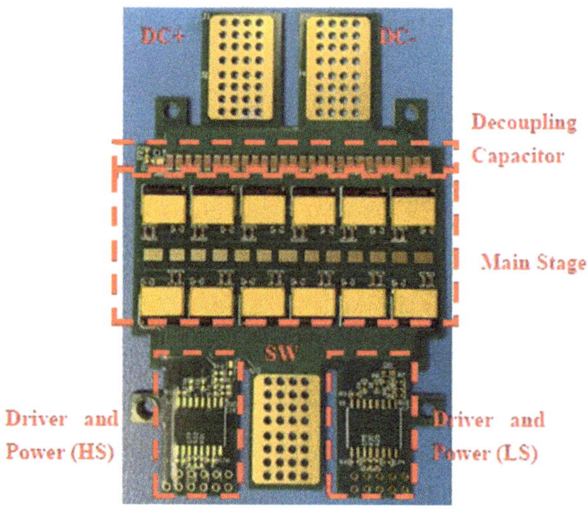

Table 5.5 Developed 650 V GaN module specifications

Parameters	Numbers	Conditions
$R_{ds(on)}$@25°C	2.1 mΩ	V_{GS} = 6V, T_J = 25 °C, I_{DS} = 300A
Q_G	170 nC	V_{GS} = 0V to 6V, V_{DS} = 400V
C_{OSS}	1.44 nF	V_{GS} = 0V, V_{DS} = 400V
R_G	20 Ω	To Each GaN gate
DC Docoupling Capacitor	0.58 µF	Integrated in the module
R_{JA}	103.1 mK/W	ANSYS/calcualtion

a thermal interface material (TIM). Double-side cooling can therefore be achieved from the top and bottom sides simultaneously with each GaN switch cooled from the top side. The overall dimension of the fabricated GaN power module is 64 mm × 98 mm × 4.7 mm.

Unlike conventional IGBT-based traction inverter, in which the DC capacitors and gate driving circuits are mostly placed externally to the power module, the developed high-current GaN module benefits from the PCB-based design and relocates the decoupling capacitors on the same PCB in proximity with the GaN switch. The decoupling cap is needed due to much faster dI/dt in the GaN module. With the decoupling cap, the internal power loop stray inductance along the current path is minimized by minimizing the area covered by the main current. The tested power loop inductance is 0.52nH. This is estimated based on the measured oscillation frequency between power loop inductance and decoupling capacitors.

The example GaN inverter is designed for double-sided liquid cooling, as illustrated in Fig. 5.53. However, the GaN switch used in this work from GaN system comes in a non-isolated package, necessitating the addition of an insulating layer. In the depicted example (Fig. 5.53), thermal interface materials (TIMs) are employed on both sides. These TIMs provide galvanic isolation between the devices and the liquid-cooled heatsink while maintaining excellent thermal performance. The overall thermal resistance on each side comprises three components:

- GaN device junction-to-case thermal resistance (R_{jc})
- TIM thermal resistance (R_{TIM})
- Heatsink-to-ambient (liquid flow) thermal resistance (R_{heatsink})

By calculating the total thermal resistance from the junction to the liquid, we arrive at a junction-to-liquid thermal resistance of 103 mK/W as shown in Table 5.6. With its result, the DC current rating, assuming a liquid temperature of 70 °C and a maximum device junction temperature of 150 °C, is 378A.

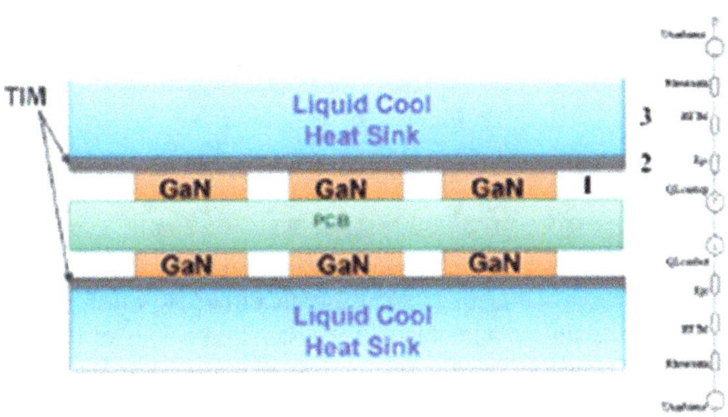

Fig. 5.53 Double-side cooling structure for the GaN inverter

Table 5.6 Junction-to-liquid thermal resistance calculation

Layer	Material	Thickness (mm)	λ (W/mK)	R_{th} (mK/W)
1	Al			5.6[a]
2	TIM	0.7	11.5	75
3	GaN	0.5	27.1	22.5
	Total			103.1

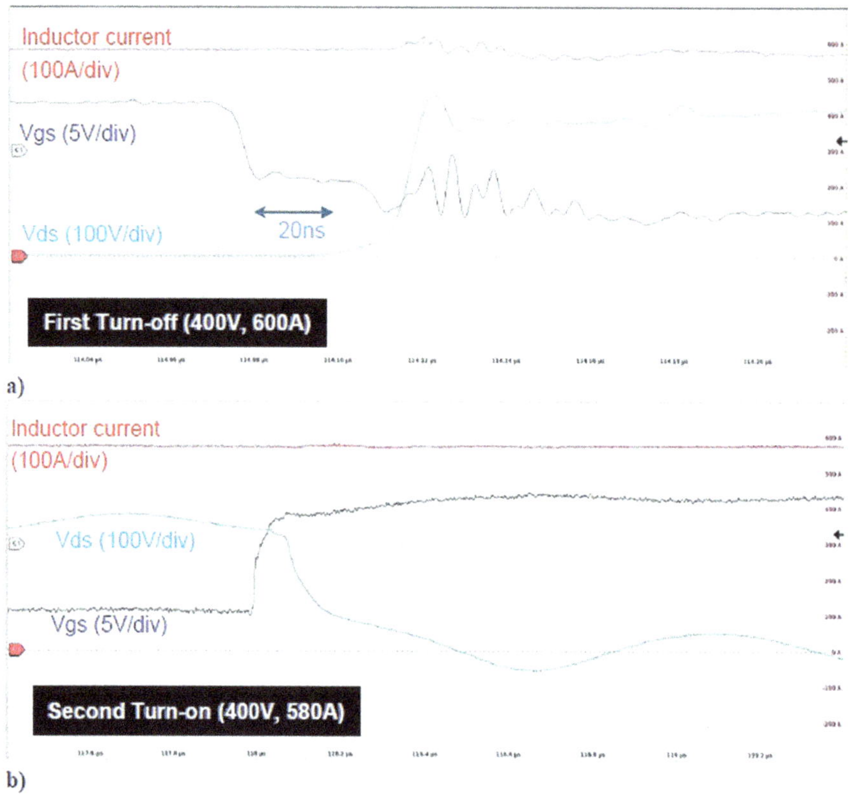

Fig. 5.54 GaN half-bridge module double pulse test at 400 V/600A. (**a**) turn-off; (**b**) turn-on

Three of the developed half-bridge modules are utilized to create the three-phase traction inverter. Prior to subjecting the inverter to a full-power test, the half-bridge module underwent testing in a double-power tester. The results of the turn-on and turn-off performance at 400 V/600A are depicted in Fig. 5.54. In this test, very fast and clean switching was achieved. The device's turn-off time was less than 40 ns. The clean waveform observed during testing validates the feasibility of our design methodology for achieving very high power and high-current performance using discrete GaN power devices.

5.7.4 High-Voltage Application Example: 800 V EV On-Board Charger (OBC)

Although EVs offer a wide variety of advantages, they still have some yet-to-be-solved concerns, and above all is the so-called range anxiety. Some automobile manufacturers utilize larger and heavier battery packs, which enable customers to drive further between charges. Another approach is accelerating the charging speed or minimizing the charging time. One of the significant developments that can remarkably enhance charging speed is the introduction of 800 V electrical systems. The 800 V architecture offers several advantages, such as lower weight, higher thermal efficiency, and, above all, a significantly higher charging speed. For the same amount of current, the 800 V EVs can be charged at two times the power hence cutting down the charging time by 50%. Porsche Taycan was the first car to use the 800 V architecture, and it is also being adopted by other cars, such as the Hyundai Ioniq5, Kia EV6, Audi e-tron GT.

Power conversion system used in 800 V EVs will require power devices with 1200 V or high-voltage ratings. This is one area that SiC has substantial advantages over GaN due to the availability of numerous vendors that offer 1200 V SiC MOSFET. The blocking voltages of commercially available GaN switches are very limited—650 volts at the moment. To address this limitation, a three-level converter provides the opportunity to apply 650 V GaN switches in 800 V applications. Multilevel structures are suitable topologies that not only enable designers to use GaN devices for high-voltage applications but also offer additional benefits, such as smaller filter size, lower dv/dt, and reduced electromagnetic interference (EMI).

Among all known three-level topologies, the FCML (flying capacitor multilevel) topology offers a versatile structure that can be replicated across multiple levels, allowing for the use of switches with lower blocking voltages and higher FOM in high-voltage applications. This replication also facilitates achieving very high power density. However, increasing the number of levels introduces complexities such as hardware design, converter control, and voltage balancing of flying capacitors. This complexity often results in higher costs for the converter. As the current trend in the powertrain architecture of EVs shifts toward 800 V systems, a three-level FCML structure can be deemed an appealing choice. It facilitates the utilization 650 V GaN switches. During the operation, the voltage stress on each device is 400 V while the battery voltage can be 800 V.

An 800 V OBC can be realized using the three-level FCML as shown in Fig. 5.55. The front end is a single-phase three-level FCML AC/DC converter or power factor correction (PFC) circuit, while the DC/DC stage is a three-level resonant converter. The design example of the PFC is discussed below as an example on how to achieve high voltage by using the FCML. The design example is conducted by using 650 V GaN HEMT V22TC65S1A from VisIC.

In the design of the three-level FCML based on GaN, increasing switching frequency is desirable to achieve higher power density. Therefore, minimizing loop parasitic inductance is very important. In the FCML, there are multiple loops since

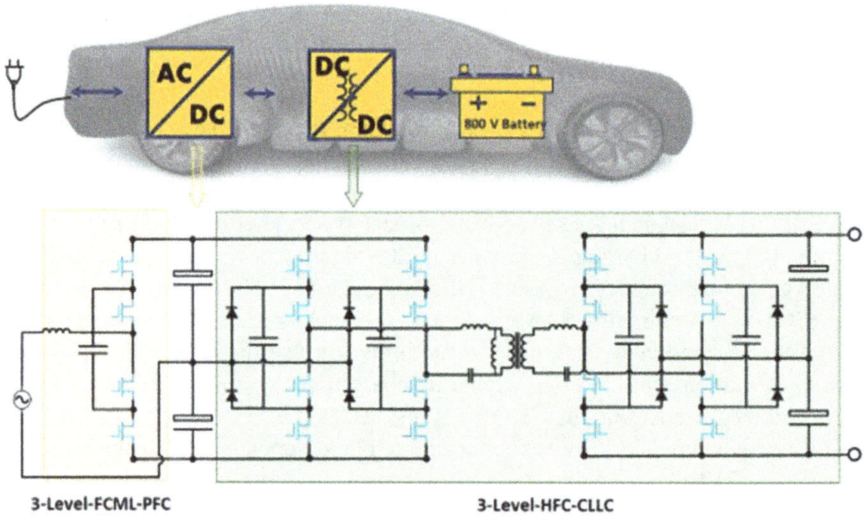

Fig. 5.55 Three-level FCML can be used as the front-end power factor correction (PFC) circuit for a high-power OBC in 800 V EVs. The second stage is a three-level DC/DC converter using 650 V GaN

there are four power devices involved in each phase leg. This aspect is substantially different from two-level converters; hence, careful design considerations are needed to obtain excellent switching performance. While increasing the gate resistance can reduce the transition rate and alleviate this issue, it will significantly increase switching losses, leading to reduced converter efficiency and decreased power density. In the three-level FCML phase leg shown in Fig. 5.56a, we have two loops, an inner loop and outer loop. The inner loop includes S_2, S_3, and C_{fly} and shows 2.51 *nH* loop inductance based on the layout shown in Fig. 5.56b. The outer loop, also shown in Fig. 5.56a with the purple line, includes S_1, C_{fly}, S_4, and C_{in} showing 2.9 *nH*. This reduces switching ringing and device voltage overshoot and provides very high turn-on speed which substantially reduces the switching losses. The actual layout is shown in Fig. 5.56b. In this layout, switches S1 and S4 are positioned on one side, while switches S2 and S3 are located on the other side. This configuration helps minimize the commutation loop distance.

The developed three-level FCML PFC hardware is depicted in Fig. 5.57a, along with its dimensions. For this design, we selected a VisIC GaN device with a junction-to-case thermal resistance of 0.3 K/W. This device features an isolated package, allowing direct attachment of a compact heatsink to all four devices. To evaluate thermal behavior, we conducted a thermal test by applying a constant power loss to all four devices in the reverse direction and measuring their corresponding temperatures. Figure 5.57b displays the thermal image of the backside of the printed circuit board (PCB) during this test. Based on the captured data, we calculated the thermal resistance for each device. Specifically, the junction-to-ambient thermal resistances

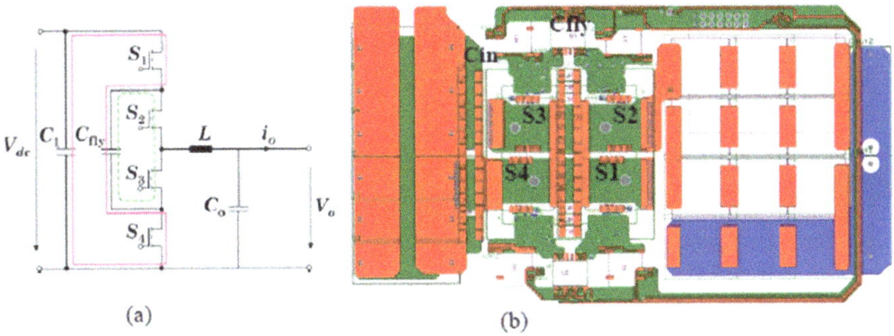

Fig. 5.56 Commutation inductance within the three-level FCML

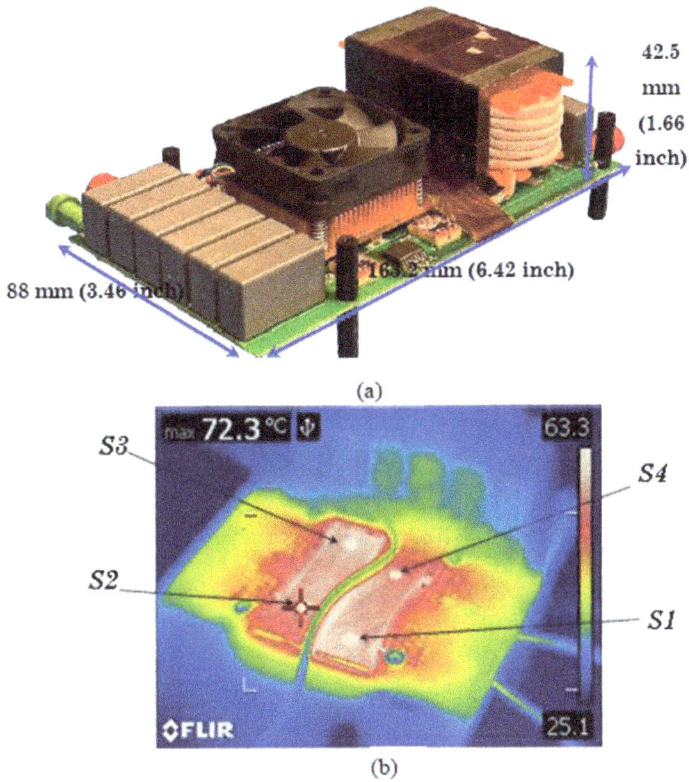

Fig. 5.57 (**a**) Hardware of the developed GaN three-level FCML converter; (**b**) thermal test result

for switches one through four are as follows: 3.4 K/W, 3.42 K/W, 3.3 K/W, and 3.8 K/W, respectively.

Using this thermal resistance data and considering the projected losses (as shown in Fig. 5.58b), we can confidently rate the PFC circuit at 10 kW. This rating enables the creation of a compact OBC suitable for 800 V electric vehicles.

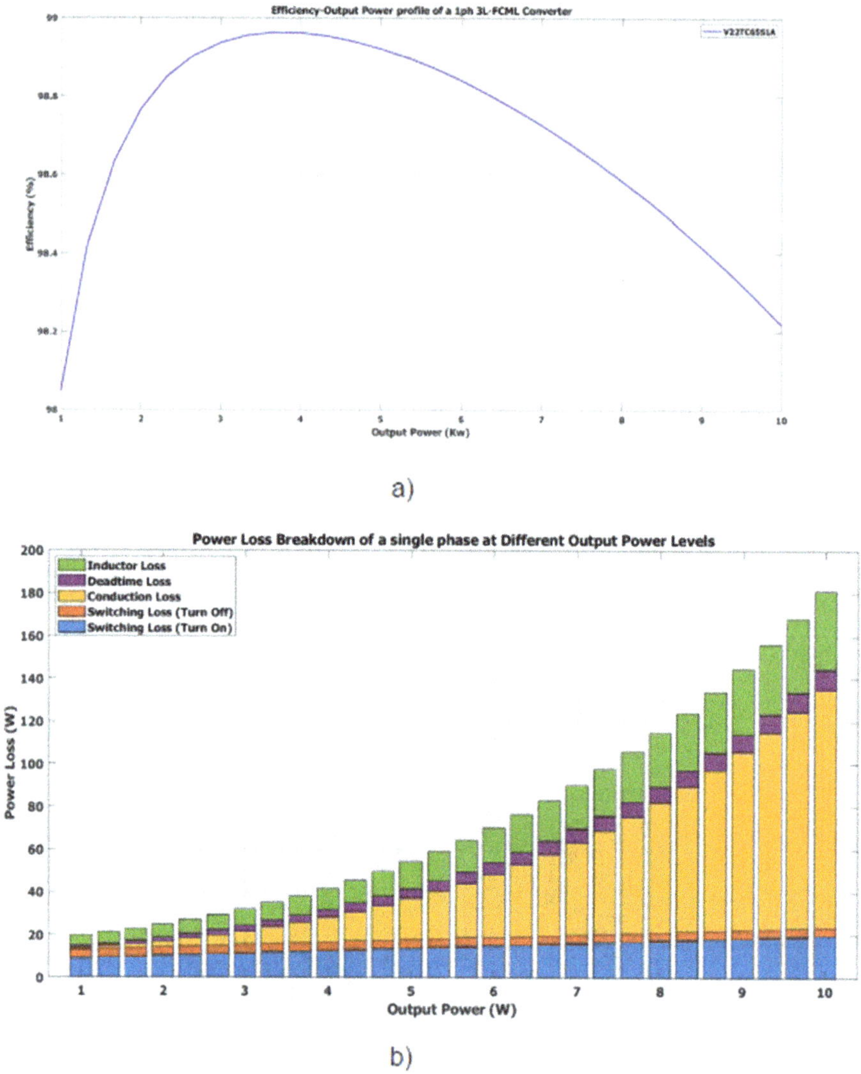

Fig. 5.58 Calculated efficiency (**a**) and loss breakdown (**b**) of the three-level FCML PFC

Although experimental data for the PFC are not available at the time of writing, we have calculated the PFC efficiency. Figure 5.58a illustrates the predicted peak efficiency, which closely approaches 99%, while the full-load efficiency (at 10 kW) stands at 98.2%. Additionally, we partially validated this higher efficiency by operating the converter in a DC-DC configuration, achieving a peak efficiency of 98.9% at 10 kW. Remarkably, at this power level, the power density reaches an impressive 271.2 W/in^3.

5.8 Foundry Service: Part 1

Maurizio Di Paolo Emilio

During my recent trip to Taiwan, a hub of technological innovation, I had the privilege of meeting representatives from three trailblazing companies that are sculpting the future of the semiconductor industry.

Episil Group, headquartered in Taiwan, is at the forefront of this burgeoning sector. I had an insightful discussion with JH Shyu, chairman of Episil Technologies, who detailed the company's focus on specialized foundry services. These services, ranging from crystalline to silicone processes, are meticulously tailored to match each client's unique requirements. Shyu emphasized the challenges faced when manufacturing with SiC and GaN, particularly given their unique attributes like hardness, transparency, and susceptibility to contamination. As I gathered from our conversation, Episil's intimate understanding of their clients' needs sets them apart in delivering precise solutions.

In another enlightening dialog, I conversed with John Wei, President of Vanguard International Semiconductor (VIS), and Shyh-Chiang Shen, GaN Program Director at VIS. The duo painted a vivid picture of the trajectory of gallium nitride (GaN) and silicon carbide (SiC) in the electronics and power systems domain. They shared that although GaN and SiC promise groundbreaking applications, their adoption faced initial reluctance from the traditionally conservative industry. Our discussion further delved into the industry's reservations, from the evolution of packaging technologies to the ongoing debates about defect density in GaN and SiC. The progressive mindset of VIS was evident in their nuanced approach to these challenges.

Lastly, my visit led me to Wavetek Microelectronics, where CTO Barry Lin shared his perspective on the compound semiconductor landscape. The versatility of compound semiconductors, as Lin pointed out, is evident in their broad applications, be it in electronics, photonics, or beyond. Lin emphasized Wavetek's unique capability to manage both aluminum and gold-based back-end processes, highlighting the company's agility in navigating the challenges of compound semiconductor manufacturing.

As I journeyed through Taiwan, meeting these industry leaders, it became evident that companies like Episil Group, VIS, and Wavetek are not just navigating the challenges of the semiconductor domain but are shaping its future. The dedication, innovation, and adaptability of these firms are paving the way for a technologically advanced future. The trip was not only informative but reaffirmed my belief in the potential and vitality of the semiconductor industry.

5.8.1 Episil

In the realm of electronics, wide bandgap (WBG) semiconductors, notably silicon carbide (SiC) and gallium nitride (GaN), are trailblazing advancements. With a more substantial critical electric field than their silicon counterparts, these materials excel in power electronics, paving the way for enhanced energy efficiency and diverse applications.

The Episil Group is at the forefront of this revolution, specializing in compound foundry services, especially in SiC, GaN, and silicon technologies. Both SiC and GaN have recently gained prominence due to their superior wide bandgap attributes. Such properties empower them to surpass silicon, especially in high-voltage and high-frequency applications. Silicon carbide is recognized for its excellent thermal conductivity, making it apt for high-power applications. In contrast, GaN is celebrated for its high performance and radiation resistance, suiting space-related applications like satellites.

The term "wide bandgap semiconductors" stems from their expansive energy bandgap, setting them apart from typical silicon semiconductors. This broader energy bandgap allows WBG materials to function at increased temperatures and voltages, a boon for power electronics emphasizing efficiency and dependability.

However, the transition to WBG materials is not without challenges. The electronic industry's immense investments in silicon-centric manufacturing are a testament to the evolution of technology. Adopting wide bandgap materials necessitates distinctive production methods and equipment, an endeavor that can be both pricey and prolonged. Issues like thermal expansion coefficient discrepancies can induce complications during manufacturing, leading to wafer deformities and subsequently, increased production costs. Further complicating matters is the inherent difficulty in producing GaN and SiC due to their intricate crystal structures, which pose challenges in crafting large, flawless wafers.

Currently, Episil's strategy revolves around the trench structure of SiC devices, aiming to minimize size and Rds(on).

5.8.1.1 Applications Driving Demand

One of the driving forces behind the demand for WBG semiconductors is the urgent need for decarbonization and energy efficiency across industries. Power electronics, utilizing the exceptional capabilities of WBG semiconductors, are playing a pivotal role in achieving these goals.

The applications are far-reaching. In the realm of power systems, WBG semiconductors are integral to energy generation and storage through solar and wind systems. They enable efficient electrolysis for green hydrogen production, power electric vehicles, drive electric trains and buses, facilitate high-efficiency heat pumps for residential needs, and power industrial applications such as uninterruptible power supplies (UPS) and motor drives.

"We have positioned well in these applications since two years ago. As our products have been widely used in renewable areas, the EV market has been developed for some time with ISO 16949 and VDA 6.3 qualified in our manufacturing lines," said Shyu.

5.8.1.2 Manufacturing Challenges and Progress

The manufacturing process for these semiconductors involves intricate steps, including the growth of epitaxial layers and the fabrication of devices. However, several challenges must be addressed to ensure high-quality epitaxial films and functional devices. Due to the challenges in material quality and wafer processing, the yield of defect-free devices from a single wafer can be lower compared to silicon manufacturing. Low yields drive up the cost per functional device.

5.8.1.3 Epitaxial Growth Process

The epitaxial growth process is critical for obtaining high-quality semiconductor films. Epitaxy involves depositing a crystalline layer on a substrate, allowing the deposited atoms to align with the substrate's crystal structure. In the context of SiC and GaN, this process is key to achieving the desired electronic properties.

5.8.1.4 Challenges in Epitaxial Growth of GaN Films

Lattice Mismatching and Stress Behavior: GaN has a different lattice constant than silicon, leading to lattice mismatching during epitaxial growth on silicon substrates. This results in strain and defects in the GaN film, impacting device performance and reliability.

Defect Control: Defects in epitaxial layers, such as dislocations and stacking faults, can degrade material quality and device performance. GaN's susceptibility to defects poses a challenge to achieving high-performance devices.

Uniformity of Thickness and Doping: Achieving uniform thickness and doping concentration in each layer of the epitaxial film is crucial for device functionality. Non-uniformity can lead to variations in device characteristics.

5.8.1.5 Addressing Challenges

Companies in the semiconductor industry employ various strategies to overcome these challenges and achieve high-quality epitaxial films.

For GaN Epitaxy

Lattice Mismatching and Stress Behavior: Some approaches involve using buffer layers or adjusting growth conditions to reduce strain and improve the quality of GaN films on silicon substrates.

Defect Control: Advanced growth techniques, like metal-organic chemical vapor deposition (MOCVD), can be tailored to suppress defect formation. In-situ monitoring and real-time feedback enable adjustments during growth to enhance material quality.

Uniformity of Thickness and Doping: Similar to SiC, GaN epitaxy benefits from precise process control and advanced monitoring methods to ensure uniform layer thickness and doping concentrations.

Scaling up WBG semiconductor production to meet growing demand poses challenges due to high material costs and specific manufacturing requirements. SiC substrates, a significant cost component, are being addressed through innovative technologies like laser cutting and cold splitting.

In the production arena, SiC has seen remarkable progress. Advances in technology like laser cutting and cold splitting have been explored to reduce costs and improve efficiency. Moving from 150 mm to 200 mm substrates is also helping cut costs per die. "While the 200 mm GaN can save cost by effective usage of wafer area, however, the 200 mm migration has the challenges of warpage (due to thermal mismatch), the alternative substrate or carrier can be costly as well," said Shyu.

The competition with dominant silicon technologies, particularly in the low-voltage market segment, presents hurdles. GaN's strength lies in middle voltage ranges (around 650 volts), where its performance surpasses other options. However, establishing GaN as a credible alternative for high-voltage operations remains a challenge. Additionally, ensuring GaN's reliability for critical applications is a priority for industry players.

"The decision to switch between materials depends on various factors, including cost, performance, and availability. Companies are evaluating the best approach based on their unique situations. Using GaN on GaN is technically easier, but the cost of GaN substrates is higher compared to silicon. On GaN, different from SiC on the substrate, the difficulty is on the wafer fabrication to mitigate the mismatching of Si and GaN temperature coefficient which is severe during the migration from 150mm to 200mm," said Shyu.

The manufacturing of wide bandgap semiconductors involves intricate epitaxial growth processes that are essential for high-performance devices. The challenges of lattice mismatching, defect control, and uniformity of thickness and doping are being tackled through advanced growth techniques, process optimization, and real-time monitoring. As technology advances, semiconductor companies continue to refine their approaches, paving the way for even more efficient and reliable wide bandgap semiconductor devices. "Real-time SPC (Statistical Process Control) control of parameters of equipment and process and experienced know-how to optimize the process and reactor conditions," said Shyu.

He added that "SiC is on the higher cost of substrate. Recently, the expertise has been working on new technologies like laser cut, and cold split to enhance productivity and reduce cost. By the end, the migration to a bigger size of the substrate, like today from 150 mm to 200 mm, can reduce the cost per die. I believe the Silicon manufacturing engineers can well exercise their experiences of 150 mm to 200 mm migration in manufacturing due to similar process architecture," said Shyu.

5.8.1.6 Global Semiconductor Landscape and Economic Ramifications

The supply chain for wide bandgap (WBG) semiconductors is influenced by geopolitical dynamics. Export restrictions on materials, especially by major players like China and the USA, have considerable implications. With WBG semiconductors playing a pivotal role in technological growth and economic expansion, there is an emphasis on reducing costs and amplifying production capabilities.

Recent geopolitical maneuvers have placed constraints on the flow of critical materials. For instance, China has limited gallium exports, a move that may necessitate alternative sourcing in the West. The USA, for its part, has imposed restrictions on GaN-on-SiC technology transfer due to national security implications.

5.8.1.7 The Future of WBG Semiconductors

As the world leans more into renewable energy and electric vehicles, the demand for WBG semiconductors, including SiC and GaN, skyrockets. Companies are strategizing to benefit from this trend, emphasizing consistent manufacturing and quality control. However, creating universal standards remains challenging given the varied applications of WBG semiconductors.

5.8.1.8 Factors Impacting WBG Semiconductor Costs

One of the chief barriers to the broader adoption of WBG semiconductors has been their cost, with SiC substrates being particularly pricey. To counter this, the industry is emphasizing diversifying suppliers and promoting competition. As more suppliers venture into this space, prices are expected to moderate, making these advanced semiconductors more affordable.

With electric vehicles becoming more common, there is an escalating demand for energy-efficient WBG devices. SiC and GaN play pivotal roles in enhancing energy efficiency in EVs and their charging infrastructure.

5.8.2 VIS

Manufacturers with existing silicon infrastructure and expertise benefit from maintaining compatibility with silicon technology when transitioning to GaN to harness existing resources. Epitaxial growth is a notable challenge in both GaN and SiC production, and substrate choice, like silicon or sapphire, is crucial. GaN on silicon makes use of existing infrastructure but has a limited voltage handling capacity, typically up to 650 V. In comparison, GaN on QST substrates enables thicker epitaxial layers, suitable for higher voltage applications, potentially reaching 1200 V or even 2200 volts (2.2 kV). Devices built on this substrate can achieve a 2.2 kV breakdown voltage. For 1.2 kV devices, a 1.8 kV breakdown voltage overhead is ideal. To manufacture GaN on silicon devices with higher voltage ratings, advanced fabrication techniques and materials are necessary. Challenges related to wafer size, uniformity, and yield need addressing. If the process is not optimized, leading to low yields or high costs, the technology might not be viable commercially. In the 650-volt range, there is stiff competition in the power electronics sector, with silicon carbide (SiC) being a significant competitor. GaN on silicon must present unique advantages over SiC for a 1200-volt rating to gain market traction. Many experts, considering the material's intrinsic properties, believe GaN on silicon can achieve this. Researchers remain hopeful about GaN's potential. In the future, the adoption of materials like gallium oxide (Ga_2O_3) will be slow, as the power electronics industry is traditionally conservative. Ga2O3 must overcome doubt and establish its reliability for widespread acceptance. GaN's direct growth on silicon substrates is often deemed cost-effective and simpler. Substrate choice impacts cost, with silicon carbide substrates being pricier than silicon, a major consideration in cost-sensitive power electronics. In specific applications like RF ICs, GaN on SiC, despite cost and complexity, is favored for its superior performance. Existing silicon infrastructure makes GaN on silicon a popular choice in many scenarios, but this limits the substrate thickness and its application to lower voltages, with 650 volts being the highest noted. The key points highlight the pros and cons of GaN on silicon vs. GaN on QST, focusing on voltage capabilities and substrate preparation.

5.8.2.1 Challenges

During the fabrication of gallium nitride (GaN) on silicon substrates, substrate fracturing poses the greatest difficulty. Despite using sturdier silicon substrates, companies are still contending with breakage issues, leading to a search for alternative substrates. Breakage remains a significant obstacle, influencing the production and dependability of GaN devices.

Regarding scalability and cost-effectiveness, the transition to larger wafer diameters (e.g., 8 inches and 12 inches) is viewed as productively advantageous. Nonetheless, this transition presents its own set of obstacles, including the management of stress and the adaptation of existing technologies and toolkits. Companies transitioning from 6-inch to 8-inch wafers may encounter issues with material quality, tool compatibility, and the need to develop new processes.

In this context, some companies are investing in 8-inch fabs to capitalize on the advantages of larger wafers from the inception, but this requires a protracted and difficult development process. GaN industry manufacturing challenges are complex and multifarious, requiring careful consideration of substrate selection, scalability, and technological adaptation.

The greatest obstacle to the widespread adoption of GaN is resolving reliability concerns, particularly those associated with dynamic Rds(on) and surface entrapment states. Standardization efforts and ongoing investigation are essential to overcoming these obstacles. As GaN technology matures, its acceptance in the electronics industry will increase.

5.8.2.2 Other Considerations

Other different aspects of semiconductor technology and packaging in the context of power electronics are discussed below.

Substrate Choice: Silicon carbide and sapphire are used as substrate materials for semiconductor devices, but their availability and scalability can be limiting. QST substrate, particularly in 12-inch size, is seen as a more flexible option for scaling up production.

Packaging Technology: Taiwan is considered a hub for semiconductor packaging solutions, providing a turnkey solution to introduce customers to suitable packaging vendors as they do not want to be tied to a specific packaging technology.

Standardization: There is a lack of standardization in power electronics, particularly in terms of application and system requirements. While there are some standards for device reliability, the absence of standardized packaging and application requirements can lead to higher costs.

Market Size: One challenge in achieving standardization is the relatively small market size for power electronics. The market needs to grow before standardization efforts can gain momentum.

Cost Considerations: The cost of compound semiconductors like silicon carbide is currently higher than traditional silicon components, which can be a barrier to their adoption.

5.8.2.3 Applications and Next Steps

Wide bandgap semiconductors are set to play pivotal roles in sectors like electric vehicles and solar energy. SiC enjoys a growing market share in EVs, while GaN's compactness and efficiency make it suitable for integrated chargers in EVs. In solar energy, GaN's application is still under exploration, especially for high-power infrastructure. GaN's potential is recognized in power delivery (PD) chargers due to its dynamic on-state resistance properties. According to the experts, the dynamic on-state resistance properties of GaN make it ideal for fast-charging applications because these applications do not require very stringent device specifications and they can be implemented in a soft-switching fashion. As our understanding and control of these dynamic issues improve, we can expect to see GaN technology expanding into more applications. Additionally, GaN's performance attributes make it a promising candidate for space electronics.

The decision between SiC, GaN, or other materials remains a topic of debate. SiC has benefited from more extensive investment, but GaN's versatility extends beyond power electronics. For GaN to thrive, substantial investment and development are essential.

Training a competent workforce is of the essence. Educational institutions must equip students with knowledge in wide bandgap materials, power electronics, and related areas. Partnerships between the industry and academia are necessary to equip the next generation with the required skills and awareness about the significance of semiconductor technology.

References

1. A. Lidow, M. De Rooij, J. Strydom, D. Reusch, J. Glaser, *GaN Transistors for Efficient Power Conversion*, 3rd edn. (John Wiley & Sons, 2019). ISBN: 978-1119594147
2. EPC2152 – 80 V, 10 A Integrated ePower™ Stage, Efficient Power Conversion Preliminary Datasheet, Jan. 2020, [online]. Available: https://epc-co.com/epc/products/gan-fets-and-ics/epc2152
3. J. Wang, E.A. Jones, M. de Rooij, Optimal eGaN FET Scaling for Minimal Power Loss in High Step-down Ratio Half Bridge Converters, in *PCIM Europe 2019; International Exhibition and Conference for Power Electronics, Intelligent Motion, Renewable Energy and Energy Management*, (2019), pp. 1–6
4. Y. Zhang, *Automotive Buck/Reverse-Boost Converter with GaN for Efficient 48V Power Distribution* (Bodo's Power Systems, 2021), pp. 36–38
5. M. de Rooij, A. Negahdari, Beyond 4 kW/in^3 power-density for 48 V to 12 V conversion using eGaN FETs in an LLC DC-DC Bus Converter, in *PCIM Europe 2022; International Exhibition and Conference for Power Electronics, Intelligent Motion, Renewable Energy and Energy Management*, (2022), pp. 1–9. https://doi.org/10.30420/565822013
6. M. de Rooij, J. Wang, A. Negahdari, Y. Zhang, A 1 kW eGaN FET-based LLC resonant converter in them 1/8th power brick size for 48 V server applications, in *PCIM Europe Digital Days 2021; International Exhibition and Conference for Power Electronics, Intelligent Motion, Renewable Energy and Energy Management*, (2021), pp. 1–8
7. M.H. Ahmed, F.C. Lee, Q. Li, Two-stage 48-V VRM with intermediate bus voltage optimization for data centers. IEEE J. Emerg. Sel. Topics Power Electron. **9**(1), 702–715 (2021). https://doi.org/10.1109/JESTPE.2020.2976107
8. D. Huang, S. Ji, F.C. Lee, LLC resonant converter with matrix transformer. IEEE Trans. Power Electron. **29**(8), 4339–4347 (2014). https://doi.org/10.1109/TPEL.2013.2292676
9. Efficient Power Conversion Corporation, EPC9159: 1 kW, 48V/12V LLC Converter Reference Design Board, 2023, [online]. Available: https://epc-co.com/epc/products/demo-boards/epc9159
10. J. Zhao, K. Yeates, Y. Han, Analysis of high efficiency DC/DC converter processing partial input/output power, in *14th IEEE Workshop on Control and Modeling for Power Electronics (COMPEL)*, June 2013, pp. 1–8
11. J.-I. Itoh, T. Fujii, A new approach for high efficiency buck-boost DC/DC converters using series compensation, in *IEEE Power Electronic. Specialists Conference*, June 2008, pp. 2109–2114
12. Efficient Power Conversion Corporation, EPC2619 datasheet, 2023, [online]. Available: https://epc-co.com/epc/Portals/0/epc/documents/datasheets/epc2619_datasheet.pdf

13. Efficient Power Conversion Corporation, EPC2067 datasheet, 2021, [online]. Available: https://epc-co.com/epc/Portals/0/epc/documents/datasheets/epc2067_datasheet.pdf
14. uPI SEMI, uP1966E datasheet, 2021, [online]. Available: https://www.upi-semi.com/files/2279/19724ef5-d308-11eb-8d5f-d30ecdd7d08d.pdf
15. Texas Instruments, LMG1020 datasheet, 2018, [online]. Available: https://www.ti.com/lit/ds/symlink/lmg1020.pdf?ts=1664290417482.pdf
16. Kemet, U2J Dielectric, [online]. Available: https://content.kemet.com/datasheets/KEM_C1086_U2J.pdf
17. S. Colino, *Chasing the Speed of Light* (Bodo's Power Systems, 2022), pp. 22–24
18. Efficient Power Conversion Corporation, EPC2619 datasheet, [online]. Available: EPC2619_datasheet.pdf (epc-co.com), 2024
19. Efficient Power Conversion Corporation, EPC2204 datasheet, [online]. Available: EPC2204_datasheet.pdf (epc-co.com), 2022
20. Infineon, ISC080N10NM6 Datasheet, [online]. Available: Datasheet ISC080N10NM6 (infineon.com), Feb. 2023
21. S. Colino, *Improving Performance While Reducing Size and Cost with Monolithic GaN Integration* (Bodo's Power Systems, June 2023), pp. 22–25
22. A. Lidow, J. Glaser, GaN-based solutions for cost-effective direct and indirect time-of-flight lidar transmitters are changing the way we live, in *International Power Electronics Conference (IPEC-Himeji 2022- ECCE Asia) IPEC*, May 2022
23. Efficient Power Conversion Corporation, EPC21601 datasheet, [online]. Available: EPC21601_datasheet.pdf (epc-co.com), Dec. 2022
24. Efficient Power Conversion Corporation, EPC9154 development board, [online]. Available: EPC9154 – 40 V, 10 A High Current Pulsed Laser Diode Driver (epc-co.com), 2022
25. B.J. Baliga, Power semiconductor device figure of merit for high frequency applications. IEEE Electron Device Lett. **10**, 455 (1989)
26. Efficient Power Conversion Corporation, EPC9173 Quick Start Guide, 2022, [online]. Available: https://epc-co.com/epc/documents/guides/EPC9173_qsg.pdf
27. Efficient Power Conversion Corporation, (2023). EPC9186 quick start guide, 2023, [online]. Available: https://epc-co.com/epc/Portals/0/epc/documents/guides/EPC9186_qsg.pdf
28. A. Lidow (ed.), *GaN Power Devices and Applications* (Power Conversion Publication, El Segundo, 2022)
29. D. Cittanti, V. Mallemaci, F. Mandrile, S. Rubino, R. Bojoi, A. Boglietti, PWM-induced losses in electrical machines: An impedance-based estimation method, in *24th International Conference on Electrical Machines and Systems (ICEMS)*, November 2021
30. Efficient Power Conversion Corporation, EPC2152 – 80 V, 10 A Integrated ePower™ Stage, Preliminary Datasheet, Jan. 2020, [online]. Available: https://epc-co.com/epc/documents/datasheets/EPC2152_datasheet.pdf
31. Efficient Power Conversion Corporation, EPC23102 datasheet, Preliminary Datasheet, 2023, [online]. Available: https://epc-co.com/epc/Portals/0/epc/documents/datasheets/epc23102_datasheet.pdf
32. Efficient Power Conversion Corporation, 2023. EPC23101 datasheet, Preliminary Datasheet,2023, [online]. Available: https://epc-co.com/epc/Portals/0/epc/documents/datasheets/epc23101_datasheet.pdf
33. Application Note, *Block Commutation Vs. FOC in Power Tool Motor Control* (Infineon Technologies AG, 2019)
34. M. Wattenberg, E.A. Jones, J. Sanchez, A low-profile GaN-based integrated motor drive for 48V FOC applications, in *PCIM Europe Digital Days 2021; International Exhibition and Conference for Power Electronics, Intelligent Motion, Renewable Energy and Energy Management*, Online, 2021, pp. 1–8
35. M. Wattenberg, O.G. Lorenz, J. Sanchez, A multi-kilowatt low-profile GaN inverter for light electric vehicles and high-power tools, in *2023 IEEE Applied Power Electronics Conference and Exposition (APEC)*, Orlando, FL, USA, 2023, pp. 1443–1450, https://doi.org/10.1109/APEC43580.2023.10131197

36. G.C. Messenger, M.S. Ash, *The Effects of Radiation on Electronic Systems* (Van Nostrand Reinhold Company, New York, 1986)
37. T.P. Ma, P.V. Dressendorfer, *Ionizing Radiation Effects in MOS Devices and Circuits* (John Wiley and Sons, Inc, New York, 1989)
38. A. Lidow, J.B. Witcher, K. Smalley, Enhancement mode gallium nitride (eGaN®) FET characteristics under long-term stress, in *GOMAC Tech Conference*, Orlando Florida (March 2011)
39. G. Sonia, F. Brunner, A. Denker, et al., Proton and heavy ion irradiation effects on AlGaN/GaN HFET devices. IEEE Trans. Nucl. Sci. **53**(6) (2006)
40. S.J. Pearton, F. Ren, E. Patrick, et al., Review—Ionizing radiation damage effects on GaN devices. ECS J. Solid State Sci. Technol. **5**(2), Q35–Q60 (2016)
41. S. Bazzoli, S. Girard, V. Ferlet-Cavrois, et al., SEE sensitivity of a COTS GaN transistor and silicon MOSFETs, in *9th European Conference on Radiation and Its Effects on Components and Systems, RADECS* 2007 (2007)
42. A. Lidow, K. Smalley, Radiation tolerant enhancement mode gallium nitride (eGaN®) FET characteristics, GOMAC Tech Conference, Las Vegas, Nevada, March 2012
43. A. Lidow, J. Strydom, M. Rearwin, Radiation tolerant enhancement mode gallium nitride (eGaN®) FETs for high-frequency DC-DC conversion, GOMAC Tech Conference, Charleston, South Carolina, April 2014
44. J. Brandt, A. Lidow, R. Strittmater, B. Sun, M. Zafrani, S. Zhang, Radiation results for modern GaN-on-Si power transistors, in *IEEE Nuclear and Space Radiation Effects Conference (NSREC), Utah, United States* (2022)
45. L.Z. Scheick, Recent Gallium Nitride power HEMT Single Event Testing Results, Poster W-6, in *IEEE Nuclear and Space Radiation Effects Conference (NSREC), Portland, United States* (2016)
46. S. Kuboyama, A. Maru, H. Shindou, et al., Single-event damages caused by heavy ions observed in AlGaN/GaN HEMTs. IEEE Trans. Nucl. Sci. **58**(6) (2011)
47. High Power Factor Preregulator for Off-Line Power Supplies, Lloyd H. Dixon, Jr., 1988 Unitrode (now TI)
48. Design of CCM Power Factor Correction Boost Converter, 2022 Lazar Rozenblat, https://www.powerfactor.us/ccm-pfc-boost.html
49. 2500 W full-bridge totem-pole power factor correction using CoolGaN™, Infineon application note AN_201702_PL52_011

Dr. Michael de Rooij is currently Vice President of Applications Engineering at Efficient Power Conversion Corporation (EPC) headquartered in El Segundo, CA, U.S.A. He received the B. Eng., M. Eng., and, D. Eng. degrees from The University of Johannesburg fka Rand Afrikaans University, South Africa, in 1992, 1994, and 1998, respectively, all in electronics and electrical engineering with major in power electronics. He worked as a post-doctoral researcher at the Center for Power Electronics (CPES), Virginia Tech in 2001. Dr. de Rooij has authored and co-authored over 50 papers at various conferences and journals and has been granted over 58 U.S. and international patents. He is author and co-author of 7 books. His experience covers highly resonant loosely coupled wireless power, high-density DC to DC converters, GaN IC development, RF power amplifiers, solid-state high-frequency power converters and devices, utility applications of power electronics, uninterruptible power supplies, integration of power electronic converters, power electronic packaging, induction heating, photovoltaic converters, magnetic resonance imaging (MRI) systems, and gate drivers with protection features. Dr. de Rooij is senior member of the IEEE.

Alejandro Pozo Arribas has been a Senior Applications Engineer at EPC since August 2022. He is responsible for the design and development of demonstration boards and reference designs using GaN devices in DC/DC converters. Prior to August 2022, and since 2017, he was involved in product reliability and device characterization of EPC's GaN devices. He received his Ph.D. in Electrical Engineering from the Illinois Institute of Technology in 2017 for his work in base drivers for SiC BJTs.

Stephen Colino is the Vice President of Strategic Technical Sales at EPC. He received his BE in Electrical and Biomedical Engineering from Vanderbilt University and his MBA from Adelphi University. He has 39 years of analog and power experience at EPC, Texas Instruments, Ciclon Semiconductor, Vishay-Siliconix, International Rectifier, and ILC Data Device.

John Glaser serves as Director of Applications Engineering at Efficient Power Conversion Corporation (EPC). He earned a BSEE from the University of Illinois and Ph.D. from the University of Arizona. John has designed 800 MHz power amplifiers for Motorola and researched TWT HV converters for Hughes Missile Systems. From 1998 to 2014, he was at GE Global Research where he worked on a broad range of power conversion applications with frequencies ranging from 60 Hz to 128 MHz and power levels ranging from 1 W to 100 kW. Applications covered a broad spectrum including lighting, cooking appliances, defense, medical imaging, and aviation. At EPC since 2014, he develops applications, circuits, and methods to maximize the benefit and drive adoption of GaN FETs. Dr. Glaser is an author of over 50 technical papers and articles, three books, an inventor with over 50 granted patents worldwide, and is an IEEE Senior Member.

Marco Palma is the Director of Motor Drives Systems and Applications at EPC. Marco has over 20 years of experience in motor control power electronics ranging from switches to gate drivers, controllers, and algorithms. For International Rectifier, Marco worked on the smallest 13 kW fully integrated and programmable motor drive power modules in 2002, the first industrial sensor-less FOC controller IC in 2004, and the smallest 100 W, fully integrated fan drive module in 2009. For Infineon Technologies, he worked on the smallest programmable 100 W fully integrated motor control power module in 2018. He is the author of several articles and patents in the field of motor control. Marco received an MSEE from Politecnico di Torino and an MBA from Bocconi Milano.

Martin Wattenberg received the Bachelor's and Master's degree from Bremen University in Mechatronics and Reutlingen University in Electrical Engineering in 2012 and 2014, respectively. From 2014 to 2020, he has been involved in the research and development of energy storage systems with a strong focus on bidirectional DC/DC conversion. In 2021, he graduated with a Ph.D. from the Technical University TU Dortmund. In 2020, he joined in Infineon Technologies Austria as a System Application Engineer for wide bandgap power devices (SiC and GaN) where he covers high-frequency DC/AC conversion in telecom and server as well as innovative motor drive systems using WBG.

Max Zafrani is the Chief Technology Officer at EPC Space. Max has over 30 years in the power semiconductor and Hi-Rel industry. Max was the co-founder and CTO of Freebird Semiconductor concentrating on radiation effects of GaN. Max started his career at RCA-Harris in 1985 and held multiple positions in diffusion, test, and development of power rectifiers and IGBTs. He also worked at CPClare now Littelfuse in the development of solid-state relay and SCRs. Prior to starting Freebird Semiconductor, he worked at International Rectifier for 15 years as the Senior Director of Engineering for the Product Development of Radiation Hardened MOSFETs. Max received his Bachelor of Science degree in Electrical Engineering from Wilkes University. He has been granted three U.S. patents in the area of power semiconductor.

Alex Lidow is the CEO and co-founder of Efficient Power Conversion Corporation (EPC). Prior to founding EPC, Dr. Lidow was CEO of International Rectifier Corporation. A co-inventor of the HEXFET power MOSFET, Dr. Lidow holds many patents in power semiconductor technology and has authored numerous publications on related subjects, including co-authoring the first textbook on GaN transistors, GaN Transistors for Efficient Power Conversion, now in its third edition published by John Wiley and Sons. Lidow earned his Bachelor of Science degree from Caltech and his Ph.D. from Stanford.

Robert Strittmatter is Vice President of Strategic Research and Development at Efficient Power Conversion where he has been involved in GaN-based product reliability and device characterization since 2010. Prior to joining EPC, he worked as Director of Device Development at GEOST where his work experience covered a broad range of electro-optic systems development, primarily for DOD customers. Experience encompassed all phases of system development, from modeling and design to implementation and final testing. He received his Ph.D. in Semiconductor Physics from the California Institute of Technology.

Tony Marini is the President of dtm Associates, Engineering Consultants. Tony Marini has over 40 years of experience in the areas of electronics, semiconductors, and microelectronics. Tony also has expertise in the areas of power electronics, system reliability prediction and failure modes, and EMC/EMI design and problem remediation. He has published numerous technical papers and has been a presenter at technical conferences and symposia. Tony earned his BSEE and MSEE from Worcester Polytechnic Institute (Worcester, MA), with concentration in semiconductor physics and power electronics.

Scott Durkin has over 30 years' experience of all aspects of Electronics Design in Automotive and Industrial applications. He holds a BSc (Hons) in Electronics and Computing from the University of Central Lancashire, Preston, UK. Scott is a Principal GaN Applications Engineer at Nexperia, joining the Applications Team based in Stockport, UK, in 2020. His role is both to design Evaluation Boards using Nexperia GaN devices and also support customers to successfully integrate Nexperia GaN devices into their final products.

Jim Honea has worked on the application of high-voltage gallium nitride switching transistors to energy-conversion systems since 2007. He is presently the director of GaN Applications Engineering for Nexperia. Prior to working with GaN, he worked in the fields of IC design, industrial control, and industrial instrumentation, with both semiconductor companies and system-level manufacturers. He received the BSEE and MSEE degrees from the University of Arkansas, and a Ph.D. in electrical engineering from the University of California, Santa Barbara.

Dr. Alex Q. Huang is the Dula D. Cockrell Centennial Chair in Engineering at the Department of Electrical and Computer Engineering at UT Austin. Dr. Huang received the Bachelor's degree in Electrical Engineering from Zhejiang University, China, in 1983 and his M.S. degree from the University of Electronic Science and Technology of China in 1986. He received his Ph.D. in Electrical Engineering from the University of Cambridge, UK, in 1992. Prior to joining UT Austin, Dr. Huang has been a faculty member at Virginia Tech (1994–2004) and NC State University (2004–2017). At NC State, Dr. Huang has established a number of internationally renowned public-private partnerships such as the NSF FREEDM ERC in 2008, NCSU's Advanced Transportation Energy Center (ATEC) in 2008 and the DOE PowerAmerica Institute in 2014.

Dr. Huang is a world-renowned expert of power semiconductor devices, power electronics, smart grid, and renewable energy systems. He has published more than 650 papers in journals and conferences and is the inventor of more than 20 U.S. patents including several patents on the Emitter turn-off (ETO) thyristor technology that received a prestigious R&D 100 award in 2003. Dr. Huang is also widely credited for his contribution in developing the Energy Internet concept and the solid-state transformer (SST)-based energy router technology. His work on the SST has been named by MIT Technology Review as one of the world's 10 most important emerging technologies in 2011. He has graduated more than 100 Ph.D. students and master students. Dr. Huang is a fellow of IEEE and the general chair of IEEE ECCE Conference in 2012. He is also a fellow of the National Academy of Inventors. Dr. Huang is the recipient of 2019 IEEE IAS Gerald Kliman Innovator Award.

Emad Nazerian (Student Member, IEEE) was born in 1992 in IRAN. He earned his MSc from the Sharif University of Technology in Tehran, Iran, in 2018. He is currently working toward a Ph.D. degree in Electrical Engineering at the Semiconductor Power Electronics Center, University of Texas at Austin, in Austin, TX, USA. His primary areas of expertise include the design and development of high power density converters for electrification of transportation, WBG-based converter design, as well as the design and analysis of magnetic elements.

Peng Han (Student Member, IEEE) was born in Beijing, China, in 1995. He received the B.S. and M.S. degrees in Electrical Engineering and Automation, in 2017 and 2020, respectively, from Tsinghua University, Beijing, China. He is currently working toward the Ph.D. degree in Electrical Engineering with University of Texas at Austin, Austin, TX, USA.

His research interests include application of wide bandgap devices, power module packaging, and traction inverter design.

Maurizio Di Paolo Emilio holds a Ph.D. in Physics and is a Telecommunications Engineer. He has worked in various international projects in the field of gravitational waves research, designing a thermal compensation system (TCS) and data acquisition and control systems, and on others about x-ray microbeams in collaboration with Columbia University, high voltage systems and space technologies for communications and motor control with ESA/INFN. TCS has been applied to the Virgo and LIGO experiments, which detected gravitational waves for the first time and earned the Nobel Prize in 2017. Since 2007, he has been a reviewer for scientific publications for academics such as Microelectronics Journal and IEEE journals. Moreover, he has collaborated with different electronic industry companies and several Italian and English blogs and magazines, such as Electronics World, Elektor, Mouser, Automazione Industriale, Electronic Design, All About Circuits, Fare Elettronica, Elettronica Oggi, and PCB Magazine, as a technical writer/editor, specializing in several topics of electronics and technology. From 2015 to 2018, he was the editor-in-chief of Firmware and Elettronica Open Source, which are technical blogs and magazines for the electronics industry. He participated in many conferences as a speaker of keynotes for different topics such as x-ray, space technologies, and power supplies. Maurizio enjoys writing and telling stories about Power Electronics, Wide-bandgap Semiconductors, Automotive, IoT, Embedded, Energy, and Quantum Computing. Maurizio has been an AspenCore content editor since 2019. He is currently editor-in-chief of Power Electronics News and EEWeb and a correspondent for EE Times. He is the host of PowerUP, a podcast about power electronics, and the promoter and organizer of the PowerUP Virtual Conference, a summit where each year great speakers talk about the power electronics design trends. Moreover, he has contributed to a number of technical and scientific articles, as well as a couple of Springer books on energy harvesting and data acquisition and control systems.

Chapter 6
Case Studies

Nihit Bajaj, Daniel Makus, Rafael Garcia, Igor Spinella, Paul Wiener, Junyang Luo, Matthias Kasper, Guilherme Bueno Mariani, Hubie Noto, Thierry Bouchet, Peter Vaughan, Rémi Comyn, Dieter Liesabeths, Dima Novo, Denis Marcon, Francesco Di Domenico, Alex Zahabizadeh, Stephen Oliver, and Maurizio Di Paolo Emilio

This chapter of this comprehensive exploration delves deep into the transformative potential of Gallium Nitride (GaN) technology across various industries and applications. Authored by industry experts and thought leaders, each section presents a unique perspective on GaN's impact for different case studies.

In the section "Advantages of using GaN in class-D audio," authored by Nihit Bajaj of Infineon Technologies, the text explores how GaN technology promises

N. Bajaj
Director, GaN Product Marketing, Infineon Technologies, San Jose, CA, USA

D. Makus
Application and Business Development Director for xEV power conversion, Infineon Technologies, Neubiberg, Germany

R. Garcia
System Architect for Onboard Chargers and DCDC Applications for Electric Vehicles, Infineon Technologies, Villach, Austria

I. Spinella
Eggtronic, Modena, MO, Italy

P. Wiener
Strategic Marketing, Infineon Technologies, San Jose, CA, USA

J. Luo
Lead Principal Engineer, System Architect for Charger/Adapter, Infineon Technologies, Singapore, Singapore

M. Kasper
Innovation Group Leader, Infineon Technologies, Villach, Austria

G. Bueno Mariani
Senior Staff Application Engineer, Infineon Technologies, Villach, Austria

H. Noto
Silanna Semiconductor, Union City, CA, USA

© The Author(s), under exclusive license to Springer Nature Switzerland AG 2024

M. Di Paolo Emilio (ed.), *GaN Technology*,
https://doi.org/10.1007/978-3-031-63238-9_6

superior sound quality, efficiency, and compactness in audio amplification systems, revolutionizing class-D amplifier designs.

Daniel Makus, Application Manager at Infineon Technologies, lays the groundwork for a comprehensive analysis of electric-vehicle charging in Sect. 6.2, focusing on wide-bandgap (WBG) technologies and the intricate EV-charging ecosystem.

Igor Spinella, CEO and founder of Eggtronic, explores the transformative potential of GaN power semiconductors in wireless charging technology in "How GaN Supports Next-Generation Wireless Power Transfer," shedding light on scaling wireless power transfer to higher wattage levels.

Authored by Paul Wiener from Infineon Technologies, "GaN-based wireless power transfer: innovations in next-generation charging technologies" introduces advancements and challenges in wireless power transfer, emphasizing GaN's role in transitioning to high-frequency systems.

"Power supply—USB-C adapters and chargers," authored by Junyang Luo and Matthias Kasper of Infineon Technologies, explores GaN's impact on USB-C chargers, paving the way for more efficient and environmentally friendly power delivery systems.

Guilherme Bueno Mariani from Infineon Technologies explores GaN's transformative potential in enhancing motor system performance in "Improving motor system efficiency with GaN inverters," highlighting advantages over conventional silicon-based solutions.

Hubie Noto, Director of Marketing at Silanna Semiconductor, addresses the demand for efficient, multi-device charging in "Addressing the Multiport Challenge

T. Bouchet
Wise Integration, Hyères, France

P. Vaughan
Director of Automotive Business Development, Power Integrations, San Jose, CA, USA

R. Comyn
Technology & Patent Analyst, KnowMade, Valbonne, France

D. Liesabeths
VisIC Technologies, Munich, Germany

D. Novo
VisIC Technologies, Nes Ziona, Israel

D. Marcon
Innoscience Europe, Leuven, Belgium

F. Di Domenico
Senior Principal System Architect, Infineon Technologies, Villach, Austria

A. Zahabizadeh
Texas Instrument's GaN Business, Dallas, TX, USA

S. Oliver
VP Corporate Marketing & Investor Relations, Navitas Semiconductor, Torrance, CA, USA

M. Di Paolo Emilio (✉)
Power Electronics News, Pescara, Italy

for Simultaneous USB Device Charging," focusing on GaN technology and USB power delivery (USB PD) specifications.

Thierry Bouchet, CEO of Wise Integration, elucidates the paradigm shift towards digital control, particularly in GaN technology, emphasizing its advantages and transformative impact on power supply efficiency and sustainability.

Authored by Peter Vaughan, Director of Automotive Business Development at Power Integrations, "Benefit of Applying GaN to Automotive µDCDC Converters" discusses integrating GaN technology into automotive micro direct current (DC) to DC converters, enabling smaller, lighter, and more efficient converter stages in electric vehicles (EVs).

In the analysis conducted by Rémi Comyn, Technology & Patent Analyst at KnowMade, the power GaN patent landscape is examined in detail in Sect. 6.10. The discussion begins with an overview of the time evolution of power GaN IP activity, highlighting shifts in inventive leadership from Japanese to Chinese players. The author delves into the competitive IP spaces in the power GaN landscape, categorizing players based on their patent portfolios

In the section authored by Dieter Liesabeths and Dima Novo of VisIC Technologies, titled "Requirements for GaN Devices in Automotive Drive Train Inverter," the focus is on presenting the advancements in GaN technology for electric vehicle traction inverters. They delve into the novel D^3GaN power module technology, discussing its design, operational considerations, thermal management, and test results, emphasizing its potential for higher efficiency and reliability in automotive applications.

Denis Marcon, General Manager of Innoscience Europe, explores leveraging GaN's properties to enhance Battery Management Systems (BMS) functionalities in "Improving Battery Management Systems with GaN Technology," driving advancements in efficiency, reliability, and compactness. Also, he addresses the extreme power demands of next-generation data centers in "Powering Next-Generation Data Centers with Integrated GaN Solutions," introducing integrated GaN solutions as a potential solution to this challenge.

Authored by Francesco Di Domenico, Sect. 6.14 delves into GaN's impact on server applications within data centers, enhancing power conversion efficiency and reducing heat generation for greater energy efficiency and operational cost savings.

Alex Zahabizadeh, Project Manager at Texas Instruments' GaN business, explores GaN's potential in renewable energy solutions, highlighting advancements in solar inverters and addressing challenges in battery storage to enhance efficiency and reduce costs.

In Sect. 6.16, Stephen Oliver, VP Corporate Marketing & Investor Relations at Navitas Semiconductor, explores the pivotal role of GaN ICs in powering AI servers to over 100 kW per rack. With AI processors like NVIDIA's Grace Hopper and Blackwell driving exponential data processing demands, the need for high power-density solutions is urgent. Oliver details the evolution from legacy silicon to cutting-edge GaN technology, heralding a roadmap towards 100 kW+ per rack efficiency.

Through these diverse perspectives, readers gain insights into GaN's pivotal role in driving innovation and sustainability across various domains, paving the way for a more efficient and environmentally friendly future. The last section will analyze general aspects of foundry services.

6.1 Advantages of Using GaN in Class-D Audio

Nihit Bajaj

Many types of amplifiers are used in audio equipment today. These amplifiers have different topologies that govern their performance and quality, which gear them towards different applications.

Some of the common amplifier classes include A, B, AB, G, and H, but the most dominant class of amplifier technology today is the class-D amplifier, which has found use in almost all modern-day consumer and professional audio equipment. Class-D amplifiers are also known as pulse width modulation (PWM) amplifiers and use high-speed switches instead of linear bipolar transistors (which are found in some other amplifier classes). They also consist of a power bridge circuit and a low-pass filter.

Class-D amplifiers are switching amplifiers that have output switches that are either fully on or fully off, resulting in a lower power loss than other amplifier topologies. A class-D amplifier amplifies the voltage- and current-sourcing potential of an audio signal that has been converted into a stream of fixed voltage pulses with widths the same as the amplitude of the original signal. The high-power pulses drive the speaker, and any carrier signals are filtered out via the low-pass filter.

Class-D amplifiers can also be broken down into topology subclasses, namely, half-bridge, full-bridge, and bridge-tied-load (BTL) topologies, with half-bridge being the cheapest and BTL having the highest power output. Silicon materials have commonly been used to make class-D designs, but there has been a shift recently toward using GaN as it eliminates some of the common challenges that many class-D amplifiers experience today.

6.1.1 Problem with Today's Class-D Designs

While class-D is the dominant amplifier technology with high theoretical efficiency, most class-D designs used today have a mediocre sound quality that is below what is theoretically possible, and yet they tend to be large and bulky units.

Many modern-day designs are bulky by necessity, not choice. This is because a lot of existing systems struggle to dissipate the heat effectively, even though class-D designs typically generate only around 10% of the heat of class-AB amplifiers. This has led to the need to incorporate heatsinks into the design, which significantly increases the size and weight of the amplifiers because a high degree of retained, non-dissipated heat affects the power performance of the device.

Beyond size and thermal regulation issues, sound quality is hampered in many current systems because the PWM switching waveforms are not ideal, especially at high voltages and high frequencies. Instead of the ideal square wave, the real switching waveform experiences more overshoot, inductor recovery, and IR drops and requires wider deadtimes that affect the amplifier performance. Silicon-based metal-oxide semiconductor field-effect transistors (MOSFETs) rectify these issues and provide a stable performance to an extent, but are still far from ideal.

Ideally, an amplifier should be as close to a perfect square wave as possible with fast edges, as this increases the amount of time that a device has for accurately reproducing the audio content embedded in the pulse stream. However, even with the addition of silicon MOSFETs, today's devices suffer from noticeable distortion and variance in real-world scenarios. The signal error and distortion in modern-day amplifiers contribute to a higher total harmonic distortion plus noise and a lower signal-to-noise ratio (SNR), both of which are key contributors to the poor sound quality in class-D systems today.

Class-D amplifier performance has always suffered from a lack of "ideal" semiconductor switches, and this has led to delays in acceptance across many critical application areas. Now that GaN is being used to innovate the class-D amplifier space with better performance and switching properties than existing systems, its usage in these critical applications is being deeply explored.

6.1.2 Why GaN Technology Virtually Eliminates the Barriers

The ideal switch should transition from off to on and on to off instantly with no square wave distortion. An ideal switch should also require no energy to switch, regardless of the switching speed, and have zero electrical resistance when activated.

PWM waveforms using silicon-based switches tend to be more trapezoidal in nature, which causes more distortion in the sound and hence, require a longer dead time (when both switches are turned off). The maturation of GaN technology for class-D amplifiers brings the ideal PWM switch closer to reality, as the figures of merit (FOM) of the GaN device allow square waves that are very close to ideal. The faster edge rates in GaN amplifiers reduce the amount of dead time required between high-side and low-side switching, which results in a lower distortion and a higher sound quality. In addition to this, GaN does not have any reverse recovery charge to reverse bias the body diode, leading to higher power efficiencies, the elimination of any ringing, and a reduction in noise.

Compared to existing silicon-based amplifiers, the square waves are also more accurate and require less energy. GaN also has a higher electron mobility than silicon, which results in a lower drain-source on resistance ($R_{DS(on)}$) and a higher maximum drain-source voltage ($V_{DS(MAX)}$) per unit area, leading to higher power efficiencies and rail voltages in the same size or leading to lighter and smaller amplifier devices for the same power output. Additionally, the lower $R_{DS(on)}$ per area results in a significantly lower drain-source capacitance (C_{oss}) that reduces power dissipation from switching losses.

Compared to silicon MOSFETs, GaN-based amplifiers have a power-stage efficiency that measures 95–96%, which is 6–10 percentage points higher than Si-based class-D power stages.

The mechanism that underpins switching in GaN amplifiers is also superior to silicon when it comes to handling the third-quadrant commutation current, which is the ability of the device to conduct reverse current during deadtime. The performance of GaN switches means that no helper diodes are required in these architectures, once again reducing the number of components compared to other amplifiers. Another component that is omitted from GaN amplifiers compared to other designs is heat sinks. This is because GaN has a much better thermal regulation and a 6% efficiency improvement with GaN, resulting in a 50% reduction in thermal loss, eliminating the need for heavy and bulky heatsinks.

GaN's superior figures of merit relative to silicon also ensure low switching losses, especially at high switching frequencies. Higher switching frequencies, in turn, allow smaller passive components in the system, saving PCB area, weight, and cost.

Overall, the key aspect beyond performance is that GaN amplifiers are inherently lighter and smaller because they need fewer components to run and perform more efficiently than existing silicon architectures. Similar levels of high-frequency switching are only possible in other devices through the introduction of bulky capacitors and inductors or at the cost of higher power losses, so GaN provides value in many areas for class-D amplifiers.

6.1.3 Why Select Infineon's GaN Switches

The details mentioned so far clearly highlight the potential of GaN in class-D amplifiers.

Infineon offers different designs to help leverage the best sound out of speakers.

The comprehensive offering of Infineon includes amplifiers and turnkey amplifier modules that are based on half-bridge and BTL topologies, they include the following examples:

- Reference designs and evaluation kits
- 2-channel class-D amplifier
- GaN-based power-factor correction (PFC)/LLC power supply design kits

The solution from Infineon consists of a GaN-based switched-mode power supply (SMPS) with PFC and a stereo 200 W/8 Ω class-D amplifier. The performance of the reference design is ideal for system designers to choose between open-loop and closed-loop designs.

For the open loop design, the power supply becomes an important component and the SMPS pairs well for these designs. The class-D amplifiers provide a better audio performance, efficiency, and increased bandwidth, while the companion SMPS also increases efficiency while eliminating the need for a heatsink or a cooling fan.

The systems from Infineon use best-in-class GaN High-electron-mobility-transistor (HEMTs) to achieve maximum performance while providing optimal voltage and current options for audio amplification. They come with top-side and bottom-side thermal solutions, standard packaging for PCBA fabrication, and assembly techniques. They can also be used with any controllers with single, dual, or high-side/low-side drivers, offering a high degree of design freedom and scalability for engineers who want to take advantage of the benefits offered by GaN.

6.1.4 Conclusion

Many amplifiers exist in today's marketplace, but while many amplifier classes target specific applications, class-D amplifiers are versatile, and have become the system of choice across most audio technologies. This is because they possess the highest power efficiencies and the lowest noise, providing the best sound quality across all amplifier technologies.

However, many of the current amplifiers are designed using silicon technology, which is heavy, has high thermal losses, and has a lot of distortion. This means that their switching characteristics are far away from the square wave expected in an ideal switch. GaN bridges this gap and offers much better switching characteristics and properties compared to silicon, including a reduced thermal loss, higher power efficiency, lower weight, switching behavior that almost mimics square wave characteristics, and a superior sound quality [5].

6.2 Electric-Vehicle Charging

Daniel Makus and Rafael Garcia

The shift toward sustainable and efficient mobility is a paramount objective for electric vehicles. It is essential to highlight the challenges and advancements in e-mobility systems, particularly those enabled by wide-bandgap (WBG) technologies, and the comprehensive EV-charging ecosystem that supports this transition.

The EV-charging ecosystem is an intricate network that powers EVs, and consists of off-board and onboard charging systems (OBCs) (Fig. 6.1). Off-board chargers, both AC and DC, provide the necessary infrastructure for EV charging, while OBC is integral for converting this energy within the vehicle.

First, we will explore the challenges faced in e-mobility systems and solutions, emphasizing the role of WBG materials like silicon carbide (SiC) and gallium nitride (GaN). These materials are crucial for developing high-efficiency,

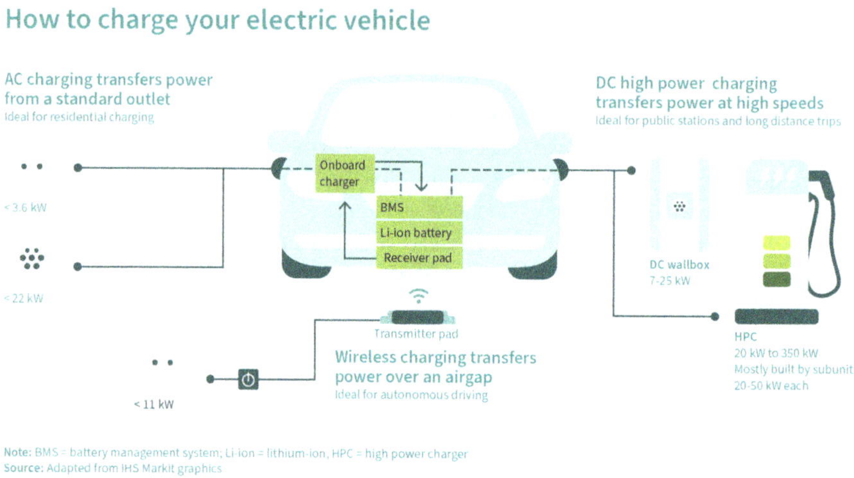

Fig. 6.1 Simplified EV-charging ecosystem

high-power-density solutions that are pivotal for EV charging systems and onboard chargers.

As we progress, we will delve deeper into the specifics of the EV-charging ecosystem, highlighting the critical role of Infineon's technologies in shaping the future of e-mobility. From pioneering WBG technologies to developing innovative charging solutions, Infineon's contributions are central in achieving a zero CO_2 future and making e-mobility work for everyone.

6.2.1 Challenges in e-Mobility Systems and Solutions with WBG

Challenges in e-mobility systems and solutions with wide-bandgap (WBG) technologies like silicon carbide (SiC) and gallium nitride (GaN) are multifaceted, primarily focusing on their integration into the broader electric vehicle (EV) architecture. These materials offer substantial benefits over traditional silicon, including higher efficiency, greater power density, and the ability to operate at higher temperatures (Fig. 6.2).

Although decreasing, the cost of WBG materials remains a barrier to their widespread use. Additionally, manufacturing complexities and the need for new thermal management solutions pose significant challenges. Integrating WBG-based systems into existing electrical architectures requires overcoming compatibility challenges, ensuring reliability under various operating conditions, and addressing the higher upfront costs of these advanced materials.

Despite these challenges, the potential for WBG technologies to revolutionize e-mobility by enabling more efficient, compact, and lightweight EVs is undeniable, driving ongoing research and development efforts.

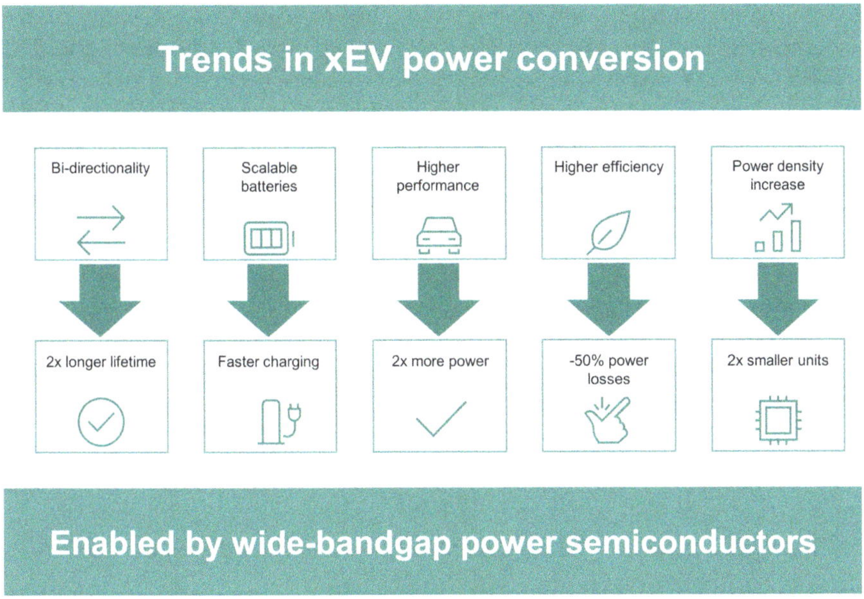

Fig. 6.2 Trends in xEV power conversion

6.2.2 The EV-Charging Ecosystem

The EV-charging ecosystem encompasses an infrastructure tailored to meet the diverse needs of electric vehicles (EVs), focusing on efficiency, accessibility, and technological advancement. This ecosystem is segmented into off-board and onboard charging systems, each having a pivotal role in the EV's operation and integration into the power grid.

Off-board chargers, including AC and DC types, are external systems providing the necessary power conversion outside the vehicle. AC chargers, typically found in homes and public spaces, rely on the vehicle's OBC to convert the AC power to DC for battery storage. On the other hand, DC chargers bypass the OBC and deliver power directly to the EV's battery, enabling rapid charging capabilities essential for long-distance travel and commercial use.

The advancements in OBC highlight the importance of compact, high-efficiency charging solutions within the vehicle. The progress in WBG technologies is instrumental in this shift as these materials offer superior performance over traditional silicon, including higher thermal conductivity, greater efficiency at high frequencies, and reduced size and weight of power conversion systems. The integration of WBG-based OBCs represents a significant step forward in achieving higher power density and efficiency, which is critical for extending the range and reducing the charging time of EVs.

The development of smart charging solutions and bidirectional charging capabilities also characterizes the evolution of the EV-charging ecosystem. Smart charging systems optimize charging schedules based on grid demand, vehicle usage patterns, and energy tariffs, enhancing the overall efficiency of the power grid. The smarter and bidirectional charging profiles will introduce the need for higher overall efficiency over the complete span of the available power classes instead of today's requirement to have best efficiency only at full load. Bidirectional chargers, enabled by sophisticated control mechanisms of WBG-based OBCs, allow EVs to return energy to the grid or power homes during peak demand, showcasing the potential of EVs as mobile energy storage solutions.

In the off-board, wall box, and portable charger's segment of the EV ecosystem, highly efficient and ultra-dense solutions become crucial. Higher efficiency enables size reduction and further improves the energy utilization in both use cases (charging and discharging). With SiC and GaN, the reduction in size of power electronics devices can reach new limits and enable new applications such as charger-in-cable or ultra-portable chargers.

6.2.3 *GaN-Enabled Innovations in Onboard and Off-board Charging Systems*

The electrification of the automotive sector is advancing rapidly, driven by innovations in power electronics, particularly through integrating GaN in onboard charging (OBC) systems. Infineon Technologies Canada Inc., formerly known as GaN Systems, has introduced a revolutionary GaN-based 11 kW/800 V OBC reference design (Fig. 6.3), marking a significant leap forward.

This design boasts a 36% higher power density and up to 15% lower bill of material (BOM) cost than its silicon-carbide counterparts. The OBC leverages a

Fig. 6.3 11 kW/800 V OBC reference design

three-level flying capacitor topology with a bridgeless totem-pole PFC structure and a dual active bridge in the AC/DC and DC/DC stages, significantly improving the efficiency and reducing the size of the charging unit. This development underscores the potential of GaN to enhance efficiency, reduce the overall weight and size of electric vehicles (EVs), and lower production costs, paving the way for more sustainable and cost-effective e-mobility solutions.

In the off-board wall-box charging segment of the EV-charging ecosystem, the development and implementation of vehicle-to-everything (V2X) systems, such as the OMRON V2X unit is a significant advancement. These systems utilize WBG devices, specifically GaN-based power solutions provided by Infineon Technologies, to offer compact and highly efficient charging solutions. The OMRON V2X unit, leveraging Infineon's CoolGaN™ technology, represents a leap forward in enabling bidirectional charging and discharging paths between renewable energy sources, the grid, and EV batteries. This approach facilitates the efficient use of renewable energy, enhances grid stability, and allows EV owners to use their vehicles as mobile power storage units.

The collaboration between Infineon and OMRON has resulted in one of Japan's smallest and lightest V2X systems, the KPEP-A series. These systems significantly reduce size and weight compared to traditional systems, coupled with 6 kW charging capability. The integration of CoolGaN™ technology has improved the power efficiency of these systems significantly, making them more than 10% more efficient at light load and around 4% more efficient at rated load. This improvement in efficiency, along with the reduction in size and weight, facilitates easier installation and maintenance, allowing for more versatile design options and installation locations.

6.2.4 Conclusion

The evolution of electric vehicles (EVs) is significantly propelled by advancements in wide-bandgap (WBG) technologies, particularly through the integration of gallium nitride (GaN) and silicon carbide (SiC) in onboard chargers (OBCs) and inverters. Infineon's role is pivotal in building these technologies that are key to efficiency, power density, and compactness, offering a promising pathway toward sustainable e-mobility solutions.

As the EV-charging ecosystem continues to evolve, the role of innovative charging solutions, including smart and bidirectional charging, becomes increasingly critical. Looking ahead, the cost-effective positioning of WBG technologies promises to accelerate the adoption of EVs, making e-mobility accessible to a broader audience and marking a significant step toward achieving a zero CO_2 future.

6.3 How GaN Supports Next-Generation Wireless Power Transfer

Igor Spinella

One of the great things about GaN power semiconductors is that their performance makes the historically impossible now possible. One example is seen in new GaN-based converters for data centers, reaching next-generation efficiency targets such as 80Plus Titanium with high power and small form factor—something that would be beyond the reach of platforms built on silicon devices.

Another opportunity, leveraging GaN's favorable characteristics, lies in scaling up wireless power transfer technologies like Qi, typically used with handhelds, to charge larger batteries with much lower energy losses than typically expected with silicon. Qi chargers are usually rated at no more than 30 W, but GaN can help extend the convenience and simplicity of wireless charging to applications well beyond this level.

The ability to charge efficiently at several hundred watts, or even over 1 kW, brings the prospect of connection-free, cable-free charging for products like laptops, e-scooters, e-bikes, and domestic robots such as autonomous vacuums and lawnmowers. Small electric vehicles (think urban delivery robots), uncrewed aerial vehicles (drones), and industrial autonomous guided vehicles can have the opportunity to charge themselves, easily and conveniently, without precision docking or manual intervention. At higher power levels, GaN has a central role in enabling more efficient wireless charging of private cars and public transport vehicles, such as buses.

There is a significant challenge associated with scaling wireless power transfer into the kilowatt range, given that typical Qi system efficiency is relatively low, at about 70%. When charging smartphones or small wearable devices, only a few watts are traded away in exchange for convenient cable-free charging. If the charging power is to be several hundred watts, or more than 1 kW, the equation becomes somewhat different. A drastic improvement in efficiency is needed – both from a marketability standpoint and to tackle thermal management and other engineering challenges. Here is where new designs that leverage the qualities of GaN can change the game.

6.3.1 Wireless Power Transfer

Established Qi designs have a complex power conversion architecture. The complete process to implement inductive charging typically involves five stages as shown in Fig. 6.4. This is one significant factor that hampers efficiency. Moreover, the efficiency is significantly reduced when the system is not operating at full load.

As Fig. 6.4 shows, an external AC/DC adapter converts the mains voltage to a DC voltage that supplies the wireless charger transmitter coil section through an

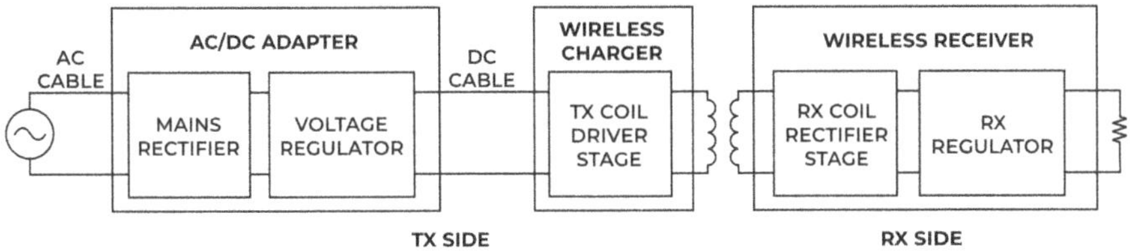

Fig. 6.4 Conventional wireless charging with multi-stage conversion

external DC cable. The coil then transfers the energy to the wireless receiver built into the device to be charged. In such a design, the typical maximum charging distance is about 5 mm and the maximum power is about 30 W.

Analyzed in detail, the first stage of the system is a mains rectifier to rectify the input line voltage. For low-power applications, this may simply be a diode bridge, or it may include power-factor correction (PFC) where mandated at higher power ratings. The next stage in the transmitter is a voltage regulator, needed to manage the power to be transmitted. This is typically a buck converter with two active elements and one high-current magnetic element.

Finally, a DC/AC coil driver stage energizes the coil. This is a resonant stage that may be a half-bridge with two active elements or a full bridge with four active elements. In both cases, operation sees a load-dependent efficiency curve that is optimal only in full-load conditions. The efficiency is degraded at lighter loads or when the coupling between receiver and transmitter is far from nominal conditions, i.e., if the device to be charged is placed at an excessive distance from the transmitter coil or is poorly aligned.

The receiver side comprises an AC/DC coil and full-bridge rectifier stage, with a DC/DC voltage regulator. This is typically a buck converter with two active elements and one magnetic element to regulate the output voltage.

From a system perspective, this traditional wireless charging solution has between 12 and 16 active devices, at least two magnetic components, two coils, an AC cable, and a DC cable. Using the same approach and technologies to create wireless charging systems for applications of several hundred watts and over 1 kW, all these components would need to be uprated, negatively impacting power density, Bill of Materials (BOM), and energy losses.

6.3.2 Simplified Architecture

Now consider a new architecture that has been designed to overcome these issues by combining GaN power components with a novel control strategy based on zero-voltage, zero-current switching (ZVS-ZCS). This effectively simplifies the circuitry, reduces the BOM and circuit size, and eliminates the energy losses associated with the large number of components needed in conventional charging systems. Leveraging this saving, combined with the inherently low switching losses

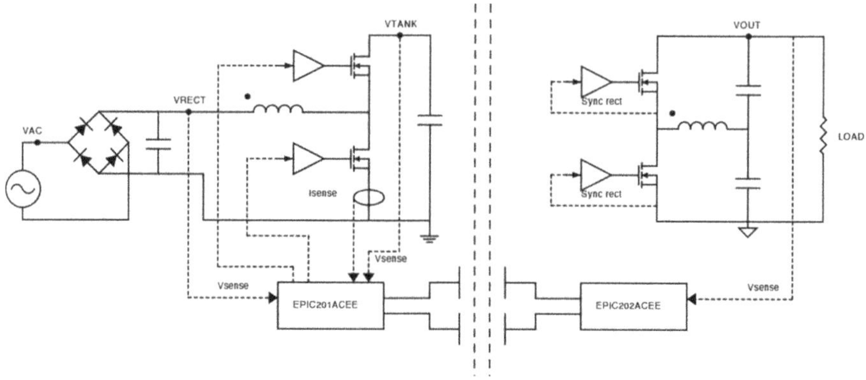

Fig. 6.5 Simplified architecture leveraging GaN and ZVS-ZCS

associated with GaN, the overall power-conversion efficiency of the new wireless power transfer system can be increased from about 70% to 96%—or even higher.

The proposed wireless power transfer architecture, developed by Eggtronic and named E^2Watt, combines a power supply and wireless charger (Fig. 6.5) and is built around a patented inductive technology. This approach reduces the number of conversion stages to two, comprising an AC/AC transmitter stage supplied directly from the mains and a single-stage AC/DC receiver half-bridge.

The AC/AC transmitter stage is modelled as a bridgeless AC converter that drives the coil using a half-bridge architecture comprising two GaN power ICs and two diodes. On the receiver side, the single-stage AC/DC half-bridge is built with two active devices and acts simultaneously as the secondary-side rectifier and non-dissipative output regulator.

The ZVS technique implemented in the transmitter controller eliminates the dependence on optimal input voltage levels, which is a weakness of typical quasi-resonant (QR) offline converters. In the US and Japan, where mains voltages are relatively low, resonant switching can operate close to 0 V and losses are small. In Europe, where line voltages are higher, a QR circuit is often unable to discharge the parasitic Field-effect transistor (FET) as the load reduces. Under these conditions, hard switching losses occur. While an active clamp flyback (ACF) converter can overcome this, it is most effective above 50% load. Below this, it operates in burst mode which degrades the efficiency. Also, an ACF converter needs a high-current, high-voltage high-side FET, which adds to the BOM cost.

In the transmitter, the effects of ZVS and near-ZCS conditions combined with robustness to input and load variations mean switching power losses are close to zero. Unlike a resonant circuit, whose efficiency is significantly reduced at load levels below full load, ZVS operation can be maintained for all load conditions. This ensures consistently high efficiency over a wide output power range (Fig. 6.6).

It is worth noting that the efficiency has the added benefit of reducing heat dissipation in all operating conditions, resulting in up to 30 °C reduction in case

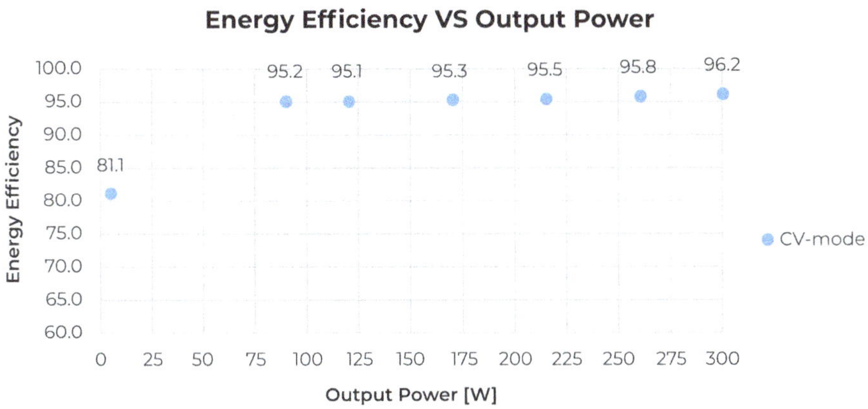

Fig. 6.6 Wireless power transfer with ZVS-ZCS raises efficiency across the load range

temperature in a 30 W design. This is important as heat is one of the design "bottlenecks" associated with the traditional wireless architectures.

Moreover, the transmitter design benefits from a rapid control loop that guarantees good voltage regulation thereby removing any requirement for additional buck converters or low-dropout regulator (LDOs) and their associated losses. This rapid control loop supports power factor control, if needed. A second control loop contributes to minimize the overall reactive power leading to further efficiency improvements.

An additional benefit compared to conventional technologies is that the receiver can be placed at a greater distance from the transmitter and still receive the full charging power. Tests of this new architecture have shown an eight-fold increase in the maximum z-axis charging distance, which can further increase user convenience.

The E^2WATT receiver operates in ZVS and additionally incorporates techniques that control the ratio between active power delivered to the load and the reactive power reflected to the primary.

Eggtronic's EPIC201ACEE control IC used in the transmitter, and the receiver's EPIC202ACEE controller, are designed for this new, high-power wireless charging architecture. While both can handle up to 10 kW of wireless power, they can also be used to build more efficient smaller chargers in the standard Qi power range.

6.3.3 *Benefits of Integrated GaN*

By incorporating GaN power semiconductors, the E^2Watt transmitter design maximizes the efficiency advantages gained through ZVS operation. GaN's ability to achieve faster switching transitions, compared to ordinary silicon metal-oxide semiconductor field-effect transistors (MOSFETs), ensures lower switching losses. Considering that E^2Watt is already ZVS in every working point, the main advantage of GaN is in the conditions where the current is not zero and the FET is turned off. Even if this is a usually smaller source of dynamic losses in comparison with losses related with non-zero-voltage-switching conditions, at high frequency the use of GaN helps to reduce current-related losses, minimizes the size of magnetic components, and ensures greater power density.

When frequency is very high, in the range of MHz (it should be noted that, in addition to inductive wireless power, Eggtronic has also developed capacitive wireless power – called EDEN – that works at more than 10 MHz), GaN can be extremely beneficial on the secondary side for high-speed and high-efficiency rectifiers. This will fuel the broader adoption of GaN beyond the primary side, high voltage-applications by supporting secondary side designs in much lower voltage applications.

One of the main challenges when designing with GaN devices can be to ensure proper gate driving voltages are applied at turn on and turn off. If the turn-on voltage is too low, the device is not fully turned on leading to excessive resistance and high losses. On the other hand, if the applied voltage is too high, the gate structure can be damaged leading to overheating at localized spots. While the same is true with silicon MOSFETs, GaN is more fragile and vulnerable to damage.

The transmitter in this design is built with GaN power ICs that integrate the circuitry needed to ensure the gate of the GaN FET is driven at the correct voltage. This greatly simplifies the circuit design by moving the complexity of the drive circuit into the IC, while also saving space and eliminating the failure points that exist in gate drivers built with discrete components.

6.3.4 Conclusion

Taking advantage of GaN's ability to minimize energy losses at high switching frequencies and with fast switching transitions, new power converter topologies and even application architectures become possible. Among these, Eggtronic's E^{2}Watt architecture boosts wireless power transfer efficiency by more than 20%, enabling high-power wireless charging with much lower energy losses than conventional silicon-based designs have been able to achieve [7].

6.4 GaN-based Wireless Power Transfer: Innovations in Next-Generation Charging Technologies

Paul Wiener

In the field of power electronics, gallium nitride (GaN) power semiconductor devices have emerged as a transformative force, bringing in a new era of Wireless Power Transfer (WPT). This chapter explores GaN's pivotal role in revolutionizing

WPT, promising to enhance system performance by switching at much higher frequencies compared to traditional solutions. As we delve into the technological advancements and challenges within WPT, we highlight the transition from the initial inductive charging methods to sophisticated, high-frequency systems enabled by GaN devices.

Through this exploration, we aim to provide insights into how GaN devices address current limitations and shape the future of power delivery. The potential of GaN-based WPT to revolutionize power delivery in industrial settings, including fields such as micromobility solutions, autonomous robotics, etc., marks a pivotal advancement in wireless charging technology. Its impact extends beyond these sectors, promising transformative benefits for consumer electronics and various other industries as well.

6.4.1 Evolution of Wireless Technologies: Cutting the Last Wire

From landlines to lightning-fast 5G technology, the twentieth century witnessed a wireless revolution in communication. 1G's bulky phones led to 2G and SMS, followed by 3G's mobile internet and 4G's streaming boom. Now, 5G promises near-instantaneous connections, reshaping our interactions [1]. Our phones are no longer tethered to an inconvenient socket in the wall.

With the demand for more data and an easier method to access it, Wi-Fi liberated wired LANs in the late 1990s, providing devices with increasing speeds and wider coverage [2]. Again, innovation freed us from another tether between our devices and a socket in the wall.

Now, the realm of power transfer is undergoing a similar transformation, shifting from tangled cords to wireless power transfer and charging. Just as communication and data witnessed a wireless revolution, a "cutting of the cord," power transmission solutions have an opportunity, in some applications, to experience an equal metamorphosis. We are transitioning from a world dominated by tangled cords and power outlets to one where wireless power transfer holds immense potential.

6.4.2 Wireless Power Transfer

At its essence, WPT is a technology that allows the transmission of electrical power without the need for a physical connection. This technology not only promises to redefine the convenience and aesthetics of charging but also opens new avenues for applications where traditional wired connections are costly, impractical, or impossible.

6.4.3 Types of WPT Technologies

Numerous WPT technologies are on the market, each at different stages of adoption and development. Key examples are:

- Inductive charging (Wireless Power Consortium (WPC)/Qi standard): Inductive charging, streamlined thanks to the Qi standard developed by the WPC, is the most widely adopted form of WPT. This method utilizes electromagnetic fields generated between two coils—a transmitter coil in the charging station and a receiver coil in the device—to transfer energy. The process begins when a device is placed on or near a charging pad, allowing for a direct transfer of power without the need for cables. Inductive charging is renowned for its convenience and safety.
- Near Field Communication charging (NFC Forum): Near Field Communication (NFC) is unique in that it can transmit power as well as data. Complimentary to Wireless Power Consortium's Qi induction charging platform, which delivers up to 15 W over a distance of 4 cm, NFC Wireless Charging Specification currently offers induction charging up to 1 W over a distance of up to 2 cm [8]. While less powerful than Qi, its compact antenna means that it can be incorporated into even the smallest devices.
- Magnetic resonance (AirFuel Alliance): Building on the principles of inductive charging, magnetic resonance technology, championed by the AirFuel Alliance, extends the capabilities of wireless charging to offer greater spatial freedom. The technology relies on both the transmitting and receiving devices being tuned to the same frequency, allowing them to resonate at this shared frequency. When the resonant frequency is matched, energy can be efficiently transferred across a gap, with significantly less sensitivity to the exact positioning of the receiver relative to the transmitter. This attribute enables devices to be charged at a distance and allows for the possibility of charging multiple devices simultaneously, without direct line-of-sight or precise alignment on a pad.

Less common WPT technologies include:

- *Capacitive coupling*: Capacitive coupling uses electric fields instead of magnetic ones for energy transfer. Ideal for short-range applications such as wearables or medical implants, it employs electrodes in both the transmitter and receiver. Upon activation, an alternating electric field forms between the electrodes, enabling power transfer through capacitive coupling.
- *Microwave power transfer*, offering the potential for long-distance energy delivery to drones or electric vehicles, with ongoing studies into its efficiency and safety.
- *Laser power transfer*, focusing energy via beams for extensive reach, could power satellites, though it is currently limited by range.
- *Acoustic power transfer*, that uses sound waves for applications in underwater settings or areas sensitive to electromagnetic interference.

- *Radio frequency (RF) power transfer*, that provides a solution for medium-range energy requirements in wearables and small electronics. It is facing challenges with efficiency.

  Challenges addressed by WPT

- *Convenience*: Eliminates the need for cables and connectors, offering a clutter-free and simplified charging experience.
- *Durability*: Reduces wear and tear on devices by removing physical connections that often fail due to repeated use.
- *Safety*: Enhances safety by minimizing electrical hazards associated with exposed connectors and cables, especially in wet or hazardous environments.
- *Innovation*: Enables new applications and designs that were previously constrained by the limitations of wired charging, such as in autonomous robots and mobility, medical implants, and fully sealed electronic devices.

As we explore the journey of wireless power technologies, it is evident that the evolution is about more than just eliminating wires; it is about enabling new opportunities for mobility, convenience, and more cost-effective solutions. The following sections will delve into the technological advancements from first-generation wireless charging to high-performing GaN-based high-frequency systems.

6.4.4 The Generations of Wireless Power Technologies

The evolution of wireless power technologies can be categorized into two generations, each marked by significant technological advancements and changing industry standards.

Generation 1: Foundations of wireless power

The first generation of wireless charging technology is predominantly characterized by the Qi standard, developed by the WPC. Qi technology was released back in 2008 and improved continuously over the years that followed [3, 4]. It employed inductive charging, requiring devices to be placed in close proximity to charging pads. Qi technology, prevalent in smartphones and wearables, faces constraints like restricted positioning, reduced efficiency over distances, and singular device charging.

Version 1.0 of the Qi standard was released in July 2010. Throughout the next 10 years, improvements were made in foreign object detection (FOD), increased charging speed, and mandatory certification requirements. Most recently, enhanced magnetic alignment methods were added to help address efficiency and heat concerns.

6.4.5 The Path to Generation 2 WPT: AirFuel, High Frequency Design, and GaN Power Semiconductor Devices

WPT has undergone a significant evolution since its inception, transitioning from the initial reliance on inductive charging methods to more advanced and flexible approaches. As seen earlier, inductive charging requires devices to be precisely aligned on charging pads, limiting the convenience and adaptability of the charging process.

Recognizing these limitations, AirFuel Alliance has shifted the focus towards magnetic resonant charging [6]. This approach not only allows for the charging of devices at a distance but also removes the necessity for precise alignment, significantly enhancing user experience.

At the core of this technological advancement is the use of high-frequency WPT, specifically centered around the 6.78 MHz frequency, coupled with the deployment of gallium nitride (GaN) power semiconductor devices. The choice of 6.78 MHz is strategic, rooted in its designation within the Industrial, Scientific, and Medical (ISM) radio bands. These bands are globally reserved for non-communication purposes, offering an optimal frequency range that minimizes interference with existing communication systems while maximizing efficiency in power transfer.

6.4.6 Wide-bandgap Devices in WPT

WBG devices, specifically those using GaN, are revolutionizing power electronics with their wider bandgap, enhancing performance at higher voltages, temperatures, and frequencies. GaN's efficiency at high frequencies enables compact designs by miniaturizing components like coils, extending power transfer ranges. GaN offers benefits such as high efficiency, compact size, low electromagnetic interference, and minimal thermal management requirements, making it ideal for WPT systems. Despite its advantages, GaN technology is in its early stages for WPT applications, presenting challenges and opportunities for ongoing research and development to unlock its full potential.

Understanding topologies

Having established the transformative potential of GaN in revolutionizing WPT, let us delve deeper into the technical aspects that unlock its power: topologies. These circuits play a crucial role in maximizing efficiency and performance, acting as the bridge between the power source and the receiving device.

The implementation of high-frequency WPT necessitates a reevaluation of circuit topologies to optimize performance. Among the popular topologies are Class D, Class E, and Class EF2, each with its own set of advantages:

- *Class D*: Known for its simplicity and ease of implementation, Class D is often used in low-power applications. However, its efficiency tends to decrease at higher power levels.

- *Class E*: Offers a higher efficiency than Class D and is well-suited for medium- to high-power applications. Its operation at higher frequencies makes it a good match for GaN transistors.
- *Class EF2*: Combines the benefits of Class E with further optimizations for even higher efficiency and better performance at high frequencies. Preferred by industry leaders for its superior efficiency and adaptability, Class EF2 is considered the best among these topologies for high-frequency WPT systems [3].

6.4.7 Challenges in the Industry

The path to broader implementation of WPT encounters two significant obstacles: the development of a specialized workforce and the misapplication of WPT where it does not address a substantial need.

- Specialized workforce development: A critical barrier to the adoption of advanced WPT technologies is the scarcity of professionals with the requisite skills. The complexity of designing, implementing, and maintaining these systems demands in-depth knowledge of power electronics, high-frequency design, and system integration. Addressing this gap requires targeted educational programs and industry-academia partnerships focused on cultivating expertise in these specialized areas.
- Application relevance: Another challenge lies in resisting the temptation to deploy WPT in scenarios where it does not solve a pressing problem or significantly enhances efficiency or convenience. The true value of WPT emerges when it is applied in contexts where traditional power delivery methods are limited, such as in some industrial applications where wired connections pose reliability and maintenance challenges. Ensuring WPT is utilized where its impact is transformative, rather than incremental, is essential for its meaningful adoption.

Previous concerns over efficiency, interoperability, cost, and regulatory standards have been addressed and the WPT industry is making significant strides. AirFuel has shown high efficiency, established standards have addressed compatibility, and advances in GaN materials have lowered costs and improved performance. Brands have launched products that comply with stringent safety and regulatory standards, demonstrating the practical viability and market readiness of WPT systems.

6.4.8 Overview of Applications

The adoption of GaN-based WPT technology spans a wide array of applications, each benefiting from the enhanced efficiency, reduced size, and improved performance that GaN offers.

Industrial and commercial applications: Enhancing efficiency

- *Industrial robots*: In manufacturing environments, GaN-based WPT allows robots to recharge autonomously, minimizing downtime and enhancing productivity. This capability is crucial for 24/7 operations where manual recharging is impractical.
- *Micromobility*: Wireless charging pads for EVs and scooters offer the promise of charging scooters and ebikes without physical plugs, simplifying the process and encouraging the adoption of electric mobility.
- *Power tools*: For construction and maintenance, wireless power tools that can charge without direct connections reduce interruptions and improve workflow efficiency.

Autonomous systems and IoT devices: Fueling innovation

- *Drones*: WPT enables drones to recharge by simply landing on a pad, crucial for applications in remote monitoring, delivery services, and inspections.
- *Medical devices*: Wireless charging offers significant benefits in medical applications, from implanted devices that can be charged through the skin to medical tools that can be sterilized without ports that harbor bacteria.
- *IoT devices*: From security cameras to environmental sensors, GaN-based WPT simplifies the deployment and maintenance of IoT devices, especially in hard-to-reach locations.

Consumer electronics: A new standard in convenience

- *Smartphones and wearables*: While Gen 1 WPT has become increasingly prevalent in the charging of smartphones, smartwatches, and earbuds, the benefits are marginally compelling compared to wired charging. As companies offer Gen 2 WPT, there is a clear leap to a seamless charging experience without the hassle of cables.
- *Laptops*: Wireless charging for laptops, enabled by GaN's high power-efficiency, will pave the way for truly mobile computing, unencumbered by the need for power outlets and charger cords.

6.4.9 Future Outlook

As we stand on the cusp of new advancements in wireless power transfer technology, the future holds promising developments that could further revolutionize how we interact with and manage power for our devices and systems. Here are some key trends and predictions for the evolution of WPT:

Trends

- Decarbonization: Increased use of micromobility as people leverage easier and more environmentally friendly means of transportation.

- Electrification: Convenience, range, and ease-of-use drive more everyday-use items to be electric.
- Automation: Robots, robotics, sensors, etc., need the flexibility of movement and location requiring to be untethered from a cable.
- Autonomy: More electrics will be smart, run on AI, and not need human intervention, requiring an easy way to be charged or receive power.

Expanding the ecosystem

- Variable Tx-Rx spacing: Advances in coil design and adaptive power control aim to enable variable transmitter-receiver spacing, accommodating diverse form factors and use cases while maintaining efficiency.
- Interoperability and standardization: To address the need for specialized skills in WPT, the industry continues to focus on fostering industry-academia partnerships for hands-on experience, and leveraging online resources for broader access to learning.
- Novel WPT technologies: Ongoing research into novel WPT methods could further expand the possibilities for certain applications, further diversifying the WPT systems.

6.4.10 Conclusion

The potential of wireless power technology to address modern challenges, such as e-mobility, electrification, automation, and the integration of smart devices, is immense. As we look to the future, it is clear that continued innovation, driven by the capabilities of GaN and other emerging technologies, will be crucial in overcoming existing challenges and harnessing the full potential of wireless charging. The journey ahead is as exciting as it is challenging, promising to transform our relationship with power in an increasingly wireless world.

6.5 Power Supply – USB-C Adapters and Chargers

Junyang Luo and Matthias Kasper

The demand for more efficient, compact, and versatile power delivery solutions has become paramount as the digital era progresses. This need has been significantly addressed by the universal adoption of USB-C connectors, known for their reversibility and high-power delivery capabilities.

This evolution has been further propelled by integrating gallium nitride (GaN) technology, an advancement in power electronics. GaN technology has led to an increase in efficiency and thus reduced surface temperatures of USB-C adapters and chargers. This caters to the ever-growing demand for rapid and efficient power delivery in compact devices.

The integration of GaN into power electronics devices has set new benchmarks for efficiency, compactness, and performance, outpacing traditional silicon-based solutions. This technological leap has enabled the development of smaller yet more powerful charging devices capable of supporting today's digital devices' fast-paced, power-intensive needs. Infineon, a leader in semiconductor solutions, has been at the forefront of this evolution, driving innovation in GaN technology to address the complexities and challenges of modern power delivery systems.

This chapter will explore the fundamental aspects of GaN-based technology in USB-C chargers, focusing on how GaN's electrical properties enhance charger performance. We cover GaN's advantages in achieving higher power densities, improved efficiency, and better thermal management.

The discussion will also include the challenges of integrating GaN technology and Infineon's role in addressing these issues. Furthermore, we look at future trends, the environmental benefits of GaN chargers, and emerging innovations in the sector, showcasing the potential for further advancements in power electronics.

6.5.1 Fundamentals of GaN-based Technology in Chargers

GaN offers several benefits over traditional silicon-based technology. GaN High Electron Mobility Transistors (HEMTs) feature superior electrical properties such as ten-fold reduced output charges (Q_{oss}), lower gate charges, and the absence of reverse recovery charges Q_{rr}. These features enable charger designs with significantly higher efficiency than Silicon-based designs. The reduction of losses translates into less heat dissipation which is crucial for designing more compact chargers since the heat can only be dissipated via the surface area of the charger by means of natural convection and radiation. A thermal simulation for a 65 W adapter underscores this point, showing that to achieve a 45 °C ΔT rise in temperature, an efficiency of 92.5% is necessary for a power density of 20 W/in^3 (Fig. 6.7).

GaN's efficiency and heat management capabilities allow for the design of smaller, more portable chargers without compromising on the power level. These chargers can deliver faster charging speeds, are more energy-efficient, and operate cooler than their silicon counterparts. This makes GaN chargers a practical choice for consumers seeking quick and efficient charging solutions and an environmentally friendlier option. Compared to silicon-based chargers, the reduced size and weight of GaN chargers further contribute to their portability and convenience, making them an excellent choice for modern, high-powered devices.

The transition from silicon to GaN technology in USB-C power adapters and chargers represents a significant advancement in power electronics. By leveraging GaN's superior electrical properties, manufacturers can meet the growing demand for compact, high-performance charging solutions. This shift enhances user experience through faster charging and reduced size and aligns with broader environmental goals by improving energy efficiency and reducing heat generation.

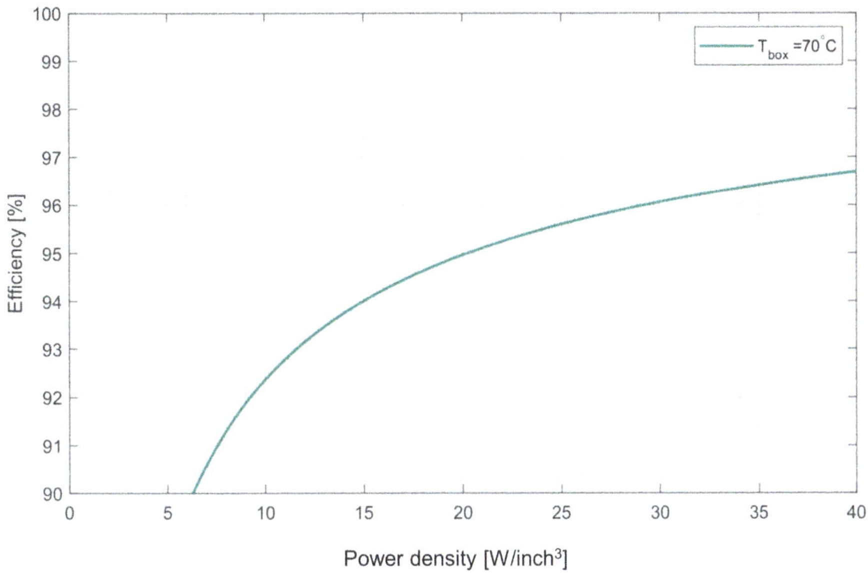

Fig. 6.7 Required efficiency for a 240 W charger as a function of power density for a given maximum surface temperature of the charger of 70 °C

6.5.2 Advantages of GaN in Charging Technology

GaN technology significantly advances charging technology, offering advantages over traditional silicon semiconductors.

Faster charging speeds with less heat: One of the key benefits of GaN is its ability to operate at higher frequencies, which leads to improved power density. GaN chargers achieve zero voltage switching (ZVS) with shorter dead times and lower currents, thanks to lower Q_{oss} charges, which minimize energy losses during switching. As a result, GaN chargers deliver more power in the same form factor and thus faster charging speeds while generating less heat, enhancing overall efficiency and safety.

High power output: GaN chargers stand out for their high-power output and the ability to charge multiple devices simultaneously without sacrificing charging speed or efficiency.

Universal compatibility: They are universally compatible with a wide range of devices, including smartphones, tablets, laptops, and potentially even electric bikes, as the support various power delivery protocols, such as USB power delivery (USB PD) and programmable power supply (PPS). Their wide output voltage range ensures that GaN chargers can efficiently cater to different devices' power needs.

Compact design: The superior figure-of-merits of GaN devices compared to Silicon superjunction devices, such as lower Q_{oss} and gate charges, and the absences of reverse recovery charges, enables the design of chargers with higher switching frequencies and smaller passive components. This leads to more compact chargers with higher power densities.

By incorporating GaN technology, manufacturers can create chargers that are not only more efficient but also significantly smaller and lighter than traditional chargers. This makes GaN chargers ideal for consumers looking for portable and powerful charging solutions.

Integrating GaN technology in chargers represents a significant leap forward in pursuing more efficient, compact, and versatile charging devices, catering to the growing demand for fast and efficient power solutions in today's technology-driven world.

6.5.3 GaN Charger Solutions and Challenges

The adoption of GaN technology in USB-C adapters and chargers offers significant benefits, such as increased efficiency, compact size, and faster charging capabilities. However, the transition to GaN technology faces several challenges that must be addressed to realize its full potential.

One of the primary challenges is the market domination by silicon-based products. Silicon has been the standard in the market for years. Despite GaN's superior efficiency and cooler operation at equivalent voltages, the existing dominance of silicon products presents a barrier to GaN's widespread adoption.

Solution providers also address the need for compact and high-power chargers by offering GaN chargers in various shapes and capacities, catering to different devices. This includes chargers with multiple output ports, providing up to 100 W capability, suitable for simultaneously charging devices such as laptops, phones, tablets, and earbuds.

Infineon Technologies has been at the forefront of addressing the challenges of adopting GaN technology in USB-C adapters and chargers. Their advanced GaN solutions aim to overcome barriers related to manufacturing complexities, cost considerations. Infineon's GaN solutions are tailored to enhance charger designs by enabling higher switching frequencies with easier implementation of zero voltage switching (ZVS), thus improving efficiency and reducing size.

6.5.4 Solutions for Single-port Chargers

Infineon proposes the active-clamp flyback (ACF) topology for single-port charger designs. The ACF topology is notable for its ability to relieve the voltage stress of primary Field-effect transistors (FET) by clamping its drain-source voltage (V_{DS})

to a capacitor using another FET. This approach reduces voltage stress and allows ZVS, leading to more efficient operation at higher frequencies. The ACF converter is suitable for higher frequency operation than the quasi-resonant (QR) flyback due to its ZVS operation and complete recovery of energy in transformer leakage inductance, which is otherwise dissipated as heat in the QR flyback RCD clamp (Fig. 6.8).

For higher power densities, Infineon introduces the hybrid flyback (HFB) topology. The HFB converter combines the benefits of resonant operation, similar to the LLC converter, with ZVS and Zero Current Switching (ZCS) operation. This topology enables high-frequency operation with high efficiency and lower RMS currents, thanks to the resonant-type current waveforms. The HFB topology also offers better voltage stress management over the FETs than ACF, thanks to its half-bridge structure with a self-voltage clamp to the bus voltage on the primary side (Fig. 6.9).

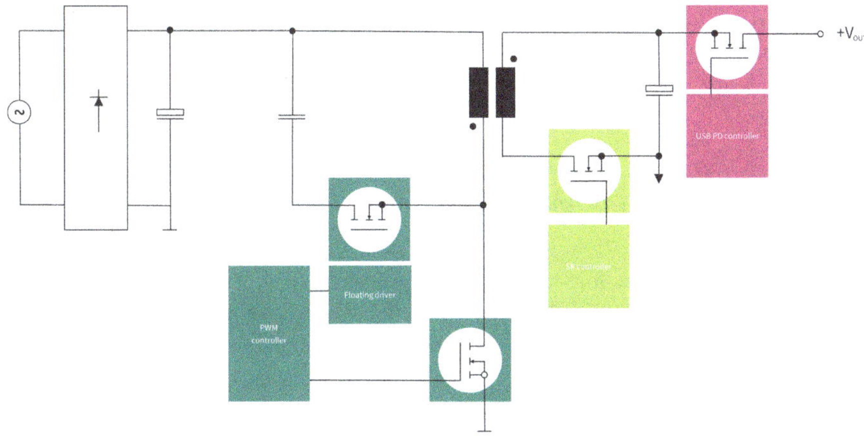

Fig. 6.8 ACF topology

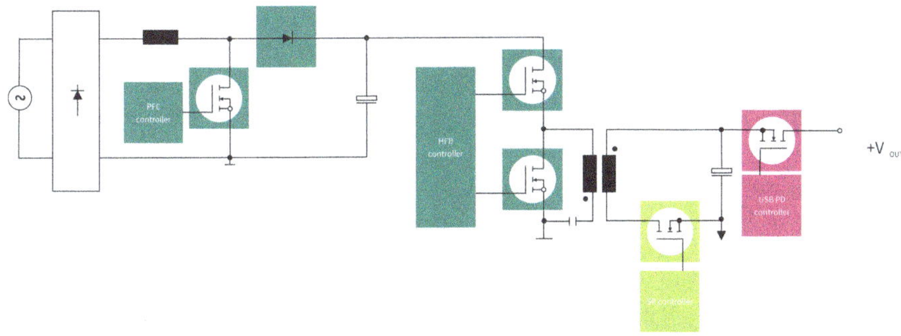

Fig. 6.9 HFB topology

6.5.5 Solutions for Dual-port Chargers

To control two USB-C ports independently, two-stage solutions have to be adopted since flyback-based topologies can only control one output port. Infineon has demonstrated a full-GaN system for a 240 W charger featuring 42 W/in^3. The system consists of an interleaved totem-pole PFC stage operating at 400 kHz (effectively 800 kHz), an unregulated series-resonant converter (DCX) at 425 kHz, and two buck stages for regulating the two USB-C ports at up to 420 kHz. In order to achieve this level of compactness with frequencies at and above 400 kHz, all stages are operating in the soft-switching mode. This is achieved for the PFC and the buck stages by variable frequency modulation depending on the operating point (Fig. 6.10).

Infineon's ongoing research focuses on optimizing GaN technology for single- and dual-port chargers. They are continually exploring new ways to improve the efficiency, cost-effectiveness, and compatibility of GaN-based solutions. By providing detailed design guidelines and evaluation boards and leveraging their CoolGaN™ technology, Infineon is paving the way for the broader adoption of GaN technology in power supply applications.

6.5.6 Future Trends and Environmental Impact

The future of GaN technology in USB-C adapters and chargers is marked by significant advancements toward higher integration levels and environmental sustainability. Innovations in GaN technology are leading to the development of more efficient, compact, and safer charging solutions by directly integrating current sense and protection features into the GaN devices. This integration is crucial for creating advanced charging solutions that enhance performance and ensure user safety through built-in protection mechanisms.

From an environmental perspective, GaN chargers offer substantial benefits. They are recognized for their enhanced energy efficiency, which translates to a reduced carbon footprint for electronic devices. GaN's ability to operate with higher efficiency means that less energy is wasted as heat, leading also to more compact chargers since less surface area is required for cooling the charger by means of natural convection and radiation. This efficiency contributes to a more sustainable use of resources and aligns with global efforts to combat climate change by reducing greenhouse gas emissions associated with power generation.

Moreover, GaN technology supports the drive towards miniaturization of electronic components, allowing for smaller, lighter devices requiring fewer manufacturing materials. This reduction in material usage decreases the environmental impact associated with the extraction and processing of raw materials. It contributes to global e-waste reduction efforts by creating longer-lasting, more durable components.

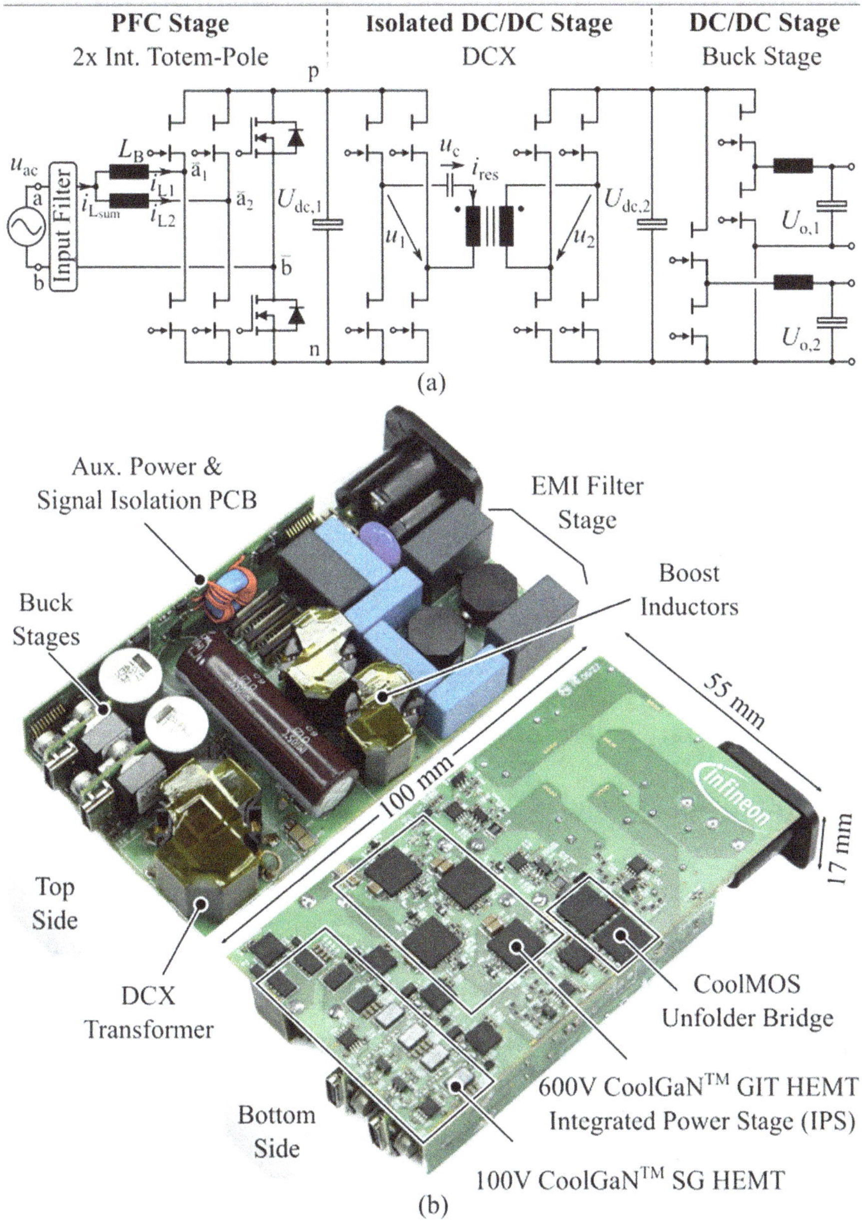

Fig. 6.10 Charger topology

The push towards a greener future is further supported by GaN's role in various industries, from consumer electronics to electric vehicles and renewable energy systems, offering a path to more energy-efficient and compact solutions. As GaN technology advances, it is expected to play a pivotal role in enabling a more sustainable, energy-efficient future for the electronics industry and beyond.

6.5.7 Key Takeaways

GaN technology has significantly impacted USB-C adapters and chargers by delivering enhanced efficiency and compact design. GaN's higher electron mobility and lower on-resistance enable devices to operate with lesser energy losses, leading to more efficient power conversion, and reduced heat generation. This efficiency leap has allowed the development of smaller and lighter chargers that are more portable and environmentally friendly due to their reduced carbon footprint and material usage.

Infineon Technologies has been instrumental in driving these innovations forward. By leveraging their GaN solutions, Infineon has addressed the challenges of integrating GaN into power supply applications, offering advanced topologies and design guidelines that enhance the performance and viability of GaN-based charging solutions.

Looking ahead, GaN technology is poised to continue its influence on the evolution of charging solutions. With ongoing research and development, we can expect to see even higher levels of integration, including smarter protection features and current sensing capabilities. This progression promises to make GaN chargers even more efficient, compact, and safe, reinforcing GaN's role in the shift toward more sustainable and innovative power electronics. The trajectory of GaN technology indicates a future where charging devices are faster and safer and contribute to the broader goals of energy efficiency and environmental sustainability [8–19].

6.6 Improving Motor System Efficiency with GaN Inverters

Guilherme Bueno Mariani

For decades, the power electronics industry has relied on conventional silicon-based devices as the backbone of all systems. But, as power electronics applications become more demanding, necessitating better efficiencies, greater power density, and higher reliability, one of the solutions proposed by the industry is to move away from silicon-based devices.

Wide-bandgap (WBG) semiconductors are a promising solution to address the new efficiency needs. The term "wide bandgap" refers to the larger energy gap these materials have between their valence and conduction bands, a characteristic that underpins their superior performance in electronic devices.

One of the most compelling attributes of WBG semiconductors is their ability to operate at temperatures, voltages, and frequencies much higher than silicon [20]. This is primarily due to their wide bandgaps, which enable devices made from these materials to withstand greater electric fields without breaking down. For instance, while silicon has a band gap of 1.12 eV, WBG materials like silicon carbide (SiC) and gallium nitride (GaN) boast bandgaps of approximately 3.3 and 3.4 eV, respectively. This inherent robustness allows for the design of power electronics and motor systems that are more efficient and power-dense, leading to smaller, lighter, and more energy-efficient systems.

Of all the WBG materials on the market, GaN is uniquely poised to revolutionize the world of motor systems. In this chapter, we explore GaN technology and how GaN-based inverters can significantly improve the efficiency of motor systems.

6.6.1 Motor System Inverters and GaN

Traditional inverters play a pivotal role in motor systems, serving as the bridge that converts direct current (DC) into alternating current (AC), thereby enabling the control of motor speed and torque in applications ranging from industrial machinery to electric vehicles. These inverters typically use silicon-based power electronics, leveraging the material's well-understood properties and manufacturing processes. However, despite silicon's ubiquity, its physical limitations in terms of efficiency, switching speeds, and thermal performance have led engineers to seek better alternatives, especially as demands for energy efficiency and compact designs intensify.

GaN technology offers a slew of advantages over traditional silicon solutions, chief among them being its high efficiency and fast-switching capabilities. GaN devices can operate at significantly higher frequencies than their silicon counterparts without suffering the same level of losses. This is due to GaN's superior electron mobility and lower on-resistance, which reduce energy dissipation during switching and conduction, enhancing overall inverter efficiency.

The impact of higher switching frequencies facilitated by GaN on motor performance and efficiency is profound. This can lead to smoother operation and reduced mechanical stress, extending the motor's lifespan. Additionally, the efficiency gains from using GaN inverters mean that less energy is wasted as heat, further improving the system's overall performance.

Beyond performance gains, GaN also opens up possibilities for size and weight reduction in motor system components. The reduced heat output of these devices allows for smaller cooling solutions, and the compact size of GaN devices enables more densely packed power electronics. Consequently, the adoption of GaN technology holds the promise of creating motor systems that are more efficient, lighter, and more compact.

6.6.2 Experimental Setup from the Case Study

A recent case study outlined an experimental setup (Fig. 6.11) devised to assess the performance of GaN inverters within motor drive systems [21, 22].

The study was based on a versatile inverter design that consisted of a modular architecture featuring three half-bridge modules, allowing for the interchangeability

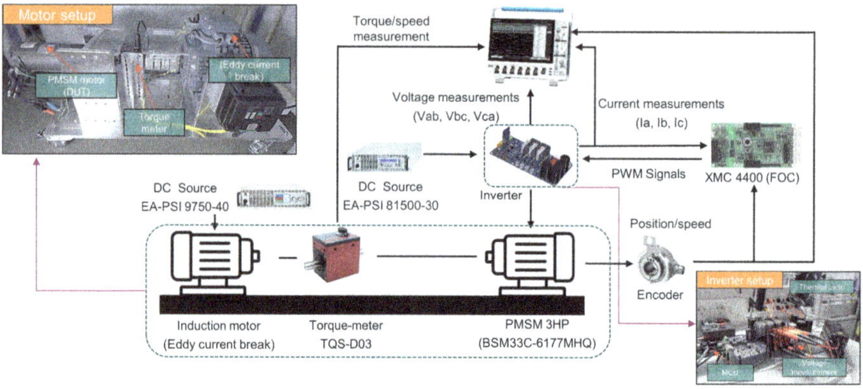

Fig. 6.11 Testbench used for investigations

between SiC and GaN devices. This flexibility allowed the researchers to explore the performance of the inverter with different WBG semiconductor technologies under identical conditions, facilitating a direct comparison of their performance characteristics. The inverter itself was designed to be capable of reaching switching frequencies up to 50 kHz, and the system's design includes power devices, dual-channel gate drivers, and an isolated power supply for adjustable bias voltage.

In the study, the machine under test is a permanent magnet synchronous machine, while an induction machine, functioning as an eddy current brake, was employed to apply a controlled loading torque, providing a stable and adjustable load for testing. The measurement material comprised of:

- Hall sensor probes for current measurements
- Differential probes for voltage measurements
- A torque meter for mechanical output readings
- A precise method for speed determination based on positional data

All data were captured and processed through an oscilloscope, ensuring high accuracy and reliability of the measurements.

6.6.3 *Results and Analysis*

The findings underscored significant efficiency improvements and high-frequency loss reductions when utilizing GaN inverters, marking a noteworthy advancement over traditional silicon-based solutions.

A key observation from the experiments was the impact of switching frequency on system efficiency. It was noted that while motor losses tended to decrease with an increase in switching frequency, the inverter losses conversely increased due to the higher number of switching events. This phenomenon highlighted a critical trade-off between inverter efficiency and motor performance, suggesting an optimal switching frequency that balances the two for maximum system efficiency.

The study also examined the overall system losses, encompassing both motor and inverter losses. It was discovered that GaN devices, operating at 50 kHz, demonstrated the lowest power loss by a significant margin compared to SiC devices, especially under conditions of 320 V and 8 A with devices rated at 600–650 V. At lower frequencies (5–10 kHz), SiC devices showed a slight advantage in terms of power loss; however, as the frequency increased to 20–50 kHz, GaN devices exhibited superior performance with significantly lower losses.

The analysis further revealed that the motor high-frequency losses at 900 RPM and a 50 kHz switching frequency were too minimal to be observed, falling within the margin of error. The most substantial high-frequency losses were recorded at 1800 RPM and a 5 kHz switching frequency, approximately around 12 W. In contrast, at the same speed but a 50 kHz switching frequency, the high-frequency losses were reduced to about 2 W, indicating a notable improvement of 10 W in high-frequency losses.

The study also investigated the interplay between load, speed, temperature, and efficiency in motor drive systems. It was found that the optimal operating conditions for maximizing system efficiency were achieved at a 20 kHz switching frequency (Fig. 6.12) when running at a nominal speed (1800 RPM) and at a 50% rated load (1.1 kW). These conditions minimized energy dissipation relative to the output power, thereby enhancing the motor's lifetime and overall system efficiency.

The study limited the switching speed (dv/dt) to 10 V/ns. This value is a normal standard used in the application to avoid degradation problems in the motor (windings, bearings, etc.) and EMI problems. This tends to reduce the advantages of GaN. However, the study shows that even with this limitation, GaN can provide a clear advantage of operating at a higher switching frequency. In the future, new machine designs should consider the possibility of operating at a faster switching frequency with a higher switching speed.

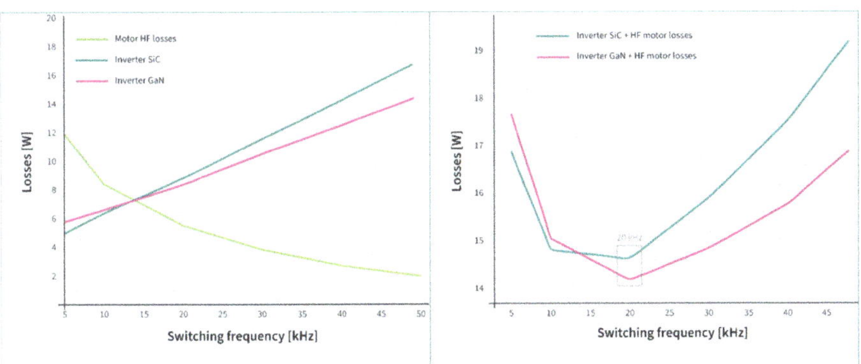

Fig. 6.12 Determining overall system losses and optimum operating switching frequency

Ultimately, the research unequivocally demonstrated that GaN inverters, particularly at higher switching frequencies, offer a pathway to significantly enhanced system efficiency in motor drives. However, the selection of the switching frequency necessitates a careful balance to optimize both inverter and motor performance, underscoring the need for a system-level evaluation when integrating WBG technologies.

6.6.4 Challenges and Considerations

While GaN has clear advantages over other technologies, implementing GaN inverters in motor systems poses challenges like thermal management and electromagnetic interference (EMI). This is largely because the high switching frequencies and efficiency of GaN devices, while beneficial for system performance, generate significant heat and electromagnetic noise that must be meticulously managed to prevent degradation of system reliability and performance.

Thermal management in GaN-based systems is paramount due to the compact and high-power nature of these devices. Efficient heat dissipation mechanisms are essential to maintain operational integrity and prolong the lifespan of the components. Similarly, EMI poses a challenge, as the fast switching speeds of GaN devices can lead to increased electromagnetic emissions, necessitating robust filtering and shielding techniques to comply with regulatory standards and ensure system stability.

The practical application of GaN inverters extends beyond addressing these challenges to considerations of system design and component selection. Engineers must carefully evaluate the characteristics of GaN devices, such as their fast-switching capabilities and thermal properties, to optimize the inverter design for efficiency and reliability. Component selection, particularly for passive elements like capacitors and inductors, becomes crucial in managing the high-frequency characteristics intrinsic to GaN-based systems.

6.6.5 Future Opportunities

As power electronics demand better efficiencies, greater power density, and higher reliability, GaN offers an exciting alternative to conventional silicon. By opening the door to new motor designs that leverage its high-frequency operation and efficiency, GaN technology can significantly improve motor performance, reduce size and weight, and increase energy efficiency. Among the evolving demands of applications like electric vehicles, renewable energy systems, and industrial machinery, GaN holds significant promise for the future.

6.7 Addressing the Multiport Challenge for Simultaneous USB Device Charging

Hubie Noto

From phones to laptops and portable gaming machines to crypto wallets, the number of portable devices that we carry and that require regular battery charging continues to grow. Consumers need convenient solutions that allow them to charge multiple devices simultaneously in a short time. A suitable charger needs to be compact, ideally smaller than a laptop adapter of a comparable rating, and able to deliver adequate power at the outputs for charging devices up to laptop size at an acceptable rate.

Small size is important if the charger is to demonstrate a significant benefit for users looking to simplify their charging setup and to travel light when on the move. GaN power technology is a critical enabler in the pursuit of a compact form factor. Its ability to operate at high switching frequencies lets designers build power converters with smaller magnetic components, while the inherently high efficiency eases thermal management. In practical terms, the gain in efficiency delivers a saving in energy losses that allows adapter dimensions to be reduced without increasing the case temperature.

The USB power delivery (USB PD) specification is another critical ingredient, needed to enable marketable multi-device chargers. USB PD enables devices to negotiate for power based on various criteria, such as their current battery state of charge, to charge significantly more quickly than previously possible. USB PD allows charging at various fixed voltages between 5 and 20 V, at a range of current levels (see Table 6.1). Proprietary charging specifications have also emerged to enable fast charging, typically optimized for the proposers' own device platforms. Qualcomm Quick Charge is a well-known example.

The 5 and 9 V voltage levels defined in USB PD are generally preferred for smartphones, to avoid exposing the batteries to excessive voltage. Charging at 9 V and 2 A (18 W), a large-capacity phone can attain a full charge in about 1 h. Some handset brands use proprietary charging settings to charge phones at rates from 30 W to as much as 65 W. A laptop may take from 1 to 2 h for a full charge, at 65 W.

Table 6.1 USB PD charging ranges

USB PD power range	Fixed voltage	Current range	Example devices
0.5–15 W	5 V	0.1–3.0 A	Headphones, crypto wallets, small accessories
15–27 W	9 V	1.67–3.0 A	Smartphones, handheld game terminals, cameras
27–45 W	15 V	1.8–3.0 A	Tablets, small laptops
45–100 W	20 V	2.25–3.0 A, 3.0–5.0 A	Larger laptops

6.7.1 Multiport Charger with GaN

Figure 6.13 shows a recently published reference design for a 65 W charger that provides three outputs and presents USB Type A and Type C ports. The AC/DC converter stage uses an active clamp flyback (ACF) topology. The flyback is chosen for its simplicity, benefiting from a minimal bill of materials and galvanic isolation due to its transformer-based design. Compared to conventional quasi-resonant (QR) flyback, the ACF achieves closer to zero-voltage switching (ZVS) than is possible with a passive clamp and dissipates almost zero energy due to the on-resistance of the clamp metal-oxide semiconductor field-effect transistor (MOSFET). The clamping MOSFET also allows return of leakage energy to the transformer that is otherwise dissipated in a conventional passive clamp. This saves about 20% of energy losses. In addition, the ACF flyback maintains higher efficiency at all loads and across a wider input volage range.

The SZ1131 ACF PWM controller shown in the diagram controls a Transphorm TP65H300G4LSG 650 V, 240 mΩ SuperGaN Gen IV FET. This fourth-generation SuperGaN technology uses advanced epi and patented design technologies to ensure very high gate robustness. This allows the devices to be used without protective external circuitry such as additional bias rails or level shifters, which enhances power density and efficiency while reducing BOM costs. The improvement achieved by replacing a silicon MOSFET with SuperGaN in the AC/DC stage is similar to the improvement achieved by moving from QR to ACF topology.

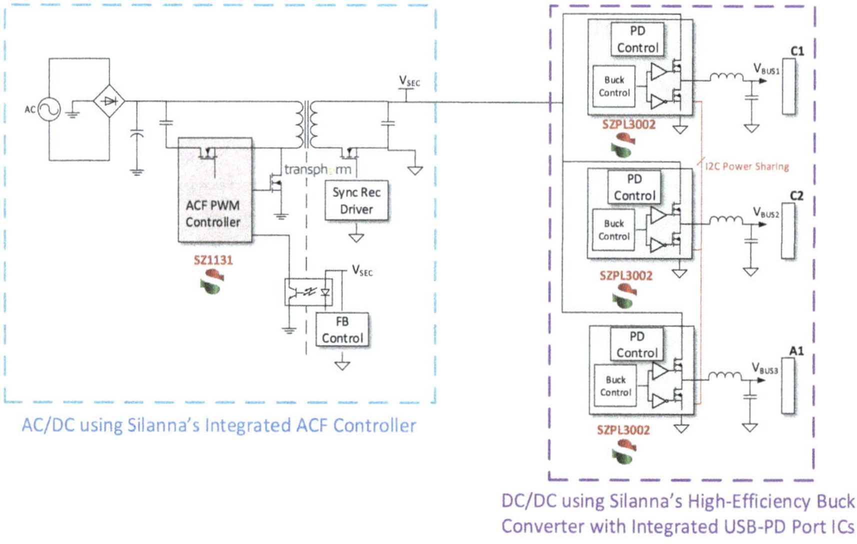

Fig. 6.13 65 W multi-port USB charger reference design

6.7.2 Adaptive Power Sharing

The charger's DC/DC stage comprises three SZPL3002 synchronous buck converters. The converter allows the switching frequency to be selected from 667 kHz to 2 MHz and produces an output voltage from 3 to 21 V. The SZPL3002 IC also integrates a USB-PD port controller that supports USB-PD R3.0 as well as the USB PPS programmable power supply specification, USB battery charging (BC1.2), and Qualcomm Quick Charge specifications.

By implementing intelligent adaptive power sharing and power rebalancing, the SZPL3002 continuously optimizes the charging rate of devices connected to the output ports. A traditional multiport power adapter without this dynamic power sharing is constrained to charge each device at an equal or pre-set rate, regardless of the device type or battery state of charge. Moreover, the adaptive power sharing algorithm lets the charger dynamically reallocate power as connected batteries become charged and devices are added or removed. In this way, each device always receives the appropriate amount of power. The power-sharing reference design shows how intelligently distributing the available power across the connected devices can enhance value for users.

When a single device is plugged into a charger built using this reference design, the PD controller integrated in the SZPL3002 communicates with the device to determine its power requirements. If the device has a low battery level, say 20%, it will receive 100% of the available power. When a second device is added, the PD controller communicates with the new device to establish its power needs. If the second device is a smartphone with a battery level of 30%, the PD controller will reduce the power allocated to the first device to 57.5 W and allocate 7.5 W to the newly connected smartphone.

Similarly, when a third device is added, the controller repeats the communication sequence to establish the new power requirements. The power allocated to the first and second channels may be further adjusted based on the battery levels of all connected devices. This ensures that the maximum output of the multiport adapter is optimally shared among the devices according to their needs.

As the battery level of the first device increases, the SZPL3002 can automatically rebalance the power to ensure that all devices continue to charge optimally. For example, the power allocated to the first device may be reduced by 10 W, and the additional 10 W may be redirected to the second device.

The PD controller built into the SZPL3002A supports both Type A and C USB outputs and permits up to four ports with adaptive power sharing. Temperature-triggered power throttling protects the device by reducing the charging current when a warning temperature is reached. Overcurrent and overvoltage protection are also built-in.

Combining the advantages of ACF operation and GaN technology in the AC/DC stage with the SZPL3002 controller's intelligent capabilities in the DC/DC stage, this reference design operates with over 93.7% peak efficiency and consumes 161 mW no-load power at 230 V AC input. It also displays a flat efficiency curve across the input-voltage ranging from 90 to 265 V AC and across a wide load range.

6.7.3 Conclusion

GaN technology not only enables existing power-converter designs to achieve next-generation efficiency meeting modern ecodesign standards but is also a catalyst for new products that can make life more sustainable as well as comfortable and convenient.

One example is multi-port device chargers, which is growing in demand to help manage modern digital lifestyles. GaN is the key to enable multi-port chargers of a form factor and weight that are attractive to end users. Fast charging standards such as USB PD and proprietary quick-charging specifications, as well as efficient, resonant power-converter topologies, are already in place. With the arrival in the market of adaptive, power-sharing PD controllers, the time is right for multi-port chargers to lighten the load on consumers and the environment.

6.8 Understanding Digital Control

Thierry Bouchet

Digital control involves representing signals, measurements, and control actions in digital form, typically as binary values (0s and 1s). This contrasts with analog control, where signals are represented in continuous, analog forms.

Digital control systems use algorithms and digital signal processing techniques to analyze data, make decisions, and generate control signals. These algorithms can be programmed and modified to adapt to different conditions, making digital control systems highly versatile. Digital control often involves the use of microcontrollers or digital signal processors (DSPs) to execute control algorithms. These devices can handle complex computations and are programmable, allowing for easy adjustments to the control strategy.

In the ever-changing realm of power supply technology, digital control, particularly in the context of Gallium Nitride (GaN), is instigating a significant shift. In contrast to traditional methods where analog signal blocks are converted into digital ones, digital GaN control strategically divides tasks between GaN ICs and digital controllers to enhance performance.

Digital control surpasses conventional approaches by simplifying intricate tasks, such as current sensing. Through the innovative use of digital GaN, a virtual current sensor is crafted through analytical estimation, reducing costs represented solely in lines of code. This breakthrough not only minimizes losses but also preserves PCB space and eliminates challenges associated with analog sensors.

The transition to digital GaN control is not just about replacing analog processes; it is a holistic transformation that impacts various facets of power supply technology. This shift allows for greater flexibility in system design, enabling engineers to optimize performance with a finer degree of control. The marriage of digital control and GaN components introduces a level of synergy that transcends the limitations of traditional analog systems.

6.8.1 Advantages of Digital Control

The benefits of digital control go beyond simplification. High-resolution pulse-width modulation, coupled with high-performance microcontroller units (MCUs), introduces solutions like zero-voltage switching, achieving extreme frequencies without increasing switching losses. This precise timing resolution, combined with GaN's efficient switching characteristics, redefines power supply efficiency.

Digital GaN does not stop at hardware optimization; it opens pathways to the cloud, contributing to the fight against climate change. Positioned at the core of power supply units and converters, digital GaN aggregates performance data, computing health-and-usage metrics with ultra-low bandwidth data flow. This intelligent link facilitates remote firmware updates and feature activations over the internet, showcasing adaptability in diverse applications, from consumer electronics to electric vehicles.

As we delve deeper into the advantages of digital control, it becomes evident that its impact extends far beyond immediate efficiency gains. The ability to harness real-time data and communicate with other devices via the cloud transforms power supply systems into intelligent, interconnected nodes within a broader network. This connectivity not only enhances user experience but also contributes to a more sustainable and adaptable power infrastructure.

One of the significant advantages of digital control is its adaptability. Engineers can modify control algorithms and parameters to optimize performance or accommodate changes in the system dynamics. This flexibility is particularly valuable in dynamic and complex environments (Fig. 6.14).

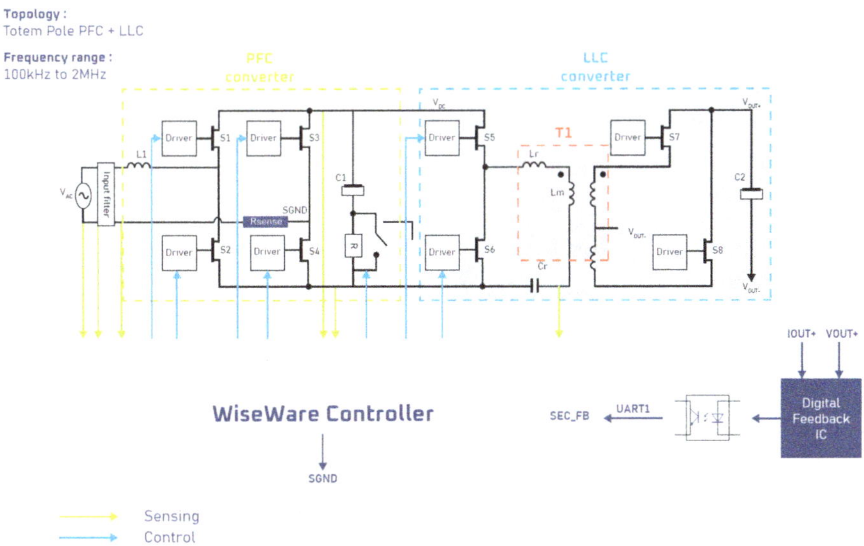

Fig. 6.14 Totem pole PFC-LLC using WiseWare digital control by Wise-integration (datasheet WIW1001ER1)

6.8.2 Innovation by Wise-integration

At the forefront of this digital revolution is Wise-integration, paving the way for transformative solutions in the power supply market. The collaboration between GaN Power Integrated Circuit (WiseGan®) and GaN Digital Control (WiseWare®) is key to this innovation (Fig. 6.15).

The GaN power IC manipulates energy pulses, while digital control efficiently manages their switching. This synergy revolutionizes the market by reducing volume while increasing energy efficiency through higher switching frequencies. What sets Wise-integration apart is the specific approach in achieving these results. Their digital control system possesses in-depth knowledge of the dynamic parameters of GaN power components, ensuring real-time control of switching for efficient energy conversion.

The drastic reduction in size without compromising efficiency is a hallmark of Wise-integration's solution. By integrating GaN technology into an IC and leveraging digital control, they maximize system-level performance. The increased operating frequency, made possible by innovations, significantly reduces the volume of passive components, optimizing overall system performance (Fig. 6.16).

Wise-integration's commitment to innovation does not stop at hardware optimization. They recognize the importance of a holistic approach that considers the entire ecosystem surrounding power supply technology. Their WiseWare® digital control system, in addition to optimizing energy conversion, facilitates seamless integration with smart grids and renewable energy sources. This forward-looking perspective positions Wise-integration as a key player in the evolution toward a greener and more sustainable power infrastructure.

In summary, Wise-integration's groundbreaking approach to digital control, coupled with GaN technology, not only redefines power supply efficiency but also contributes to a more sustainable and connected future. As we witness the transformative impact of their innovations, it becomes evident that the future of power supply technology is not just efficient but also intelligent, adaptive, and environmentally conscious.

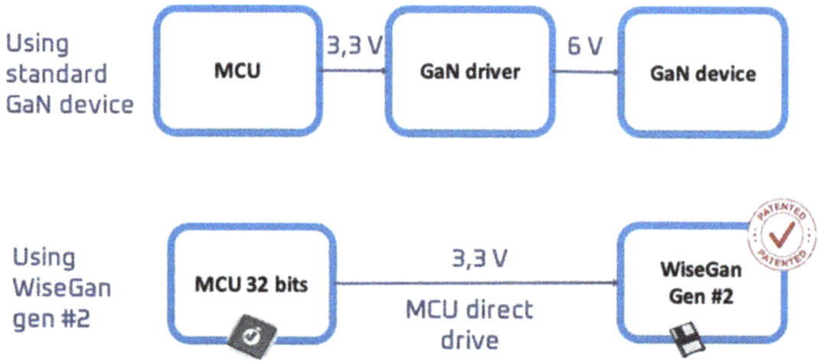

Fig. 6.15 Synergy WiseGan® WiseWare®

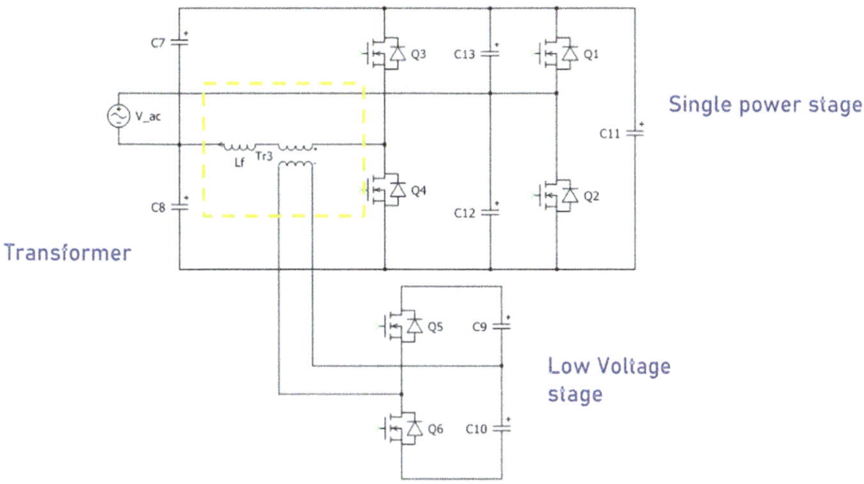

Fig. 6.16 Patented architecture using WiseWare digital control, Wise-Integration

6.9 Benefit of Applying GaN to Automotive µDCDC Converters

Peter Vaughan

The automotive industry always has a strong drive to reduce system weight and volume. The 12 V lead acid battery – still retained in some electric vehicles (EV)—is a hefty, bulky component, so industry has been looking for a way to remove it. New EV architectures employing micro DC to DC converters (µDCDC) converters based on gallium nitride (GaN) technology are enabling smaller, lighter, lower part count and more efficient converter stages, as well as the elimination of the traditional heavy car battery.

Figure 6.17 depicts a common EV architecture showing the high voltage loads and the 12 V loads. The green area is the traditional 12 V system present in any vehicle whether it is an internal combustion engine (ICE) or an EV, for infotainment, power windows, and communications functions. In the blue sector, we have the high voltage loads, the traction inverter which drives the electric motors plus functions like the cabin heater and the HVAC system. To power the high voltage loads, a high voltage battery is required (red block), which is charged via the Integrated Charging Control Unit (ICCU), sometimes referred to as "One Box," which contains the onboard charger (OBC) and 12 V supply (yellow). A unique requirement for EVs is the Auxiliary power module (APM) which feeds the 12 V system.

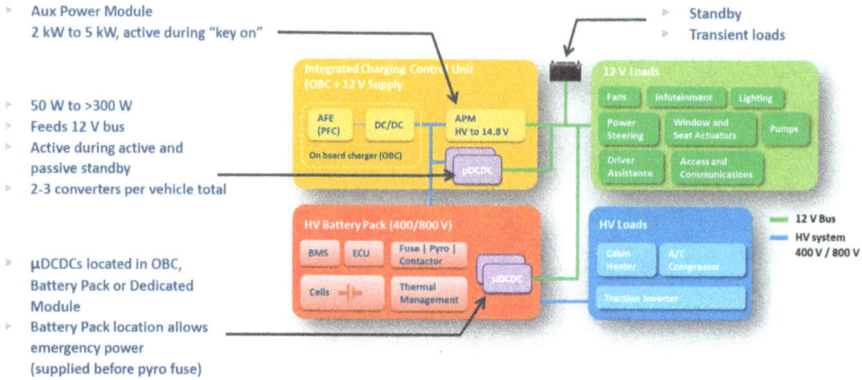

Fig. 6.17 12 V EV architecture with µDCDC

Multiple micro DC to DC converters (µDCDC) are used in the ICCU and in the high voltage battery pack. They are termed "micro" DC/DC converters because they are very much smaller than the APM. The APM typically provides 14.8 V at between 2 and 5 kW, supplying the 12 V system when the vehicle is active, or "Key On" to use industry terminology. When the vehicle is not running, traditionally a 12 V battery would also provide standby power and handle any transient loads that exceed the output power capability of the APM. µDCDCs are much lower power, ranging between 50 and 300 W—usually 200–300 W—and with much lower standby power consumption than the APM.

In the High Voltage (HV) battery pack, µDCDCs provide redundancy, and enable the HV battery pack to be independent of the main 12 V system. This is important because the battery pack is the most valuable component in the vehicle. Therefore, it must be able to protect itself even if the 12 V system fails, by managing its reaction to thermal conditions, and providing functional safety by its ability to disconnect the HV battery pack from the rest of the vehicle in the case of a severe fault or crash.

We mentioned the auto industry's ongoing desire to reduce component count, size, weight, and volume. With an ICE, the valve regulated lead acid (VRLA) battery typically has a capacity of around 1 kWh. But, of course, in an EV, there is no high current starting requirement (no need to turn over a cold internal combustion engine).

Therefore, there is really no requirement for a very high current capability. This frees up designers to use a lithium-ion cell chemistry, dramatically reducing the size and weight. Systems are now being produced that are about 20% the size of traditional units, and which have a longer life. (In early EVs, the lead acid battery was one of the major causes of vehicles failing.)

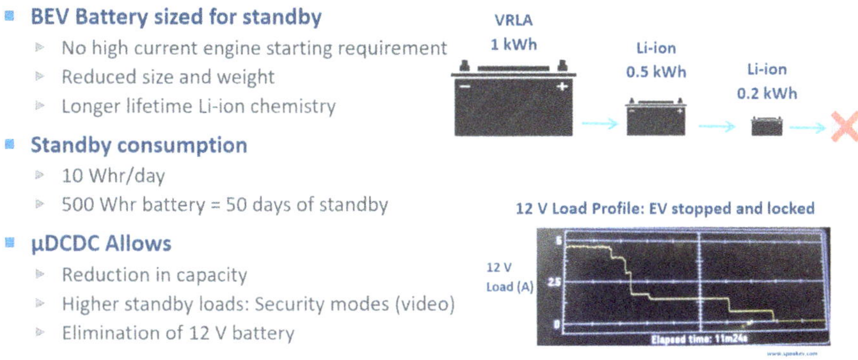

Fig. 6.18 12 V battery capacity trending to zero

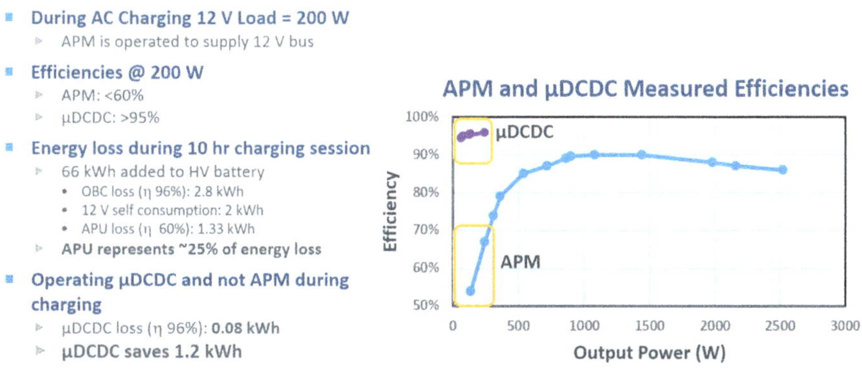

Fig. 6.19 APM and µDCDC measured efficiencies

Figure 6.18 shows that Original Equipment Manufactures (OEMs) are looking to completely removing the battery; if you do not need a starter, why have a battery at all? The 12 V load profile (bottom right Fig. 6.18) shows the load dropping as various modules inside the vehicle go to sleep after the vehicle is locked. In its lowest state, consumption is around 10 Wh/day so typical systems should have weeks of standby available. But with the proliferation of cameras for various levels of driver assistance and a desire to have security modes, rather than just a few watt hours, consumption is much higher, limiting standby to only a few hours before the battery is fully discharged. So, the ability to make the battery smaller and lighter to give more range and more internal space is rendered unfeasible.

The challenge could be solved by running the APM all the time. However, the APM is very inefficient at low power. The efficiency curve shown in Fig. 6.19 shows that at less than 100 W consumption, efficiency is only 50–60%. That is also problematic when charging the vehicle. When the vehicle is charging, the 12 V system must be running, because if the EV is only running from the 12 V battery, it will discharge before the vehicle is charged up. So, the APM must run, yet the inefficiencies of the APM will cause about 25% of the total energy loss during the charging process.

If the APU is replaced with a µDCDC that can supply the typical 200 W charging load, a very significant improvement results because the µDCDC has an efficiency of 95% or more. Under the same conditions, consumption is just 0. 8 kWh, a saving of 1.2 kWh. This has a tremendous impact on Miles Per Gallon Equivalent, a comparative figure of merit that auto makers and reviewers naturally take very seriously.

6.9.1 Why GaN?

Gallium nitride (GaN) technology is much more efficient than silicon because it has low C_{OSS} and therefore has much lower switching losses. This results in higher power density. Power Integrations consider GaN not as a product but as an enabling technology, so the company has produced a range of ICs that address specific challenges. Figure 6.20 is a reference design, DER 953, that describes a 100 W µDCDC based on Power Integrations' auto-qualified InnoSwitch3™-AQ 900 V flyback IC that uses the company's PowiGaN™ GaN technology. The device combines secondary side control, primary side control, the power device, and comprehensive protection in one package resulting in a very compact 100 W converter that can operate across the 150–500 V input range.

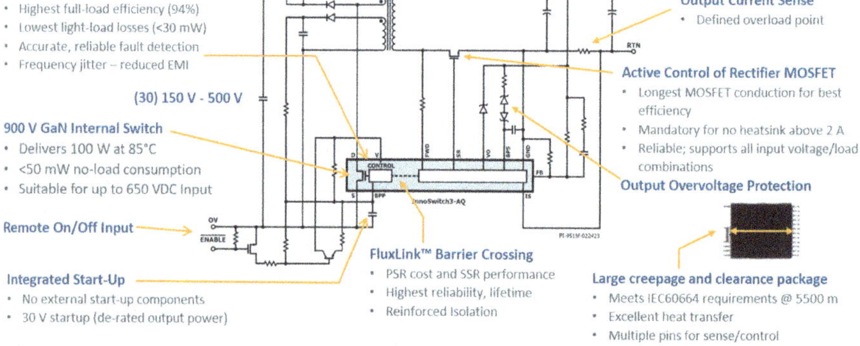

Fig. 6.20 100 W GaN-based automotive flyback (DER-953)

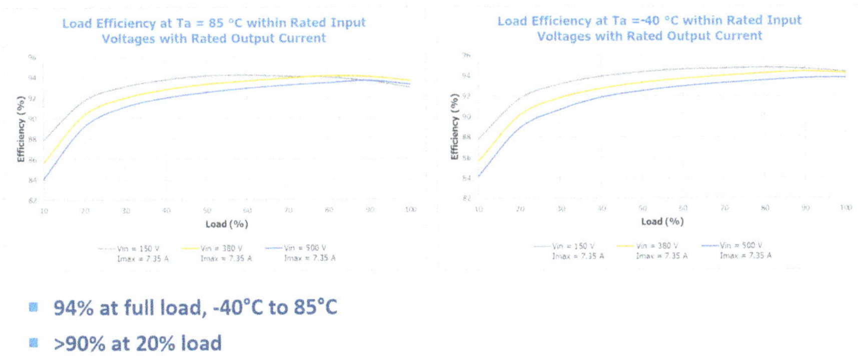

Fig. 6.21 High and constant efficiency

For comparison, simply switching from a silicon to a GaN InnoSwitch3 solution (Power Integrations offers silicon, GaN, and SiC variants of the InnoSwitch3 family) will result in a 20 °C temperature reduction, enabling more power to be delivered, or devices to run at a higher ambient temperature, or the amount of heat sinking to be reduced, depending on the design constraints. Simply, DER 953 shows a 100 W PSU that will run at 85 °C and deliver full power across the full input voltage range, with no heat sinking other than the PCB. Switching to GaN, therefore, has essentially doubled the power availability of the simple flyback approach.

There are other advantages, too, especially when considering standby. Figure 6.21 shows that the InnoSwitch3-AQ IC delivers very flat 94% efficiency at full load across its full –40 to 85 °C operating temperature range. But even at 20% load and 20 W output, efficiency is still greater than 90%. Because its light load efficiency is so good, an InnoSwitch3-AQ μDCDC can be left to run continuously. This makes the design less complex as there is no longer a requirement to have control software that decides when to turn the output on or off. Also, there is no need for an isolator to send the signal across the high-voltage to low-voltage domains to turn off the converter.

Another benefit provided by the InnoSwitch3 IC, is Power Integrations' proprietary Fluxlink™ high-speed communications technology which enables secondary side control over the isolation barrier without an optocoupler. This delivers an extremely good transient response which is important because these converters experience spend most of their time a very light load, and then may experience a transient load at close to full load. For example, on a 10-90-10% load step on a 12 V output, InnoSwitch3-AQ has a 220 mV peak-to-peak transient response; a typical competing solution might have a transient response of around 1 V. This becomes challenging for the system, because this means the 12 V is going to dip momentarily by a volt or more.

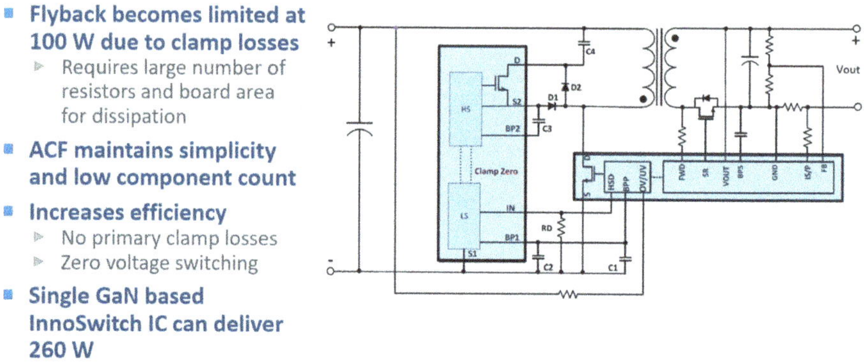

Fig. 6.22 Active Clamp Flyback (ACF) delivers higher power

6.9.2 *Higher Power?*

To achieve higher power requires a topology change. Active Clamp Flyback (ACF) uses another GaN device as the active clamp as shown in Fig. 6.22. This enables an increase in power from 100 to 260 W, again using a fairly simple topology, with a flat, high efficiency curve of around 96% across the full temperature range, with excellent light-load efficiency of around 90% at 5% load.

In comparison with competing μDCDC designs, the InnoSwitch3 IC circuit requires significantly fewer components, due to the high level of integration of the device. This not only saves space, it also increases reliability and enables the use of simple converter topologies, achieving the power levels required without having to resort to complex, expensive, and high component count two switch or LLC topologies.

6.10 Power GaN Patent Landscape

Rémi Comyn

Patents include technical, market, and legal information. Accordingly, patent analysis can provide meaningful information about the competitors, the level of their R&D investment, their technological roadmap, the future products and targeted markets, and the risks and opportunities existing in the field.

KnowMade has been researching and monitoring the intellectual property (IP) landscape across the whole power GaN value chain: from wafers to epiwafers, devices, packaging, modules, circuits, and applications. This chapter provides insights from this research. There is no mention of patent numbers that could link a patent to its owner.

6.10.1 Time Evolution of Power GaN IP Activity

Until the mid-2010s, Japanese players maintained a strong inventive leadership in the field (Fig. 6.23). Since 2015, Chinese research organizations and companies have progressively taken over the leadership in inventive activity. Eventually, in 2022, Chinese players remarkably accelerated their patent filings related to power GaN. This trend has been observed in most sectors of the semiconductor industry monitored by KnowMade. It stems from China's ambition to leapfrog the global competition, leveraging tax policies, subsidies, and loans to support the emergence of a domestic supply chain for various semiconductor technologies, including compound semiconductors. As part of this strategy, Chinese government has set up incentives encouraging patenting activities. In the nascent power GaN and power SiC industries, Chinese patenting activities successfully covered the complete supply chain with large amounts of patents. Even though the quality of such patents may be questioned, this situation brings about new challenges for global competitors in the semiconductor market looking to develop their manufacturing and business activities in China.

Importantly, Fig. 6.23 includes patent publications mentioning power electronics as the preferred application field. However, certain players do not limit the application field of their inventions, since they could be leveraged for both power and RF applications, especially when it relates to materials (wafers, epiwafers) or device processing. Lately, companies such as Innoscience, United Microelectronics (UMC), Huawei, or Intel have protected significant numbers of such generic GaN electronics inventions. In order to assess their impact on the global IP competition, KnowMade published elsewhere a patent landscape analysis which includes both generic and power GaN electronics patents [23, 24].

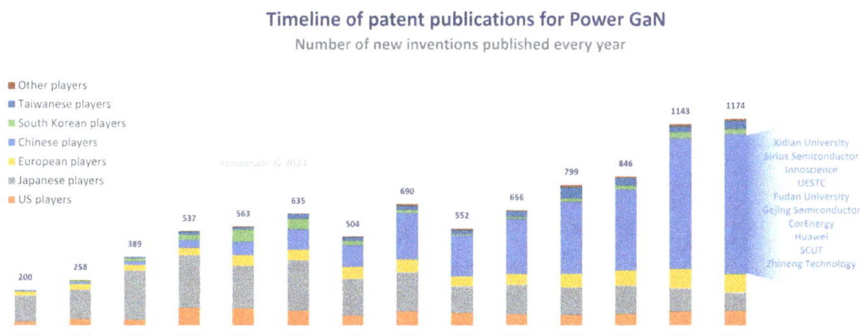

Fig. 6.23 Inventive activity in the power GaN IP landscape since 2010

6.10.2 *Competitive IP Spaces in the Power GaN Landscape*

As patents provide an exclusive right to control the use of an invention, a patent assignee may leverage them to stop or limit the activity of its competitors (manufacturing, using, selling products resulting from the invention). Importantly, this right is limited to a given geographic area. Therefore, patent applicants usually file several patents to protect an invention in the main markets targeted by the resulting products. Eventually, the budget related to a patent portfolio may become significant for the company with the increase in the number of patents, due to application fees (for pending patent applications) and maintenance fees (every year once the patent is granted). Therefore, patents not only relate to an R&D investment, but also to the investments supporting a business strategy.

A straightforward way to describe and start analyzing the current IP competition consists of comparing the number of granted patents and the number of pending patent applications assigned to each player, as shown in Figs. 6.24 and 6.25 for instance. Depending on their position in these graphs, IP players may be split into different categories:

- IP leaders (in the top right corner): They own the highest numbers of granted patents and are assigned to numerous pending patent applications, thus combining

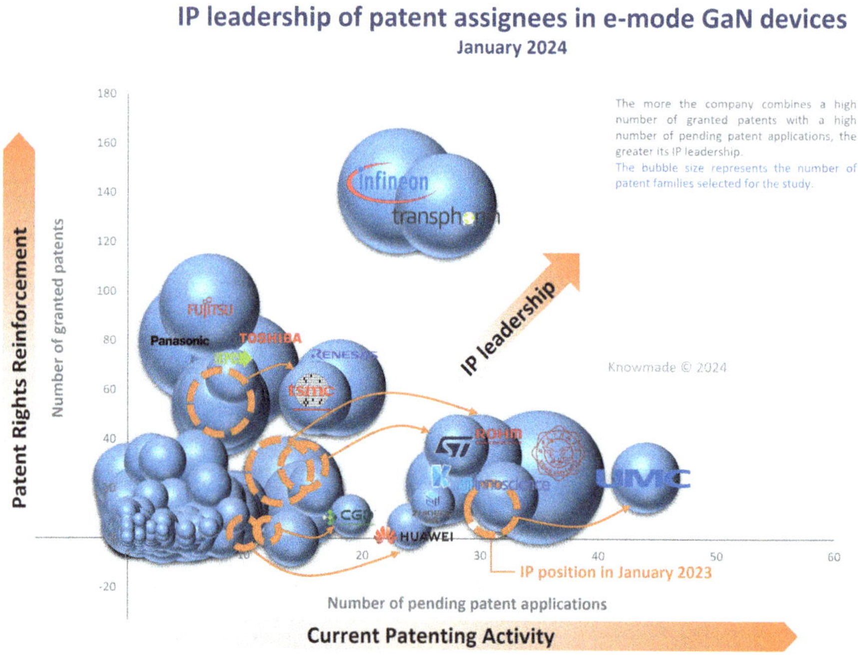

Fig. 6.24 Evolution of the IP leadership of patent assignees in Enhancement-mode (E-mode) power GaN devices from January 2023 to January 2024

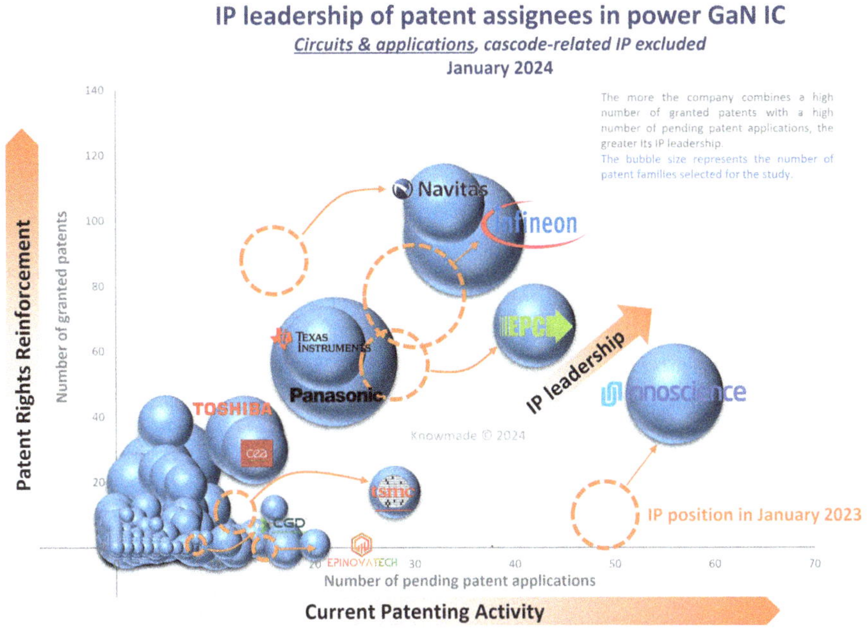

Fig. 6.25 Evolution of the IP leadership of patent assignees in power GaN IC from January 2023 to January 2024

a strong IP position and a significant potential for further strengthening this position.

- IP challengers (in the bottom right corner): They own limited numbers of granted patents compared to IP leaders, but they are assigned to the largest numbers of pending patent applications, which provide them with the potential to close the gap with leading IP players in terms of granted patents.
- Well-established IP players (in the top left corner): They own high numbers of granted patents but are currently assigned to limited numbers of pending patent applications. They maintain their IP rights live but are less active in terms of inventive activity.
- Non-competitors/newcomers (in the bottom left corner): They own limited numbers of granted patents and pending patent applications.

This approach was applied to the main segments of the power GaN value chain: wafers and epiwafers, devices, packaging and modules, circuits, and applications. It highlights an ecosystem emerging to support the fast-growing power GaN market as well as contrasted levels of IP competition between the segments.

Starting from the very upward part of the supply chain, historically dominated by well-established IP players Infineon and Fujitsu, new players from the epiwafer market are now aiming to develop their patent portfolios worldwide, such as

Sumitomo Chemical and Enkris Semiconductor. In parallel, new players such as Qromis and Shin-Etsu entered this IP space, developing innovative engineered substrates for manufacturing large-area and thick GaN RF & power epiwafers. Although wafers and epiwafers have remained a competitive space for power GaN IP, most players have rather focused their activities on devices on the one hand (Fig. 6.24), and on circuits on the other hand (Fig. 6.25).

For GaN power devices, patent applicants have focused on the protection of methods and structures that enable enhancement-mode (E-mode) operation of GaN transistors, a critical feature for switches used in power electronics. Figure 6.24 highlights the strong IP position of Infineon and Transphorm in this space, assigned to many granted patents. Their IP leadership is further strengthened by existing licensing agreements with Panasonic and Fujitsu, respectively, both well-established IP players for E-mode GaN devices. Therefore, Transphorm's acquisition by Renesas, if approved, will reshuffle the cards in the power GaN patent landscape, turning Renesas into an IP leader for E-mode GaN power devices.

Transphorm's IP activity, focused on E-mode GaN devices, strongly contrasts with its marketing activities promoting the cascode GaN device, which combines a depletion-mode GaN device in cascode configuration with an E-mode Si device. Yet, the US company anticipated early that IP would be critical to be able to safely manufacture and sell E-mode GaN devices, which eventually were not retained in Transphorm's market strategy. Still the concern about IP risks has been largely shared by other market players and became even more pregnant when EPC leveraged four of its E-mode GaN patents to sue Innoscience in July 2023. As shown in Fig. 6.24, in 2023, several players were already accelerating their patent filings in this space, including STMicroelectronics, ROHM Semiconductor, and UMC. In addition, TSMC, which is serving multiple power GaN market players, also increased its IP activity related to E-mode GaN devices in 2023.

The market adoption of E-mode GaN devices has been supported by the development of specific driving circuitry and operation methods that alleviate remaining technological issues, especially the gate vulnerability of E-mode GaN devices, and/or enable the use of such devices into new applications. It led to the development of GaN power integrated circuits (IC), integrating GaN power transistors with gate drive and protection circuitry (e.g., regulator, ESD protection diode, level shifting, bootstrap, under-voltage lockout, current sensing). Accordingly, this IP space has become increasingly competitive in recent years under the leadership of early market players (Navitas, EPC, Infineon). In 2023, most of them significantly improved their IP position in this space (Fig. 6.25). In addition, the main IP challenger, Innoscience, notably narrowed the gap in terms of granted patents with leading IP players in power GaN IC. Power GaN IC has become a very dynamic space with several players emerging in the IP competition in 2023, especially TSMC, and startup companies such as Cambridge GaN Devices (CGD) and EpinovaTech. Eventually, newcomers were identified in 2023 in the GaN IC patent landscape, such as CEA-Leti spin-off startup Wise-Integration.

6.10.3 *Emerging IP Trends for Power GaN Technology*

The IP for power GaN packages and modules still seems to be at its infancy, but the recent patenting activities suggest that this situation is not here to last. So far, modules and packaging IP activities have been relatively limited with few inventions focusing specifically on power GaN technology (Fig. 6.26), indicating that most market players have relied on existing know-how and IP developed for Si and SiC power semiconductors. Yet the number of such inventions sharply increased in 2022 with several market players actively filing patent applications worldwide (Navitas Semiconductor, Texas Instruments, Infineon/GaN Systems, Innoscience, ROHM Semiconductor, Delta Electronics/Ancora Semiconductors). They disclosed new packages developed specifically for GaN technology and dealing with technical issues such as thermal management and parasitic inductances. This evolution in IP activities attests that the development of new module and packaging technologies has become critical to support the adoption of power GaN technology in a wider range of applications. Therefore, the competition in this IP space is poised to evolve fast in the next few years, with many companies entering the field in recent years, especially from China. From this perspective, the acquisition of GaN Systems by Infineon in 2023, two players exhibiting a significant IP activity in this space, is an interesting move in the IP competition for GaN packages.

Another important emerging IP trend concerns the development of vertical GaN devices for power applications. Such devices have the potential to overcome breakdown voltage and current capacity limitations of lateral devices, while alleviating thermal issues. Although more than 1000 patent families (inventions) related to

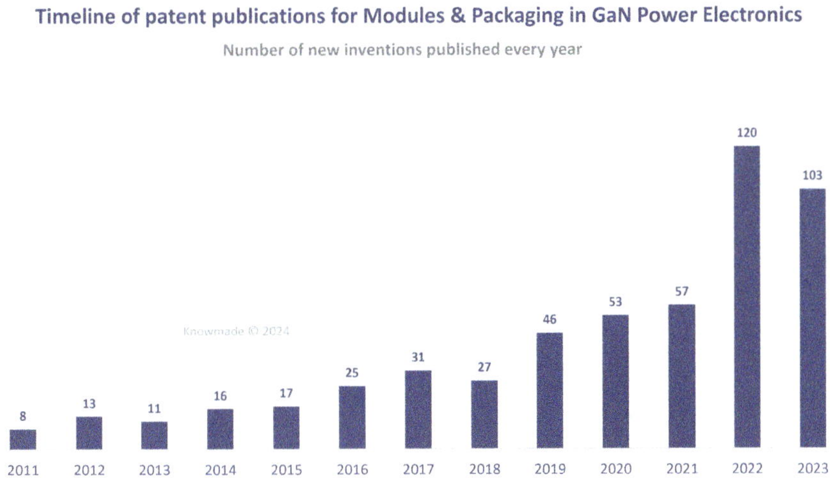

Fig. 6.26 Timeline of patent publications in the power GaN packaging and modules IP landscape: nearly half of all these inventions have been published in the last couple of years. (Note: generic wide band gap inventions have been excluded from this timeline)

vertical GaN technology have been disclosed since the 2000s, the IP competition has been remarkably moderate so far (Fig. 6.27) [25].

This situation reflects a relatively limited investment in vertical GaN device technology, especially when compared with vertical Silicon Carbide (SiC) technology. SiC technology has been successfully commercialized in recent years and would be in direct competition with vertical GaN in power applications. Overall, more than 6000 inventions have been disclosed for vertical SiC power devices (compared to 1000 inventions for vertical GaN), pointing out a gap in terms of R&D investments between vertical SiC and vertical GaN technologies. This comparison is relevant as both technologies have shared similar technical issues (crystal growth, defects, doping, etching, device manufacturing, etc.), making particularly challenging for them to reach the maturity required in high-power and high-temperature applications.

Interestingly, many players, including incumbent automotive companies, are still investing in vertical GaN technology (Fig. 6.27). Indeed, several IP leaders (Fuji Electric, NexGen, and Toyota Motor) and relatively new IP players (Denso and Bosch) are accelerating their patent filings. Accordingly, the vertical GaN patent landscape is expected to become increasingly competitive in the next decade. Importantly, as this technology proves itself, several main innovators in vertical

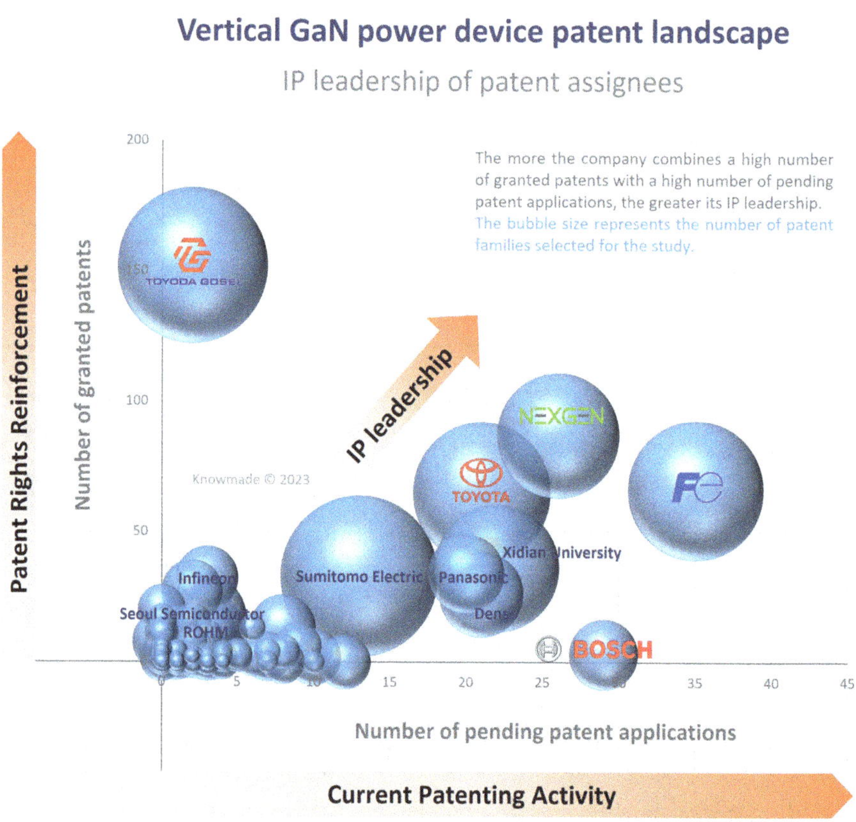

Fig. 6.27 IP leadership of patent assignees in vertical GaN power devices [30]

GaN might resume their IP activities to prepare for the industrialization and commercialization of the vertical GaN power devices (ROHM, Seoul Semiconductor, and Sumitomo Electric).

6.11 Requirements to GaN Devices for Automotive Drive Train Inverter

Dieter Liesabeths and Dima Novo

The market of battery electric vehicle (BEV) will increase until the end of the decade in 2030 to more than 60% of all newly registered cars. With approximately 60 Mio BEV produced every year, the demand on electricity will drastically increase and the industry is requesting for higher efficiency cars to extend range, reduce energy consumption, and therefore decrease CO_2 pollution. With the introduction of wide-bandgap materials (WBG) like silicon carbide (SiC) and gallium nitride (GaN), the industry is shifting away from conventional silicon-based power drive train inverter to improve efficiency and therefore, reduce energy consumption. WBG materials are offering a 3–10% better efficiency compared to traditional silicon-based inverters, as they operate especially in light-load condition in a resistor curve compared with silicon transistors in a diode curve. The introduction of GaN based power module and demonstrator will be discussed in this study as the next evolution pass after SiC toward a greener planet and a better conservation of energy.

6.11.1 Introduction

In this work, we will present our Direct Drive Depletion-Mode GaN HEMT (D^3GaN) power module technology demonstrators for EV traction inverters. Until today, GaN based power transistors are mainly used in low power application as normally-off Enhancement-mode (E-Mode) or normally-on Depletion-mode (D-Mode) with silicon metal-oxide semiconductor field-effect transistors (MOSFET) to make it normally-off, for power ranging from 50 W to few single digit kW applications. The natural choice of a GAN product is a D-Mode type of transistor, as it delivers the highest current for given die area. The developer choice is here a High-electron-mobility-transistor (HEMT) which in comparison to a Si or SiC MOSFET does not include a body diode and have almost zero switching losses and no reverse recovery charge (Qrr) induced by a parasitic body diode at the MOSFET structure. With the introduction of the D^3GAN technology, it was firs time presented that D-Mode GAN can be a door opener for high-power application as well, targeting the BEV drive train inverter beyond 100 kW.

The novel and patented D^3GAN has the highest reliability against transient voltages due to its robust gate structure with a threshold voltage of over 8 V which is very important in a high inductive load environment like motor inverter. Due to its HEMT structure it makes it an almost ideal switch for high frequency operation with zero losses and fast turn on and off times which enables switching frequencies of 10 KZ to over 100 KHz, even todays BEV inverter will just switch at low frequencies of 8–12 KHz and do not make use of full potential of the GAN technology.

The gate reliability is a critical requirement for automotive high-power applications. In the challenge of keeping simultaneously advantages of GaN and reliable operation at high power, D-mode GaN devices show a significant advantage over E-mode GaN devices with low threshold voltages of only 1.5 V. In the D^{3}GAN technology, the gate driver is connected directly to the GaN gate which allows utilizing the fast turn-on and turn-off times and slew-rate control. D3GaN topology is radically different from cascode topology.

First, we present and discuss the technology demonstrator VM022 half-Bridge power module based on D^{3}GaN 8 mOhm discrete devices, which was developed to showcase high power operation of four parallel devices in a traction inverter application. In this work, we will present the electrical and mechanical aspects of the module design and the efficiency test results performed on the dyno closed-loop test bench according to the World Harmonized Light Vehicles Test Procedure (WLTP). Second, we will present the production target transfer molded module that is currently in development, showcasing high power density and improved thermal resistance.

6.11.2 VM022 D^{3}GaN Power Module Design

Figure 6.28 shows the VM022 650 V, 2 mOhm D^{3}GaN half-bridge Power module. The key performance parameters are shown in Table 6.2.

VM022 is a PCB-based half-bridge module, with four parallel D^{3}GaN 8mOhm discrete devices for high and low sides, a total of eight GaN devices. The driving board with connectors to a control board is placed on top of the module. Gate drivers are located above the devices to reduce the gate loop length and inductance. The module block diagram is shown in Fig. 6.29.

To further reduce the power loop inductance, decoupling capacitors are integrated on the module PCB.

Fig. 6.28 VM022 half-bridge side view

Table 6.2 VM022 key performance parameters

Parameter	Value	Unit
V_{DS}	650	V
$R_{DS(ON)}$	2	mΩ
Q_G	800	nC
$I_{D,pulse}$	1520	A
I_D	460	A

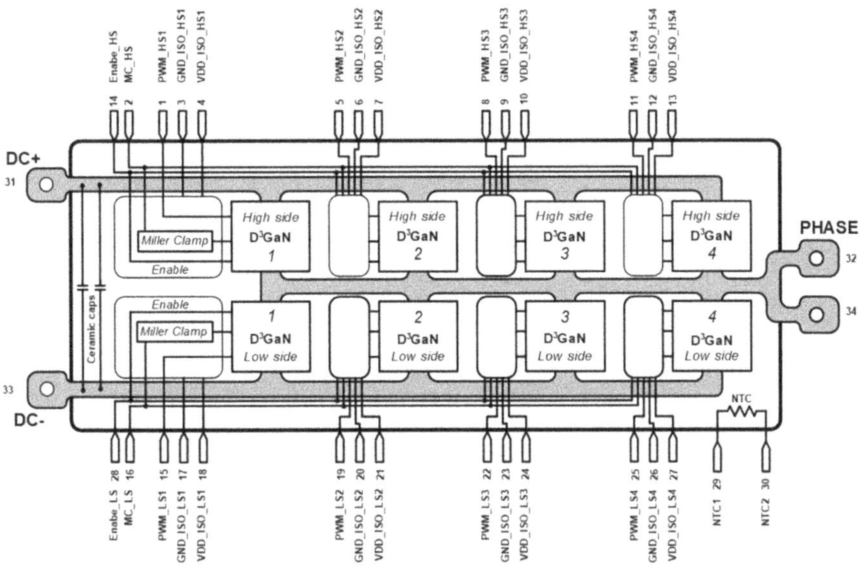

Fig. 6.29 VM022 block diagram

6.11.3 Power and Gate Loop Design Considerations

Power and gate loop inductances have a critical impact on system operation. Reducing the parasitic inductances is even more important in applications utilizing GaN devices due to their high switching speeds, leading to high di/dt. Much attention was put into reducing power and gate loop inductances, both in the power PCB layout, control pin's location, and driver board PCB layout. The measured power loop inductance is approximately 3.9 nH. To minimize mismatch in time during turn-on between parallel power devices, special attention was put into the design of the power and driver PCBs, ensuring that the control signals will commutate the devices simultaneously. In power loop design, we need to make sure that the parasitic inductances per device are equal in value to create even current distribution during switching events. GaN devices have a positive Ron/Temp coefficient, i.e., Ron increases with an increase in temperature. This effect helps to "naturally" balance current between parallel devices during on state. Reduction of power loop parasitic inductances is done by two methods described in [26, 27]: (1) reducing the

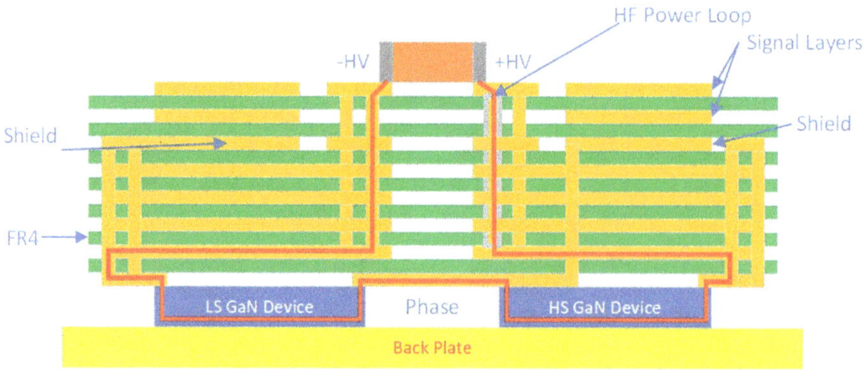

Fig. 6.30 VM022 power PCB stack-up

area enclosed by the power loop and (2) using field cancelation techniques. In Fig. 6.30 you can see the power PCB stack-up. The High Frequency current path is shown in red, where (1) and (2) are implemented.

6.11.4 *Operation of Parallel Devices*

For testing the parallel operation of the GaN devices, we have developed a special technique utilizing the LV MOSFET integrated in the D^{3}GaN platform. By measuring the voltage drop on the MOSFET of parallel devices during turn-on, we can verify that the devices turn on simultaneously (current flow begins at the same time). In Figs. 6.31, 6.32, and 6.33, the voltage spike on the parasitic inductance of LV MOSFET during turn-on is shown comparing pairs of the four parallel devices. In these waveforms, it is easy to see that current flow starts simultaneously on the devices ($\Delta t = 2.807$ nS, at its max, on device #1, the farthest from reference device #4) (Blue is the DUT, Gray is the reference device).

6.11.5 *Thermal Stack*

As described in the previous sections and figures, the VM022 power module uses a multilayer power PCB for routing both signal and power traces. Soldered to one side of this PCB are the pre-packaged discrete 8 mOhm D^{3}GaN devices. These packages are electrically isolated via an internal aluminum nitride Direct bond copper (DBC) substrate and are then attached to a copper baseplate via a standard soldering process. The assembly, shown in Fig. 6.34, is considered a typical "baseplate" power module from a thermal standpoint but uses a combination of wire bonding and PCB layers for all interconnections.

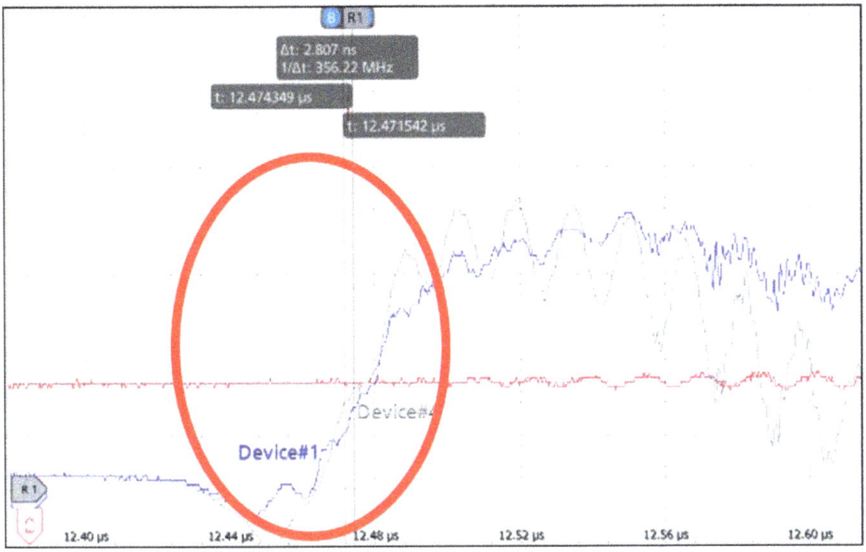

Fig. 6.31 VM022 LV MOSFET or device #1 compared to device #4

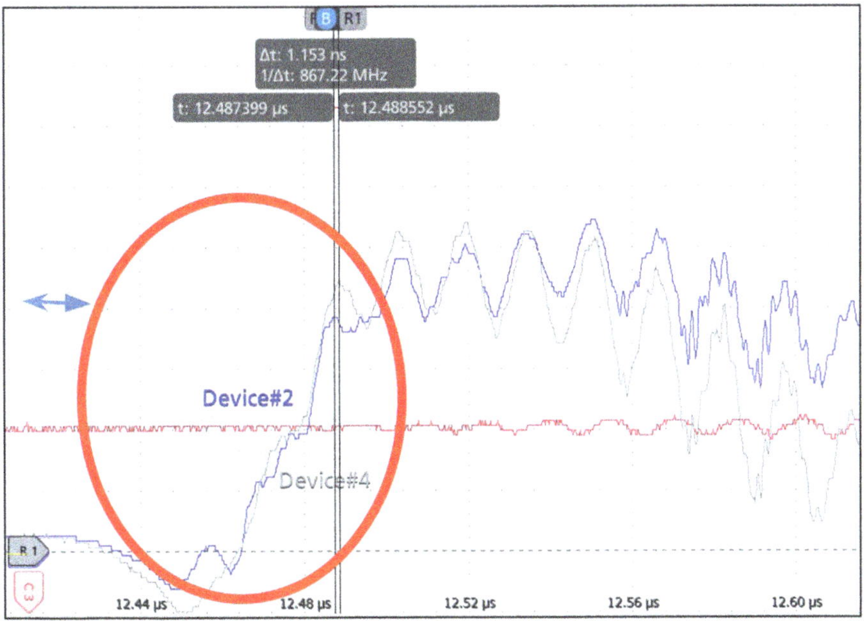

Fig. 6.32 VM022 LV MOSFET of device #2 compared to device #4

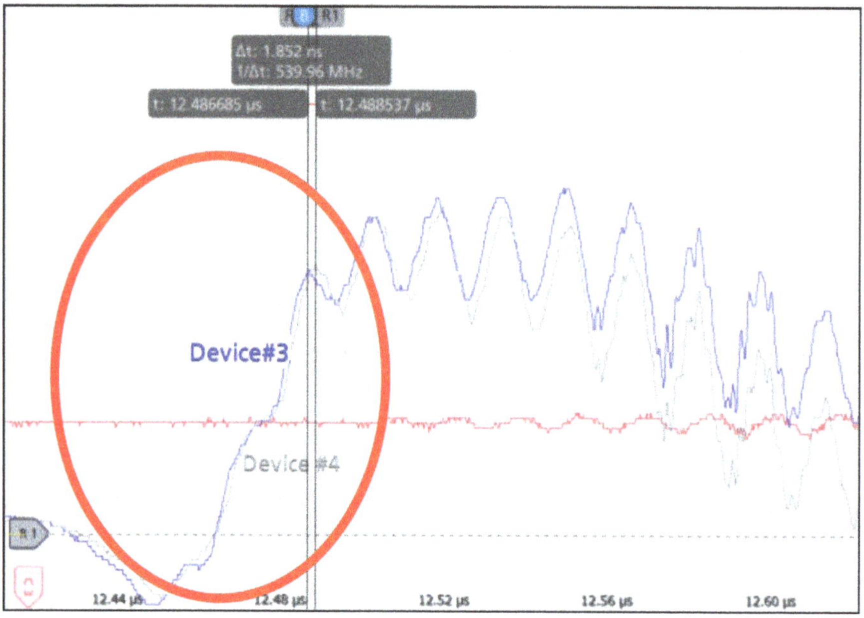

Fig. 6.33 VM022 LV MOSFET or device #3 compared to device #4

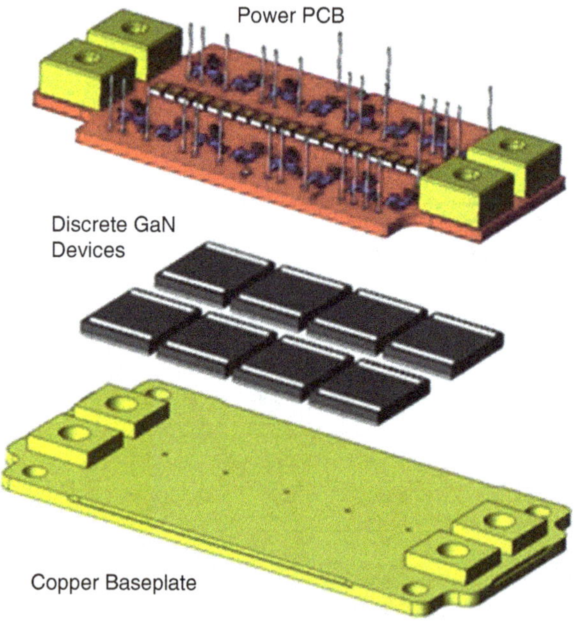

Fig. 6.34 VM022 power PCB, discrete devices, and the baseplate

The industry mainstream is to use semiconductor bare die directly attached to the DBC and interconnect through wire/ribbon bonding or clip connection. While this approach is optimal for making mass-production products, it makes the process of tuning, optimization, or debugging of the module design practically impossible. Therefore, we used a different approach which allows tuning of the layout and circuit on a "living" technology demonstrator module. The primary difference is that the pre-packaged discrete GaN devices are first assembled and tested to confirm the operation, then attached both to the power PCB on the terminal side and the baseplate on the DBC side. This provides a conventional thermal path to a cooler (typically liquid cooled), but also a secondary thermal path to the power PCB, where some heat, 10–15%, can be dissipated via the thick copper traces, encapsulation materials, and surrounding environment. Figure 6.35 depicts the complete vertical structure from die to fluid.

With this thermal stack, and with typical fluid flow rates of 8–10 LPM per inverter, we can expect a thermal resistance of device channel to fluid inlet in the range $Rth_{\text{C-FLUID}}$ 0.4–0.55 °C/W per die, depending on the details of the cooler. Some improvements can be made but these must be balanced with minimizing mechanical stresses and other reliability considerations.

6.11.6 WLTC Test Results

Worldwide harmonized Light-duty vehicles Test Cycle (WLTC) is a cycle test that is part of WLTP [28] and is a globally adopted chassis dynamometer test cycle for the determination of emissions and fuel consumption for ICU or electric efficiency for EV light-duty vehicles. The test was set up in the lab of a major EU car manufacturer. The set-up included a 3-phase inverter power stage (Fig. 6.36) based on a single VM022 module per phase, a dynamometer system with an EV PMSM machine, a DC power supply, a cooling system, and other auxiliary systems required for the operation and measurement during the experiment. The whole 3-phase power stage system, including control and measurement equipment, was designed and assembled by the VisIC team.

6.11.7 WLTP Tested Conditions

In Fig. 6.37, we show the full WLTP test conditions (conditions tested are marked in red). In this experiment, we tested up to 350 A_{RMS} phase current and 2400 rpm (electrical frequency of 160 Hz).

Red dot indicates the maximal tested conditions. The red rectangle indicates the area where several data points were collected to build a torque vs. efficiency plot (Fig. 6.38).

In this test, we achieved a maximal efficiency of 99.295% at 24.4 kW output, 2400 RPM, and 92.8 Nm.

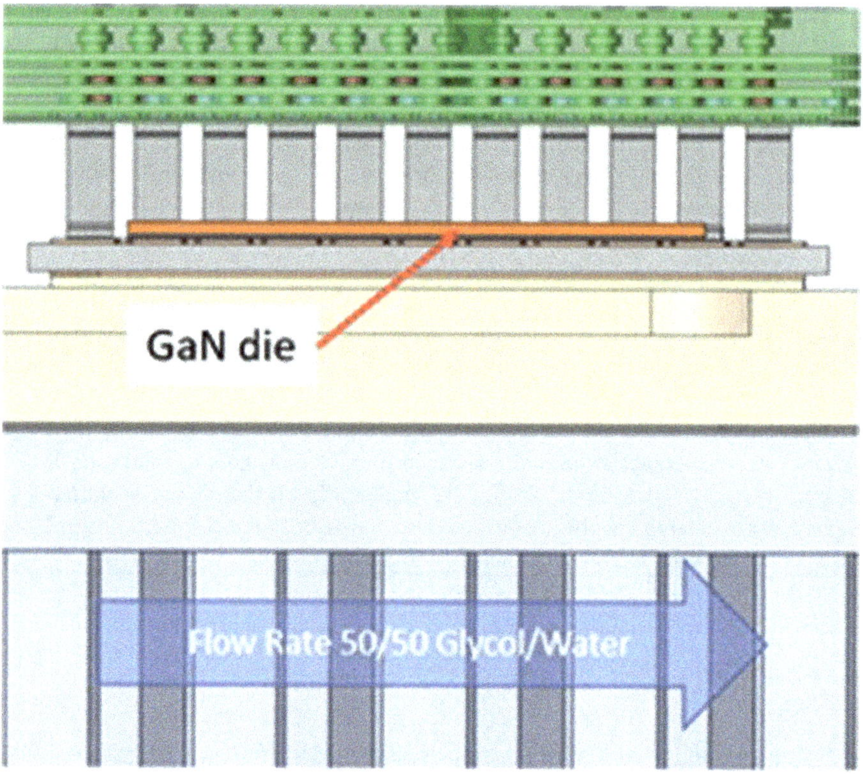

Layers from GaN die to fluid

Layer	Material	Thermal Conductivity	Thickness
		W/(m·K)	uM
Die	Si/GaN	90	270
Die Attach	Pb 90 / Sn 10	35	50
Top Metal	Copper	390	100
Ceramic	AlN	170	635
Bottom Metal	Copper	390	100
Substrate Solder	Sn 63 / Pb 37	50	250
Baseplate	Copper	390	3000
TIM	Grease	10	100
Coldplate (base)	Aluminum	180	2000

Fig. 6.35 VM022 thermal stack

Fig. 6.36 VM022 3-phase power stage test system

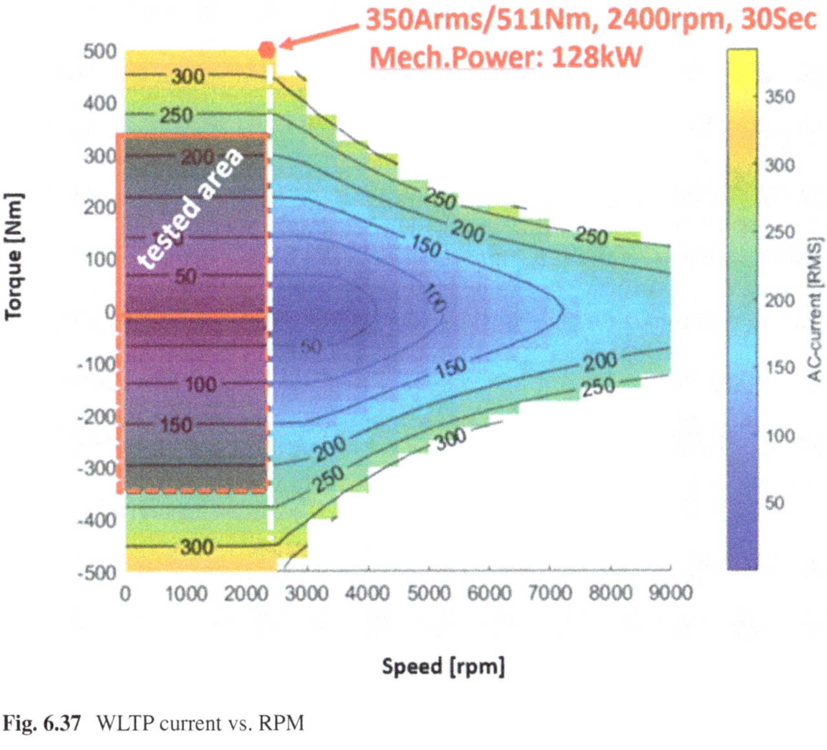

Fig. 6.37 WLTP current vs. RPM

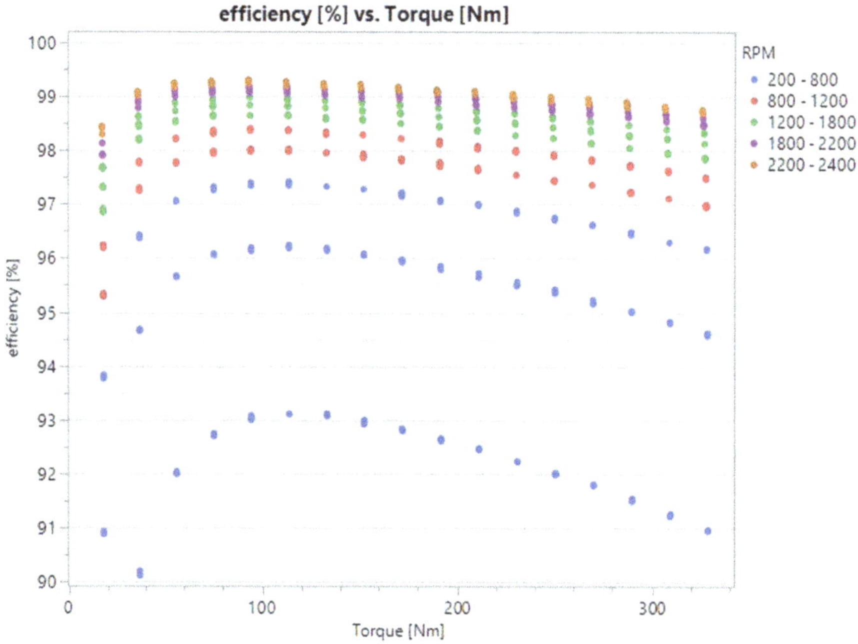

Fig. 6.38 Measured torque and efficiency

6.11.8 D^3GaN Transfer Molded Half- Bridge Module

Following the successful development and testing of the VM022 module VisIC took the next step of developing a transfer-molded half-bridge module. This module is much more compact: 50 × 30 × 4 mm (L × W × H]. Power terminals are designed for laser welding to reduce power loop inductance when connected to busbars, and the module is designed for sintering or soldering to a cold plate to reduce thermal resistance. The development is a work in progress and here we are presenting the preliminary data.

For optimal thermal efficiency as well as long-term reliability, the VM027 module structure is designed for a straightforward assembly process and transfer molding. The module is designed for a direct attachment of the Direct Bonded Copper (DBC) ceramic substrate to the cold plate, in this design a baseplate is unnecessary, lowering the cost and improving the thermal performance. Figure 6.39 shows an exploded view of this design.

One beneficial feature of this structure is that a typical stamped lead frame is not needed for the power or signal terminals. Instead, low-cost metal injection components are used and placed using standard pick-and-place processes. The novel power terminal geometry provides for low inductance and allows for laser welding to a system-level power connection. The fabricated module after molding is shown in Fig. 6.40.

Fig. 6.39 Transfer molded module exploded view

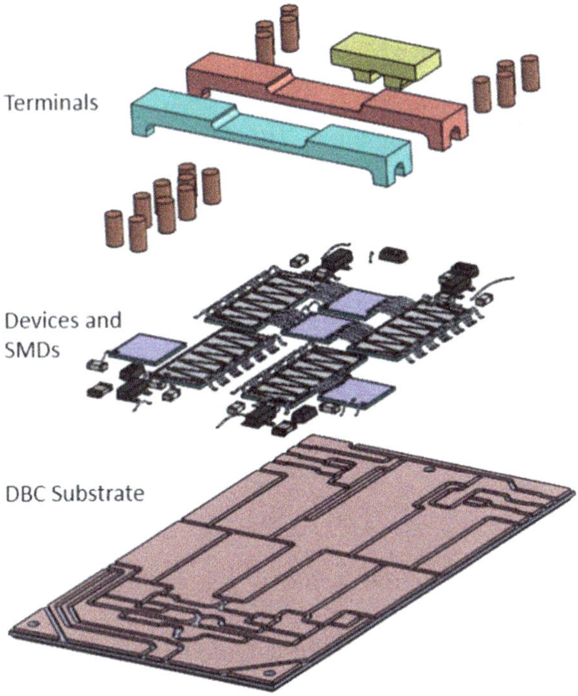

Fig. 6.40 Transfer molded module

The incentive for minimizing power and gate loop inductance is the same as for the VM022 module. The transfer molded modules present additional constraints: Smaller size and less area for a high cross-section of the power conductors. Cancelation layers are also not practical as readily available DBC has only a single layer of copper. Multi-layer DBCs have been long promised but we were not able to exercise this path. As mentioned above, power terminals enable laser welding to busbars, reducing the overall power loop inductance. Control pins are located to ensure a minimal length of control paths. ANSYS Q3D is used for power and gate loop parasitic simulations. Based on the simulations, the main power loop inductance is 2.45nH and the gate loop inductance is 6.82nH for the high side gate and 7.53nH for the low side gate.

GaN die statistical V_{th} data as measured on-wafer is shown in Fig. 6.41 (46,004 dice). Narrow V_{th} spread assists with the paralleling of dice, as the turn-on and turn-off difference at the same gate voltage is low. Real-life operation (as we did with the VM022 module) has not been performed yet.

6.11.9 Thermal Stack

This power module is a DBC-based transfer molded power module that does not utilize a typical copper baseplate, allowing for direct and thermally efficient attachment to a cooler via solder or large area Ag-sinter bonding. This design has many of the same features as other transfer molded products suitable for automotive applications, including a high-volume production assembly process and cost structure. To maintain an acceptable final flatness of the substrate and to minimize the stresses on the GaN die, special attention was given to optimize the mechanical design, as well as the molding process parameters. For the transfer molded module, this has been achieved by careful optimization of DBC layout geometry and metal thicknesses, mechanical design, die attach material, and mold design details. Figure 6.41 shows this module design. Power terminals are designed for laser welding, which has become the preferred method due to lower costs and the elimination of typical electrical contact resistances.

The substrate is silicon nitride (Si_3N_4) due to its superior mechanical strength, compared to AlN, and its increased acceptance in the power electronics industry. Without a baseplate and associated attachment layer, the thermal stack, shown in Fig. 6.42 is further simplified. Consequently, thermal resistance is also lower, with achievable $Rth_{C\text{-}FLUID}$ values between 0.3 and 0.5 C/W depending on details of the cooler and with flow rates of 8–10 LPM.

The DBC metal thickness in this design is 300 μM, much thicker than in the VM022 demonstrator, primarily because this layer must handle the full current load directly on the substrate. A benefit of this approach is more heat spreading, where there is space around the GaN device. The Si_3N_4 ceramic is thinner than AlN to compensate for the lower thermal conductivity value.

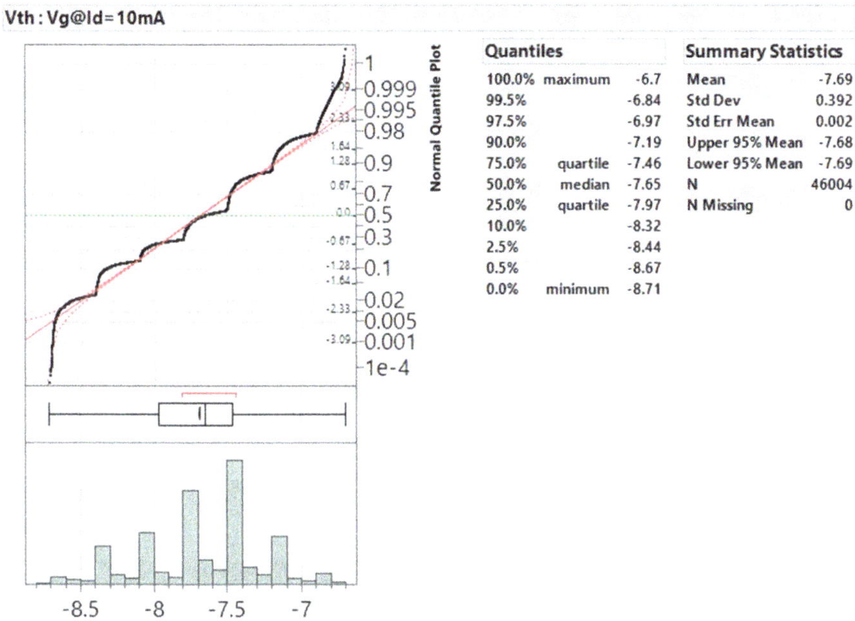

Fig. 6.41 Threshold voltage statistics

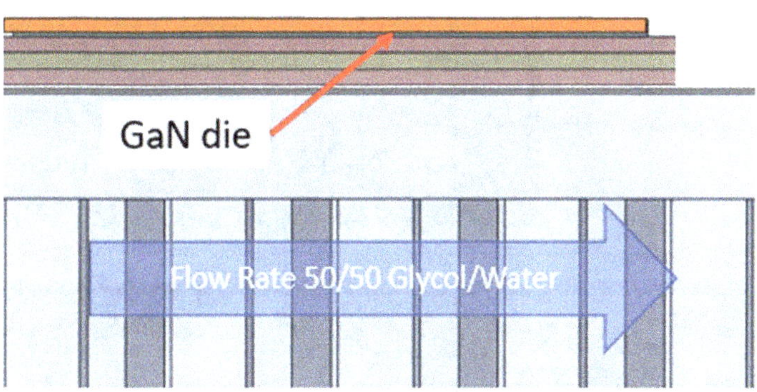

Layers from GaN die to fluid

Layer	Material	Thermal Conductivity W/(m-K)	Thickness uM
Die	Si/GaN	90	270
Die Attach	Pb 90 / Sn 10	35	50
Top Metal	Copper	390	300
Ceramic	Si3N4	90	320
Bottom Metal	Copper	390	300
Substrate Solder	Sn 63 / Pb 37	50	250
Coldplate (base)	Aluminum	180	2000

Fig. 6.42 Module thermal stack

6.11.10 Multi Pulse Test Setup and Preliminary Results

For dynamic evaluation of the module, we are using a multi-pulse test. After module fabrication and testing that the module is electrically operational, a multi-pulse test set-up is assembled using a dedicated evaluation board, developed for the module by VisIC team, and providing all the necessary control signals. A 1 GHz high bandwidth oscilloscope and high accuracy optical differential probes are used to measure drain-source and gate signals with minimal distortion. The test set up is shown in Fig. 6.43. The module is located at the bottom of the EVB.

At this early stage of the development, the purpose of the testing is to verify the stability of the switching, evaluate the power loop parasitic inductances, and measure switching energy and additional parameters that are critical for making improvements for the next revision. In Fig. 6.44, the half-bridge mid-point to ground

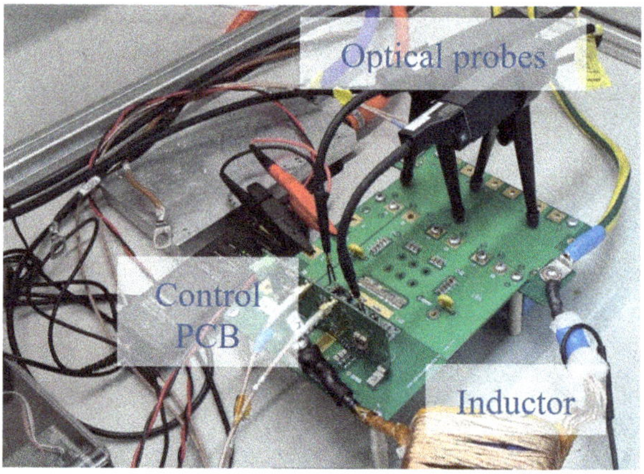

Fig. 6.43 Multi-pulse test setup

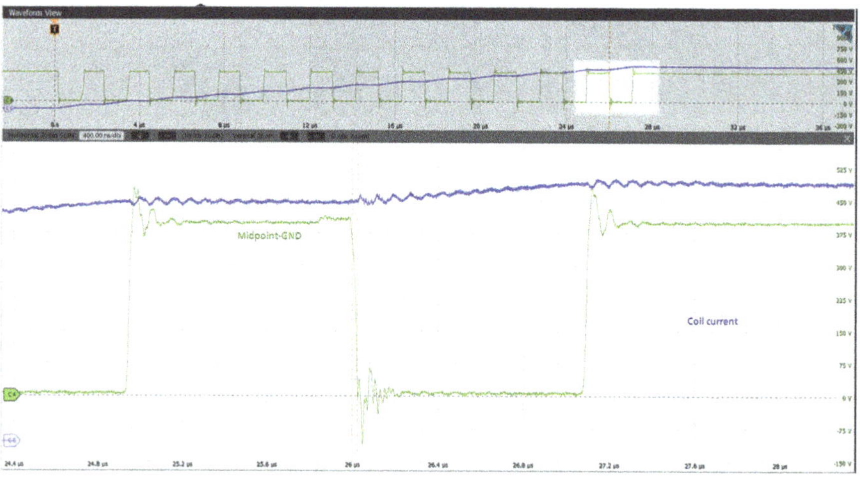

Fig. 6.44 Transfer molded HB module multi-pulse waveform

voltage and the coil current were captured. In this test, bus voltage is 400 V, and coil current was up to 200 A.

The measured power loop inductance is 8 nH (including busbars and power capacitor), this measurement indicates a high inductance of the connection between the DC link capacitor and the module.

Switching energy measurements are next in line, after finishing the initial optimization of the switching behavior.

6.11.11 *D³GaN Driving Circuit and Its Implementation with D³GaN Power Modules*

D³GaN schematic is shown in Fig. 6.45 and includes: (1) D-mode GaN die, (2) LV MOSFET die, (3) p–n diode. In this example, the GaN and P-MOSFET die are connected in series source to source. The p–n diode is connected between the GaN gate and the drain of the LV MOSFET (Fig. 6.45).

The LV MOSFET is in conductive mode all the time driver is present, and not playing a role in the switching of the GaN die. Figures 6.46 and 6.47 show the simplified gate drive and MOSFET enables circuit schematics. The gate of the MOSFET is connected to a dedicated enable circuitry. Enable signal turns the MOSFET to on state when the auxiliary power supply turns on and stays on as long as the auxiliary power is above the Undervoltage lockout (UVLO) Level. If the auxiliary power supply voltage level drops below the UVLO, the MOSFET turns OFF and allows for a safe turn-off of the GaN die.

Driving of the D³GaN:

Standard off-the-shelf gate drivers are used for driving a D³GaN and are connected to the GaN gate (sizing of the gate driver is done in the same way as for SiC or IGBT power devices). There is no need of inverting or changing the driver input PWM signal. The difference in the D³GaN driving circuit is that the GaN die Kelvin

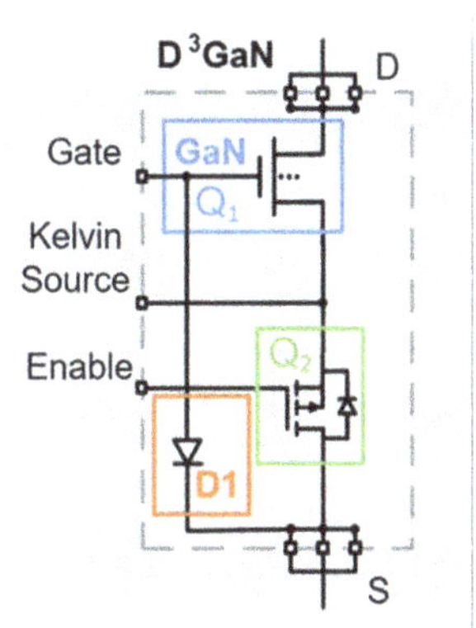

Fig. 6.45 D³GaN schematics

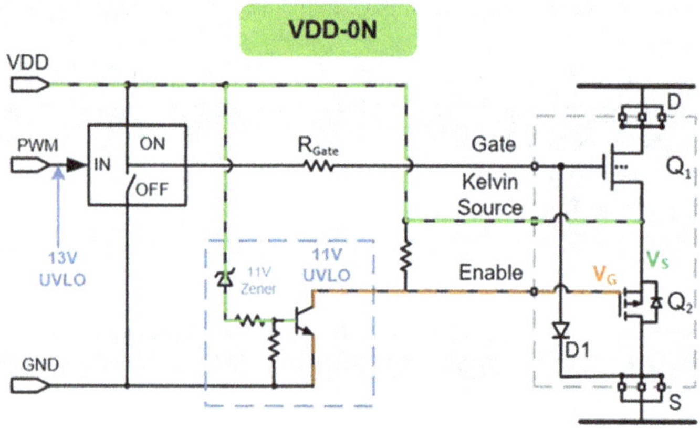

Fig. 6.46 D³GaN driving circuit schematics—on state

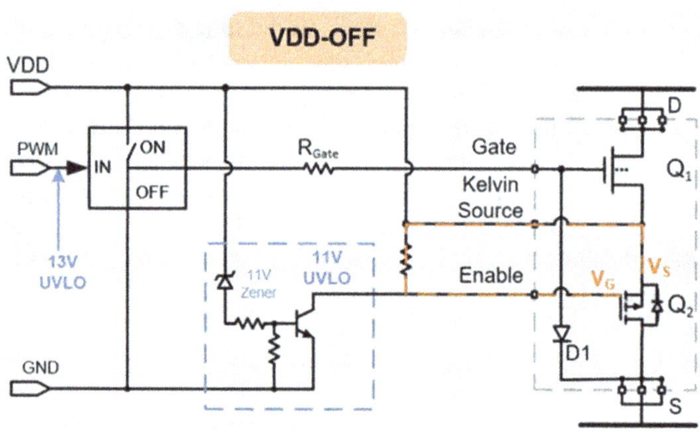

Fig. 6.47 D³GaN driving circuit schematics—off state

source is connected to the auxiliary power supply (VDD) instead of GND. This allows for switching the GaN gate between VDD and GND potentials, while GaN source is hardwired to VDD.

For turn off, the GaN V_{GS} = GND − VDD = −VDD, and on when V_{GS} = VDD − VDD = 0.

For implementing the D³GaN inside a module. it is required to have the GaN and Si MOSFET on the substrate for cooling, while all the other components can be implemented on the gate drive PCB.

D³GaN utilizes the benefits of a more robust D-mode GaN device with highly reliable gate structure without the added switching losses of the Si MOSFET.

6.12 Using GaN's Inherent Properties to Deliver Bidirectional Functionality in BMS Systems

Denis Marcon

The electrification of everything movement means that we are reducing our dependence on fossil fuels by replacing them with a number of other, cleaner alternatives—like solar, wind, hydro, or heat pumps. The most obvious application example is Electric Vehicles (EVs), where huge batteries are now being used instead of the internal combustion engine. Batteries are also deployed in many other applications as well, including e-bikes and e-scooters, small household appliances, powered hand tools, etc. This massive usage of batteries demands intelligent, safe storage systems, so battery management systems (BMS) are now being employed. New developments in gallium nitride IC technology are supporting innovations in BMS resulting in smaller, more reliable, and more powerful equipment.

We have all read stories of the dangers of EVs catching fire. Fires occur when a battery goes into thermal runaway. It is especially dangerous in high power lithium cells, as these are exothermic and the heat they generate in thermal runaway spreads from cell to cell within the battery, resulting in temperatures of over 2500 °C. (It is worth saying here, that EV fires are rare and current data from Tesla shows that approximately just five Tesla car fires occur per billion miles driven compared to 55 fires per billion miles driven among all—primarily ICE—cars. A caveat is that the majority of car fires occur with older models, and there are few older EVs on the road.) There have also been well-publicized instances of cell phones catching fire and the author has personal experience of an e-bike fire.

Thermal runaway in lithium cells occurs for several reasons. If a battery is damaged, internal short-circuits can be the cause. Also, other factors such as overcharging, multiple over-discharges followed by a charge, external short circuits, and high- and low-temperature environments can cause thermal runaway. Hence, the need for proven, reliable BMS implementation. However, of course, the BMS is still subject to pressures such as equipment volume and price.

In recent years, gallium nitride (GaN) power electronics technology has emerged as an alternative to silicon. GaN delivers higher power, higher power density, and greater efficiency. This is due to the fact that GaN has no body diode and therefore switching losses in HEMTs are much less than even the best high-performance silicon metal-oxide semiconductor field-effect transistor (MOSFET). Yet GaN is much more than a simple silicon transistor replacement, and innovative companies such as Innoscience are now expanding the functionality of GaN ICs with new parts that have increased capabilities. One example is a range of bidirectional VGaN power devices that can be employed to reduce component count in voltage blocking and current conduction circuitry for over-voltage protection (OVP) in BMS designs.

Figure 6.48 shows the traditional approach to achieve OVP, using four back-to-back silicon MOSFETs to shut down the unit when needed. The limitations of silicon technology in these applications are that low $R_{DS(on)}$ parts, which are required for

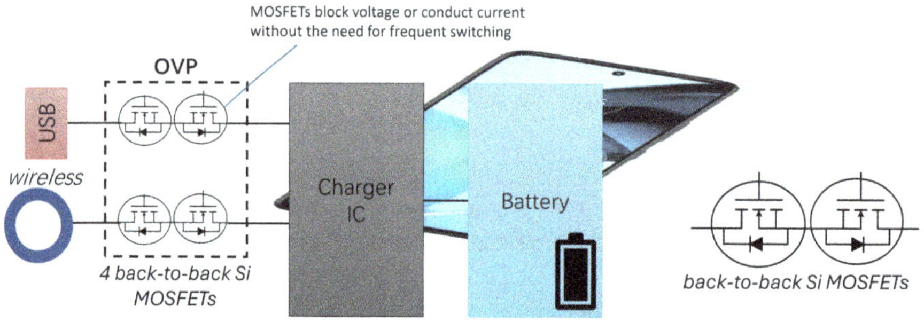

Fig. 6.48 Traditional approach to achieve OVP, using four back-to-back silicon MOSFETs

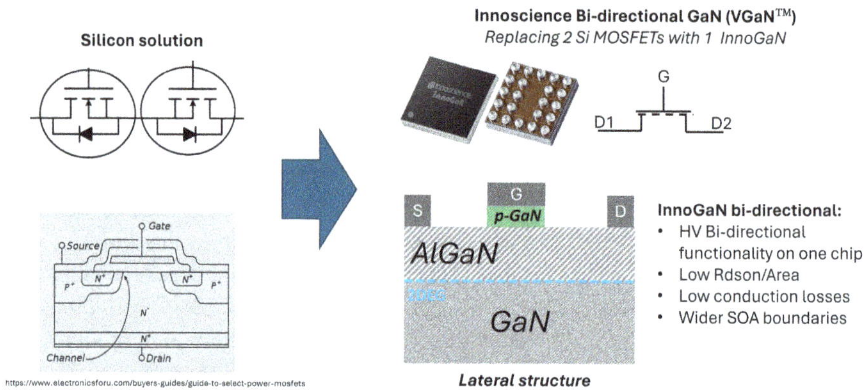

Fig. 6.49a Replacing two MOSFETs with Innoscience's bidirectional VGaN device

high efficiency, have a large footprint. They also have a limited SOA. Silicon parts cannot achieve bidirectionality due to their inherent body diode, and they are difficult to integrate on one chip.

By contrast, GaN HEMTs have no body diode, so Innoscience has produced bidirectional parts, enabling one VGaN device to replace two MOSFETs (Fig. 6.49a).

Therefore, the OVP circuit inside a BMS produced using VGaN will be at least 50% smaller and more efficient than its silicon counterpart (Fig. 6.49b) and reduces case temperature rise by 40%.

Figure 6.50 shows the traditional silicon BMS OVP solution side-by-side with the new VGaN bidirectional implementation. Measuring just 2.2 × 2.1 × 1 mm, Innoscience's 100 V VGaN part is itself smaller than the competing MOSFET (3 × 2 × 2 mm) and together with the component count reduction, the BMS PCB footprint is cut by 33%. While 40 V VGaN devices can be used with low voltage batteries (<32 V), the 100 V VGaN can be used with batteries up to 80 V, thus enabling the BMS to be significantly smaller and more cost effective.

For BMS OVP applications, devices need to be paralleled. It turns out that GaN is highly suitable for paralleling because it exhibits a stable threshold voltage over

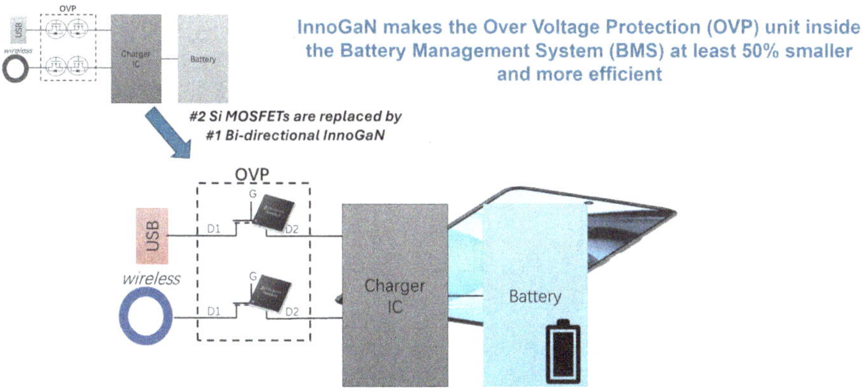

Fig. 6.49b The OVP circuit inside a BMS produced using VGaN will be at least 50% smaller and more efficient

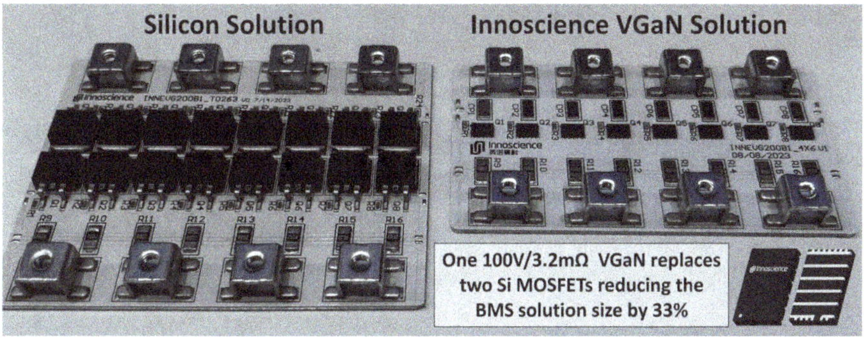

Fig. 6.50 Traditional silicon BMS OVP solution side-by-side with the new VGaN bidirectional implementation

temperature. Also, the transconductance of GaN decreases with temperature so there is no danger of the runaway effect by paralleling VGaN devices. GaN also has a positive temperature coefficient of the $R_{DS(on)}$ which, again, is helpful when paralleling devices (Fig. 6.51).

Other important characteristics are the SOA, leakage, and robustness of the solution. SOA is an important consideration for load switching applications. SOA is the combination of voltage and current over which a device can be operated without damage or a reduction in performance. Factors limiting SOA include R_{on}, the package, and thermal considerations. Improving the SOA of an Si MOSFET is challenging due to the negative temperature coefficient of the threshold voltage (V_{th}). Due to a reduced temperature dependence of V_{th} in GaN devices, VGaN can maintain a wider SOA, even at high temperature (Fig. 6.52).

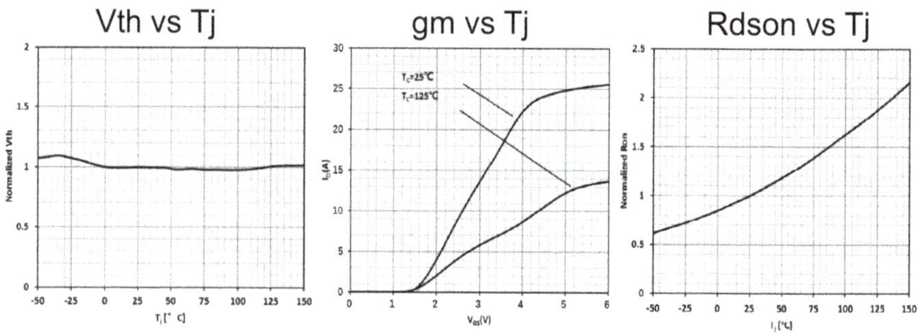

Fig. 6.51 GaN has a positive temperature coefficient of the $R_{DS(on)}$

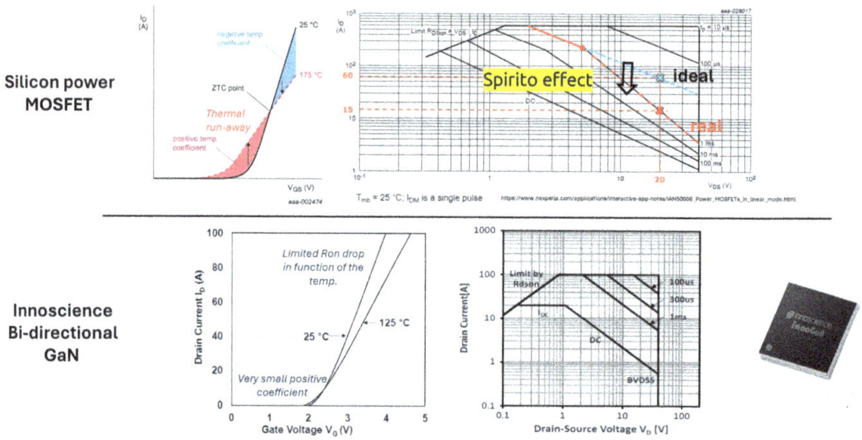

Fig. 6.52 Due to a reduced temperature dependence of V_{th} in GaN devices, VGaN can maintain a wider SOA—even at high temperature

Gate leakage is another important specification in bidirectional voltage blocking applications. Si MOSFETs have very good gate leakage performance. These devices have gate insulated from the channel by a gate oxide resulting in a sub-microampere leakage at 25 °C. Since Si MOSFET V_{th} decreases at higher temperatures, the leakage current rises at elevated temperatures. VGaN devices have an inherently different gate structure that can be visualized as two back-to-back diodes (Fig. 6.53). The upper diode is a Schottky structure with the gate metal as cathode and the pGaN as the anode. The lower diode is a junction structure with a pGaN anode and an AlGaN cathode. Without proper control, the gate leakage of a VGaN device would be higher than that of a silicon MOSFET. However, Innoscience's R&D team has developed differentiating process steps and process controls to ensure that the gate leakage is smaller than 3 μA at 85 °C throughout the device lifetime. This was a critical requirement for its adoption in smartphone handsets.

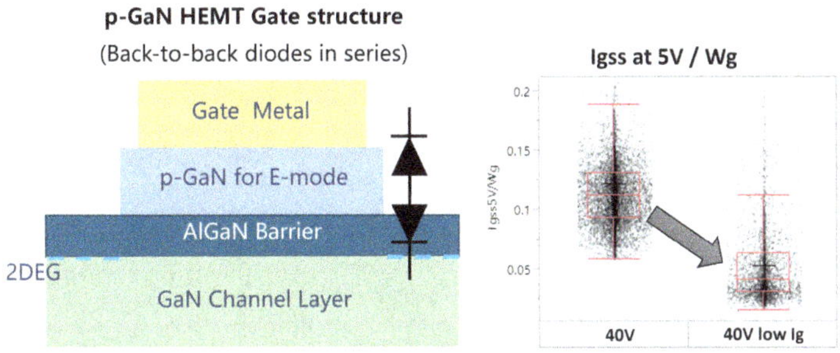

Fig. 6.53 VGaN devices have an inherently different gate structure that can be visualized as two back-to-back diodes

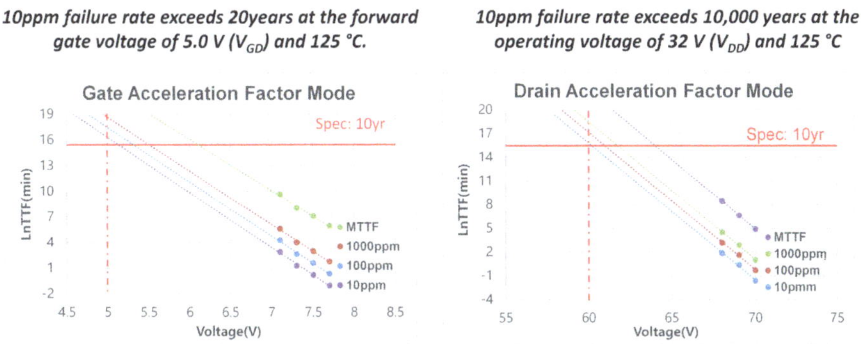

Fig. 6.54 Innoscience's VGaN bidirectional device has passed stringent JEDEC's qualification, and it shows excellent lifetime performance with a very low failure rate

Indeed, since its market introduction, Innoscience's bidirectional switch VGaN has been successfully adopted by smartphone handset makers, the first time GaN has been directly used inside a mobile phone.

Of course, to be deployed inside the mobile phone, the reliability of the new devices must be well proven: Innoscience's VGaN bidirectional device has passed stringent JEDEC's qualification and it shows excellent lifetime performance with a very low failure rate (Fig. 6.54), for example:

- over 20 years lifetime with 10 ppm failure rate at full $V_{GD} = 5$ V and 125 °C, and
- over 10,000 years lifetime with 10 ppm failure rate at V_{DD}=32 V and 125 °C.

Innoscience introduced its 40 V bidirectional GaN IC family, INN040x, in WLCSP and FCQFN packages last year. Recently, the company has launched a 100 V part, the 3.2 mΩ $R_{DS(on)}$ INV100FQ030A also in 4 × 6 mm packaging. As well as high efficiency 48 V battery management systems, the INV100FQ030A can be employed in high-side load switch applications in bidirectional converters, switching circuits in power systems, and other fields.

6.13 Integrated GaN Solutions Accommodate Extreme Power Demands of Next-gen Data Centers

Denis Marcon

By 2030, we would need 200 nuclear reactors just to run data centers! This shocking calculation—presented in a paper in Science Direct from information provided by the American Geosciences Institute and the US Department of Energy—is based on the exponential increase in the use of artificial intelligence (AI) and the computational demand that this is placing on data centers (Fig. 6.55).

The result is that data center racks are now being specified to have a capacity of 100 kW or even 200 kW, where current units have specifications averaging closer to 15 kW/rack.

The challenge is made even harder when we consider that power conversion systems are required to be much more efficient, meeting 80 plus Titanium power efficiency standards which demand levels above 90% from low level of 10% load through 100% loading, and above 96% at 50% load (Fig. 6.56). Finally, of course, since data centers operate 24/7 and must provide a guarantee of continuity of service, reliability is also critical.

Clearly a new approach is required, using new architectures and new technology.

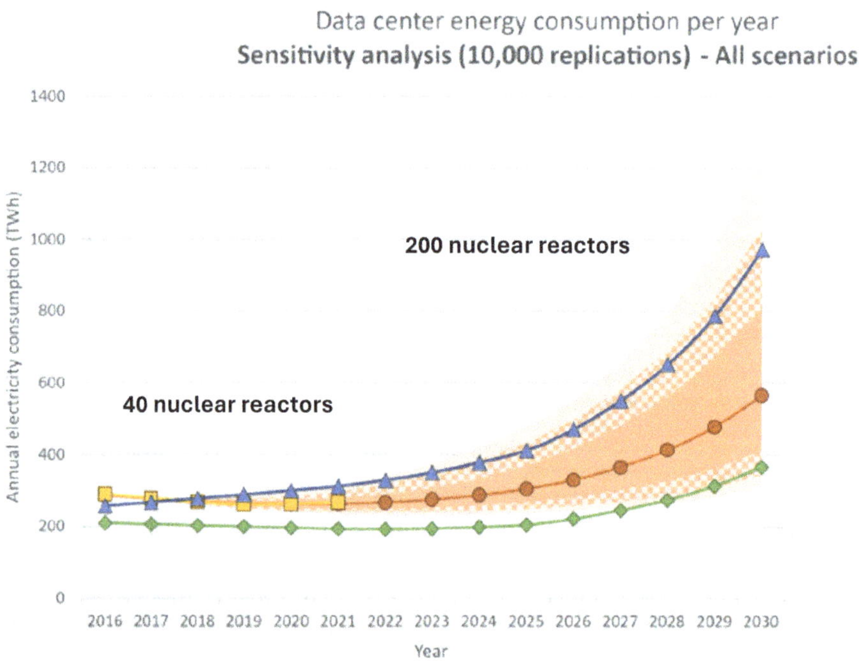

Fig. 6.55 The exponential increase in the use of AI resulting in power-hungry data centers

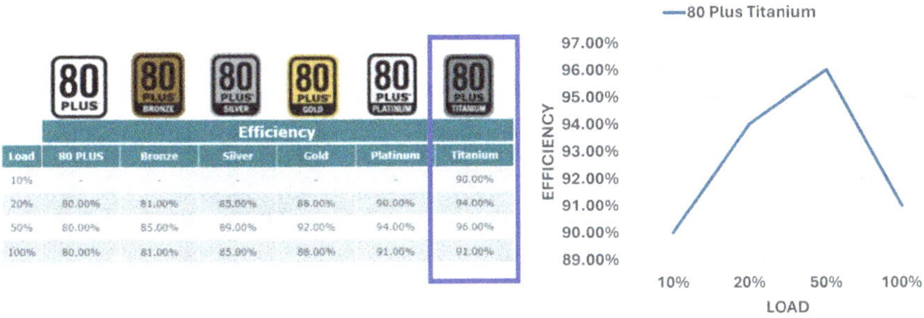

Fig. 6.56 Titanium Plus efficiency and small PSU size made possible with GaN

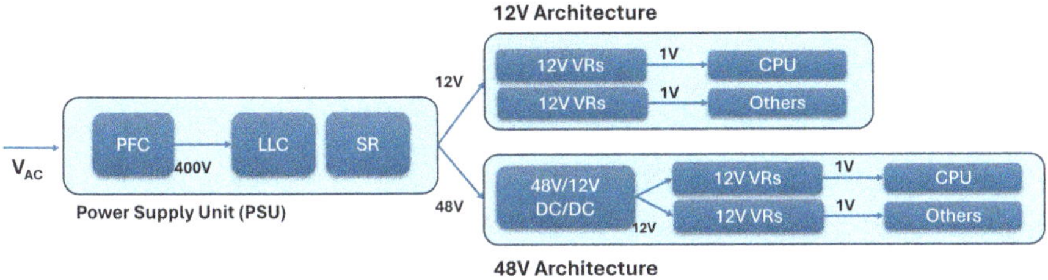

Fig. 6.57 Simplified data center power rack schematic

If we look at a simplified schematic of a data center (Fig. 6.57), first we see the power supply that converts the grid down to either 12 or 48 V. Previously 12 V was preferred, but now, to keep power losses low, there is a trend to go to a 48 V architecture (losses are proportional to the square of the current). Therefore, we need a 48–12 V conversion stage for the CPU and other functions. With silicon, designers struggle to achieve the mandated efficiency levels at high power for this stage.

However, using discrete gallium nitride (GaN)—which has no body diode and therefore minimal switching losses—in a totem pole plus PFC topology, Innoscience has shown in a 2 kW PSU demo that this first AC/DC conversion is achievable while achieving the 80 Plus Titanium efficiency rating. In contrast, a silicon solution only hits the lower platinum efficiency standard with roughly double the PSU size (Fig. 6.58). Thanks to the absence of body diode on GaN devices, a simple totem pole PFC architecture can be implemented while still reaching high levels of efficiency.

However, as we have seen, AI is fueling voracious power demands, and even a switch to GaN—with its well-publicized power density and efficiency benefits—may not be enough. Intel, e.g., is proposing a 48 V voltage regulator common module (VRCM) (Fig. 6.59) with the 48–12 W power conversion stage on a small daughter-card embedded on the server motherboard.

Such level of miniaturization could be difficult to achieve with a simpler discrete GaN solution.

To address this, companies are bringing integrated GaN power solutions to the market that allow the next level of miniaturization and performance. In Innoscience's

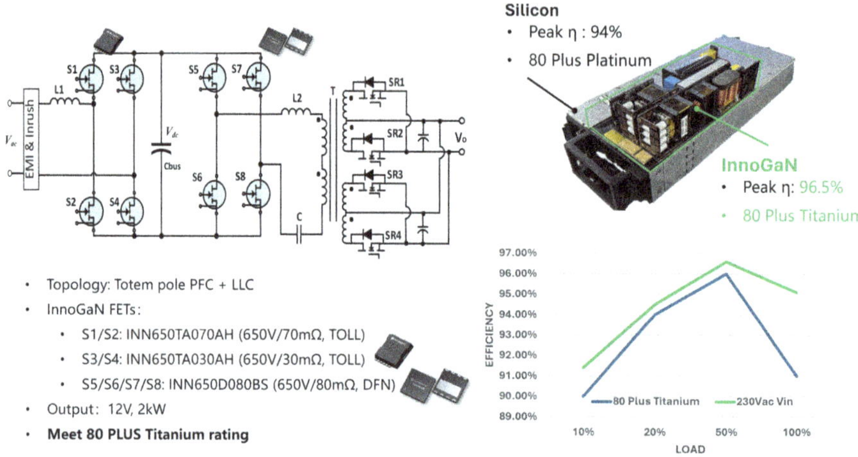

Fig. 6.58 2 kW PSU demo

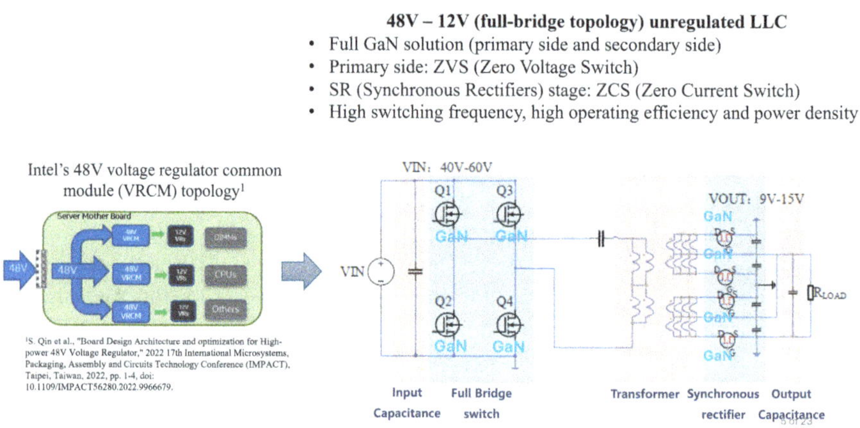

Fig. 6.59 48 V voltage regulator common module (VRCM) architecture proposed by Intel

case, this is an expanding family of SolidGaN™ devices which combine driver and GaN HEMTs or half-bridge and driver in one package, reducing power stage volume, increasing power and cutting component count for a smaller BOM. SolidGaN devices can be used not only for data center power designs but also for Buck converters, half-bridge or full bridge converters, Class D audio amplifiers, LLC converters, and power modules in applications including motor drivers, telecoms, and super computers.

Figure 6.60 highlights the benefits of the integrated approach using example circuits created using Innoscience's ISG3201 IC that combines two 100 V 3.2 mΩ enhancement mode GaN devices with a 100 V half-bridge gate driver in a 30-pin LGA package measuring just 5 × 6.5 × 1.12 mm.

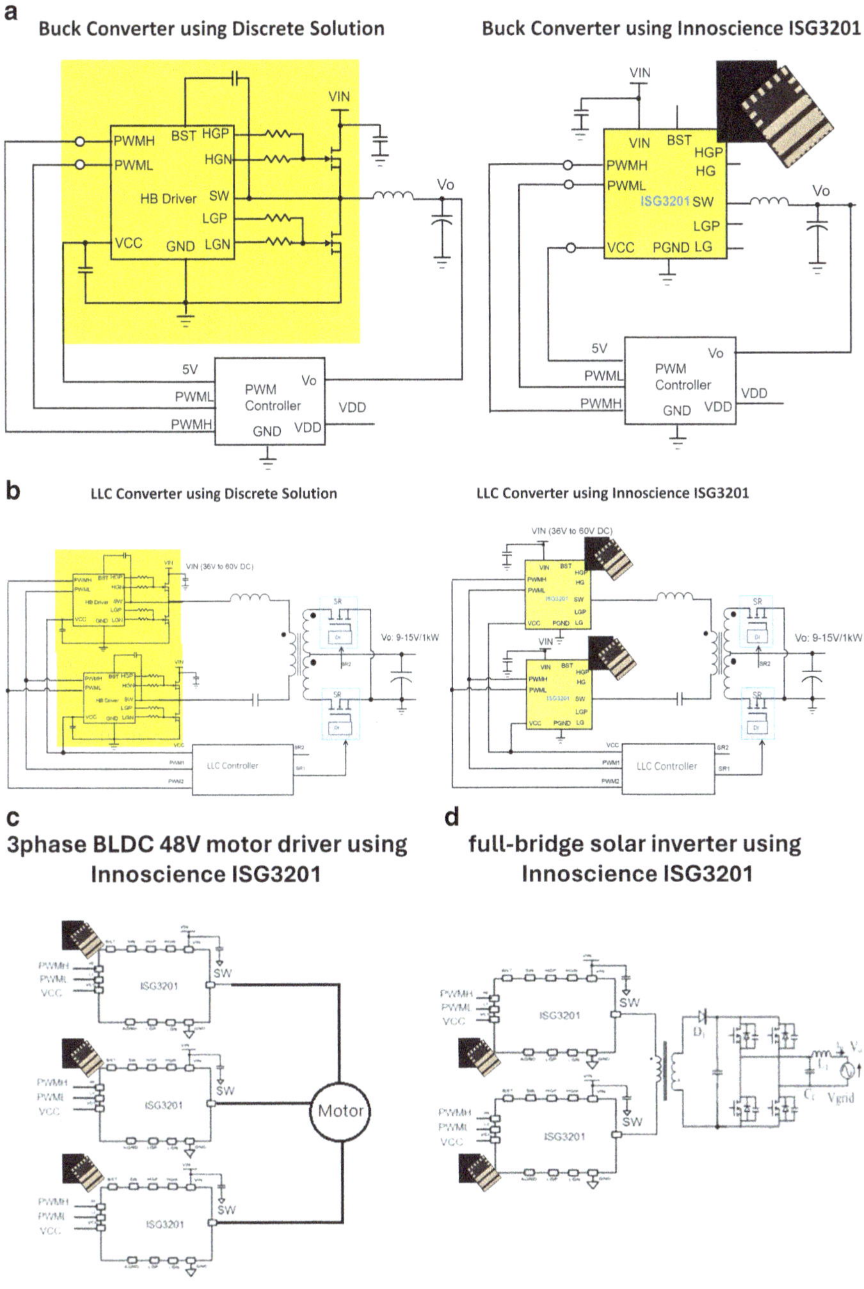

Fig. 6.60 Comparing discrete vs. integrated designs for Buck and LLC converters (**a**, **b**). Three-phase BLDC and full-bridge solar inverter schematics using the ISG3201 also shown (**c**, **d**)

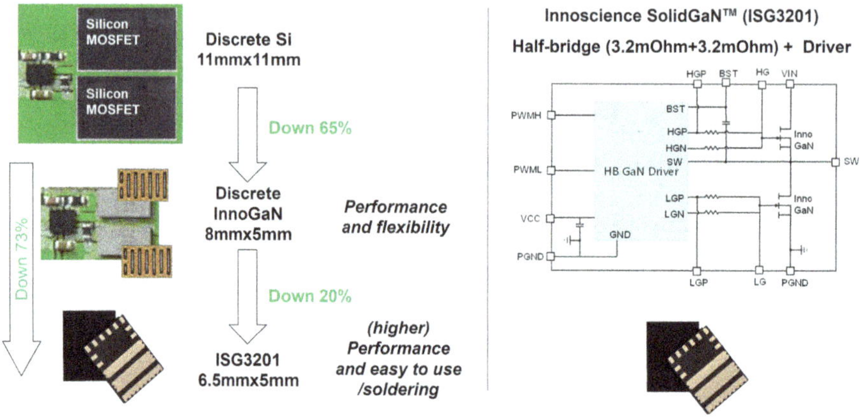

Fig. 6.61 Half-bridge and driver in one package (SolidGaN) showing size reduction

Using a discrete approach requires more external components than when implemented using an integrated solution as shown in Fig. 6.60a (Buck) and Fig. 6.60b (LLC). The simplicity of a three-phase BLDC (Fig. 6.60c) and full-bridge solar inverter (Fig. 6.60d) are also shown.

Figure 6.61 compares the reduction in size that is possible for the half-bridge circuit.

While the discrete GaN solution is 66% smaller in area than a silicon implementation, the integrated SolidGaN part is a further 19% smaller, and a massive 73% smaller than the silicon circuit. This schematic also shows the reduced BOM, as the driving resistor, Bootstrap, and the VCC capacitor are all also included in the SolidGan IC.

In addition, because all these functions are performed inside the SolidGaN LGA package, the gate loop inductance is reduced, as is the power loop inductance—typically by 40%. This reduction in parasitics leads to less ringing and reduced overshoot. This not only increases efficiency and simplifies design but also improves reliability, since overshoot is minimized to 4 V, 80% less than some alternatives (Fig. 6.62). Also, fewer clamping components are required.

The design also benefits from power stage layout. In some configurations the switch node is between V_{in} and P_{GND} which assists device construction but requires extra external components. The Innoscience layout positions the switch node at the edge, requiring only a simple decupling capacitor to be placed between V_{in} and P_{GND}, and the switch node connects to the external circuit (Fig. 6.62 right). Depending on the selected topology, many components can be saved.

Finally, manufacturing issues must be considered. The placement of fine-pitch Wafer Level Chip Scale Package (WLCSP) discretes require good process control. It also limits the amount of copper that can be used on the PCB. Although GaN typically runs cooler than silicon because it is more efficient, at high power levels, more copper to improve thermal performance is desirable, even essential. By using

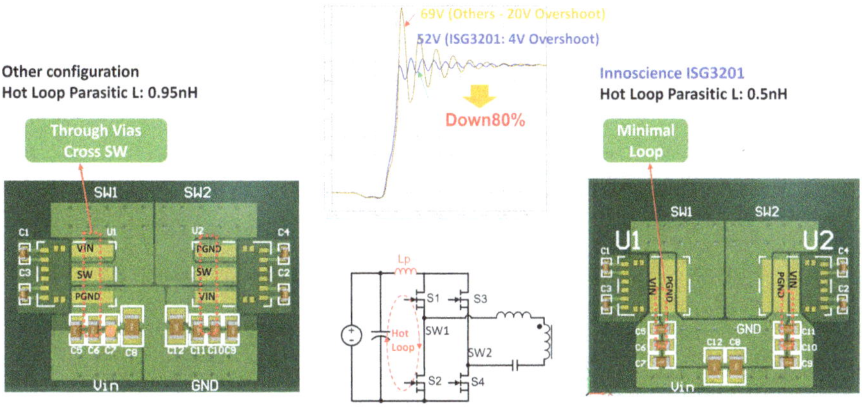

Fig. 6.62 A reduction in parasitics leads to less ringing and reduced overshoot

integrated packages with a larger pin pitch, 2 ounce copper can be used rather than one ounce which is the practical limit for fine-pitch discretes. This reduces power loss and increases efficiency, extending the practical useful power range of GaN technology. Tests have shown that the efficiency of half-bridge designs using integrated parts such as the ISG3201 is 0.3% higher than implementations using discrete GaN devices, which can, in part at least, be credited to the improved thermal construction made possible by the SolidGaN technology.

These efficiency improvements start to make a significant difference. Figure 6.63 shows an overall improvement in efficiency of 0.7% for the ISG3201 integrated half-bridge approach. Thermal plots prove that for the same temperature rise, this increase in efficiency means that 18% more power can be achieved using SolidGaN.

Let us get back to the data center challenge. To address the 48–12 V DC-DC power conversion stage inside the data center, Innoscience has developed a 1 kW 48–12 V unregulated LLC demo (Fig. 6.64) that features GaN power devices both at the primary side (100 V devices) as well as at the secondary side (40 V devices). In order to maximize the power density and simplify the circuit, the demo uses the integrated SolidGaN solution, ISG3201, with all the features and benefits described earlier.

On the secondary side of the converter, Innoscience selected a discrete 40 V InnoGaN solution, which is a cost-effective approach with efficiency and power density benefits over traditional silicon implementations.

The 1 kW 48–12 converter described measures just 50 × 30 × 9 mm (Fig. 6.65), delivering a high power density of greater than 2000 W/in^3.

Efficiency and power loss performance up to 1 kW is shown in Fig. 6.66: efficiency is higher than 98% peak efficiency across the whole 40–60 V input voltage range, confirming the promises of GaN technology in high-efficiency power conversion.

A final word concerning reliability, which is paramount in many applications, including data centers as described at the top of this article. Innoscience has strong

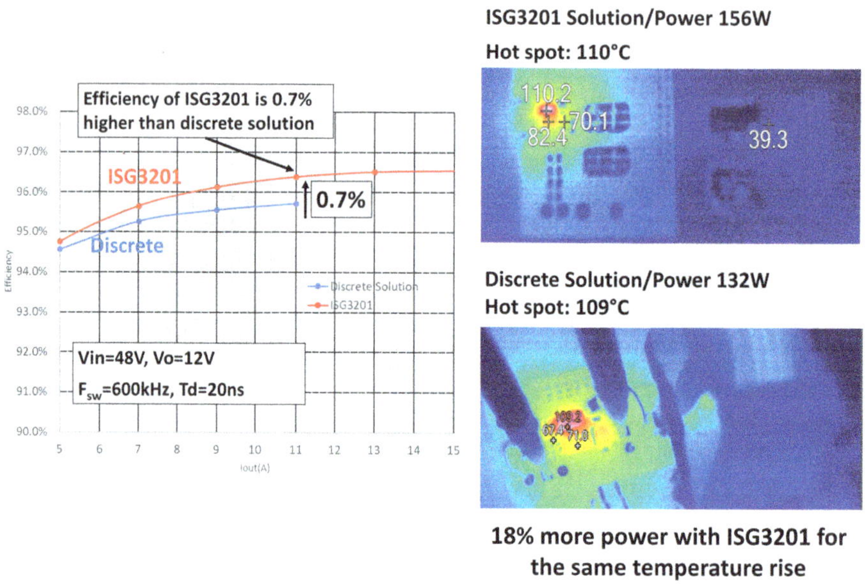

Fig. 6.63 Efficiency comparison of integrated and discrete GaN solutions

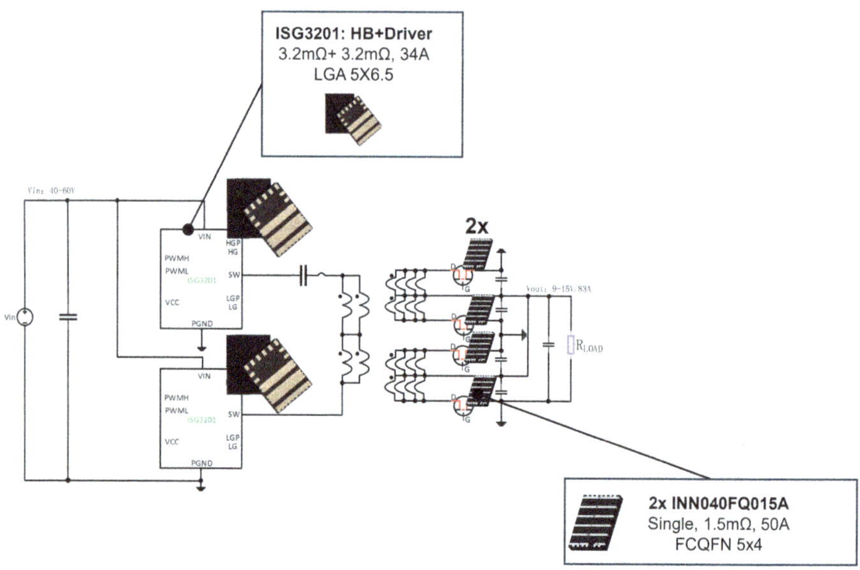

Fig. 6.64 1 kW 48–12 V unregulated LLC demo

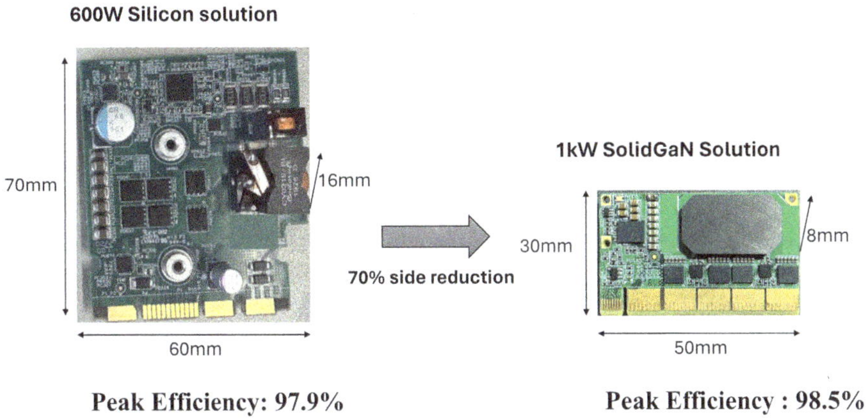

Fig. 6.65 SolidGaN vs. Silicon comparison

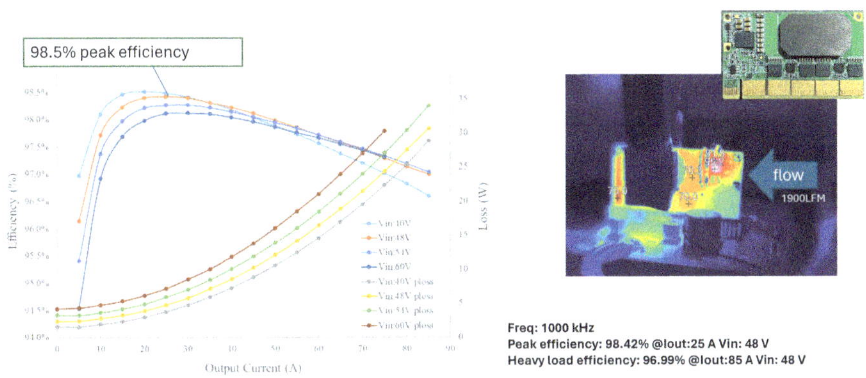

Fig. 6.66 Efficiency measurements

reliability data of its high and low-voltage GaN power devices, including end of life testing for life-time extrapolation at the ppb level, indicating that Innoscience's GaN power devices are reliable and can be safely used in long-lifetime applications in data centers for many years.

6.14 Role of Gallium Nitride (GaN) in Server Applications

Francesco Di Domenico

The growing demand of cloud-based internet services and artificial intelligence (AI) has generated a strong growth of processing power in data centers around the world. This is reflected in increasing power demand and thus a higher number of power supply units (PSUs) per rack. Data centers are also facing rising electricity and real estate prices, therefore there is a clear trend towards highly efficient and compact server power supplies. These new PSUs do not only lead to a lower power consumption of the server but also to a lower heat dissipation, reducing secondary costs such as the cooling/air conditioning of the data center.

The commercial availability of wide-band gap power semiconductors, such as GaN devices, with their significantly better figures of merit, has given power converter designers more chances to fulfil the more and more stringent requirements in terms of efficiency and power density in the most recent specifications of power supplies for data center applications.

Figure 6.67 shows three of the most relevant device key performance indicators related to $R_{DS(on)}$ (normalized) in power converter applications.

From these figures, we can derive that GaN devices have gate charge and output charge that is around one order of magnitude lower compared to their silicon counterparts. Combined with virtually zero reverse recovery charge, it enables hard commutation of reverse conducting devices. Thus, the use of GaN is preferred in topologies not suitable for legacy Si superjunction (SJ), such as CCM totem pole PFC, shown in Fig. 6.68, left.

In other words, GaN supports topologies with minimum component count and an optimization of control methods that can seamlessly change between soft switching and hard switching.

The designer of server AC/DC power supplies has three choices as next best alternatives to the use of wide-bandgap (WBG) devices:

- Single-ended topologies such as the classic boost converter as a power factor correction stage
- Strict avoidance of hard commutation through dedicated control methods such as triangular current mode (TCM) operation in totem pole PFC
- The use of cascaded converter architecture where the voltage stress is distributed to several series-connected converter stages

While single-ended topologies may not comply with efficiency targets, alternative solutions may not comply with space or cost targets. Even though cascaded solutions have demonstrated their ability to reach more than respectable efficiency and density targets, control efforts remain challenging and may limit the use of this concept.

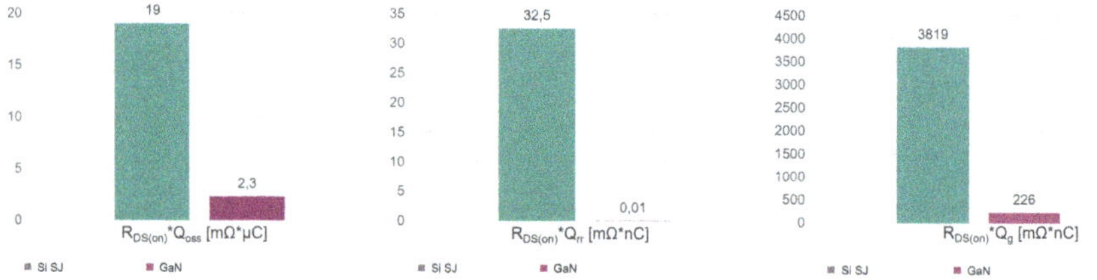

Fig. 6.67 Comparison of key figures of merit between Si superjunction (SJ) and GaN devices: Q_{OSS}, Q_{rr}, and Q_g

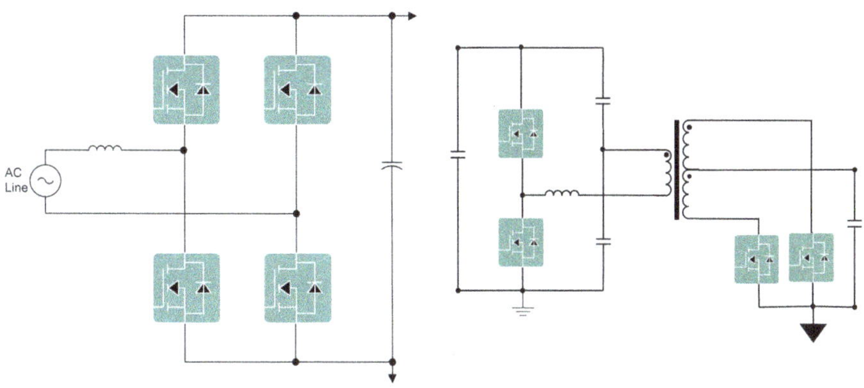

Fig. 6.68 Principle schematics of the CCM totem pole PFC (left) and half-bridge resonant LLC (right) stages of a typical PSU for data centers

The design options for highly efficient and compact server power supplies are narrowing down to silicon-based TCM operation of interleaved totem pole legs vs a CCM/TCM GaN-based totem pole stage.

The second stage of a server power supply is the isolated DC/DC stage. The GaN figures of merit shown in Fig. 6.67 make this device technology especially beneficial in isolated DC/DC resonant topologies such as the LLC converter (principal schematics on the right-hand side of Fig. 6.68).

Especially the reduction in parameters such as Q_{OSS} and Q_G, by around one order of magnitude compared to legacy silicon technologies, enables unprecedented performances in resonant topologies. Actually, it gives the chance to finally operate the LLC converter at a significantly higher switching frequency compared to traditional silicon-based solutions, with consequent reduction in the magnetic components size. Thus, benefits are definitely expected on the achieved power density.

We can summarize this general introduction, saying that the high efficiency and compactness of GaN-based systems result in lower energy losses and reduced cooling requirements, directly impacting the data center‘s operational efficiency and environmental footprint.

6.14.1 GaN in Server Power Conversion

Server power conversion plays a pivotal role in data center operations, ensuring the efficient transformation of electrical power to meet the precise needs of server hardware. This process is critical for maintaining the performance and reliability of data centers, which are increasingly required to handle vast amount of data while minimizing energy consumption and operational costs.

Gallium nitride (GaN) technology is key to enhancing server power conversion efficiency, marking a significant departure from traditional silicon-based solutions.

GaN's fast-switching properties have revolutionized server power systems, enabling operations at high switching frequencies (greater than 500 kHz), leading to an increased power density and reduced system costs.

The integration of GaN technology allows for the adoption of advanced topologies like totem pole power factor correction (PFC) and resonant converters, which operate efficiently from 500 kHz to 1 MHz. The use of GaN makes these topologies achieve efficiencies greater than 99%, a remarkable improvement over traditional methods. The high efficiency and compactness of GaN-based systems result in lower energy losses and reduced cooling requirements, directly impacting the data center's operational efficiency and environmental footprint.

The benefits of GaN in server power conversion are multifaceted. First, GaN's ability to operate in both hard-switching and soft-switching modes allow for flexible design choices, supporting continuous conduction mode (CCM), triangular conduction mode (TCM), and optimal frequency modulation (OFM) in power supply circuits.

This versatility enhances power density and efficiency in both 12 and 48 V output server power supply applications.

Consider the 12 V_{out} case. These PSUs are mostly used in the legacy server architecture, where two AC/DC power supplies provide redundancy to each other, each power supply being sufficient to cover the full power demand of the server board. The need to reduce both operational and capital expenditures (so called OPEX and CAPEX) will drive two major trends: first, local energy storage at the rack level to cut out the UPS from the power flow; second, the transition from server-based power supplies to rack-based power supplies to cut redundancy from 1+1 to n+1, thus saving cost. Both trends push to a higher output power in a given form factor. Therefore, GaN technology can play a key role there. The typical schematic of a 12 V output server PSU includes an interleaved totem pole PFC and an LLC converter with a center-tapped transformer.

Figure 6.69 shows a comparative two-dimensional optimization analysis (efficiency vs power density) between GaN and Si-based solutions for a 12 V_{out} PSU.

For a given target efficiency, exceeding the minimum requirement of the 80+ Titanium+ Standard at 50% load (96.5%), the red dotted line clearly shows the difference with a GaN 3 kW P_{out} in the same form factor as a 1.6 kW designed with a legacy Si-based solution.

On the other hand, for power density targets higher than 100 W/inch^3, the efficiency naturally starts declining for both GaN and Si solutions. This target is extremely challenging and requires a very special approach, especially in the LLC Stage. The high output current, required in the case of 12 V/3 kW rating, makes even more critical the design of the transformer and, in general, of the PSU secondary side. Matrix planar transformers with special construction geometry have to be taken into consideration for the LLC.

The emerging use of AI in computer science and the training of the corresponding neural networks has shifted the computational architectures from pure CPU-based servers to more specialized platforms with a very high degree of parallel data processing, e.g., by employing GPUs. As a result of this trend, the power consumption per rack has been recently estimated to double each year. Today, the average

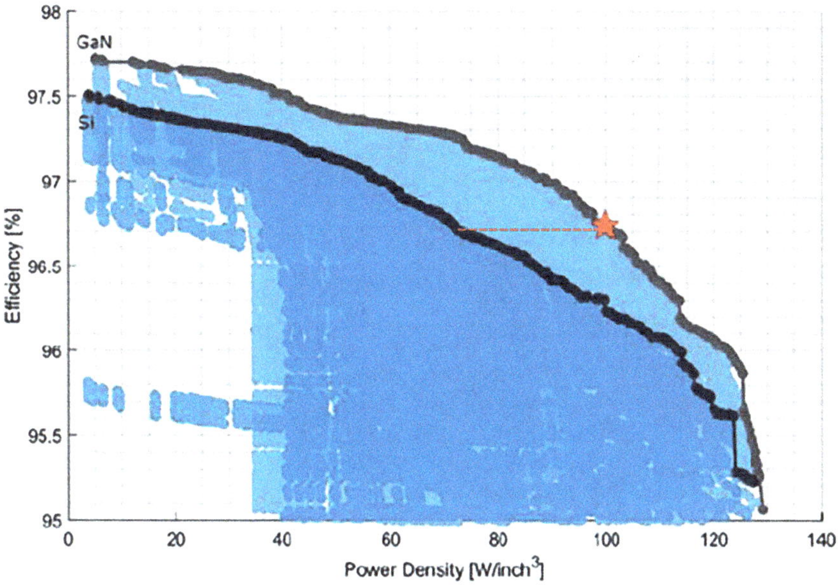

Fig. 6.69 Optimization results of a 12 V_{out} server PSU for either GaN or Si devices (efficiency optimized at 50% load) [29]

power per rack in hyperscale data centers is around 30 kW, but it is easy to predict "AI GPU racks" requiring up to 250 kW. Therefore, such a huge power level can no longer be economically handled by the classic 12 V distribution rails. As a solution, power distribution within the rack at 48 V levels instead of 12 V will be more and more common in the future. Since the 48 V server power supplies are mainly targeting applications optimized for total cost of ownership (TCO), the performance of GaN in these systems is evaluated purely from an efficiency perspective.

As a basis for the comparison, consider a 48 V/3 kW PSU, using an interleaved CCM or TCM totem pole PFC and a half-bridge LLC with secondary full bridge synchronous rectification. Figure 6.70 shows the two-dimensional optimization analysis of a GaN vs Si-based PSU. Differently from the analysis of a 12 V system in Fig. 6.69, note that for a 48 V output, the efficiency is the main target of the optimization. Also, in this case, GaN shows the best "Pareto Front" compared to the legacy Si-based PSU, by allowing 97.5% efficiency at 50% load at a power density exceeding 80 W/inch^3. Silicon is also able to get the same efficiency, but at around half the power density. Most importantly, by sacrificing the form factor a little, the GaN solution is in principle able to get an efficiency even higher than 98.5%, a target totally precluded to the compared Si-based PSU.

These examples underline GaN's transformative role in modern data centers, offering a pathway to meet the growing demand for data and power in an environmentally sustainable manner. As the digital landscape evolves, the adoption of GaN technology in power conversion is set to become a cornerstone of efficient, high-performance data center operations.

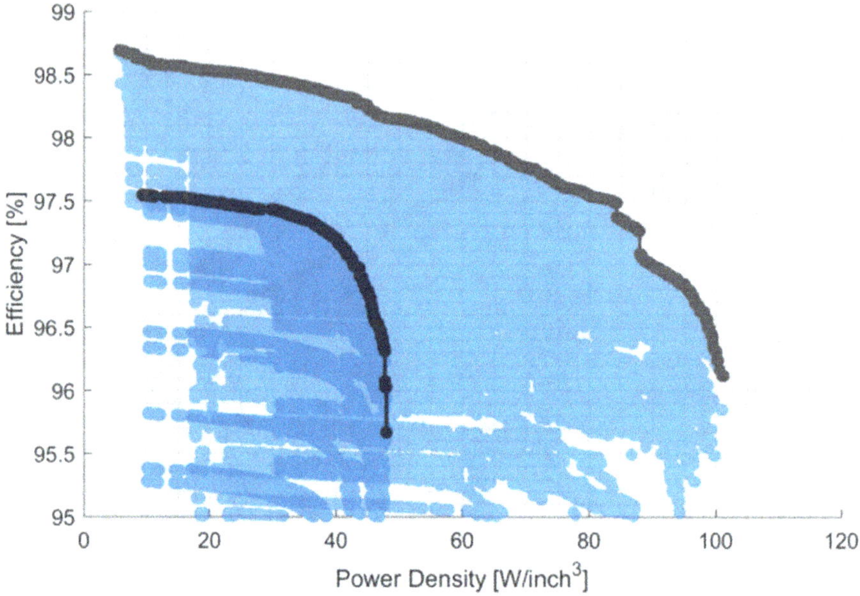

Fig. 6.70 Optimization results of a 48 V_{out} server PSU for either GaN or Si devices (efficiency optimized at 50% load) [29]

6.14.2 Cooling Solutions and Modularity of GaN-based Server PSUs

The challenges of cooling high-performance servers are increasingly complex, driven by the demand for greater computational power and the advent of technologies like generative AI, which significantly intensify power and cooling requirements. Traditional silicon-based power conversion and distribution technologies contribute to these challenges by being relatively inefficient, thus generating substantial heat that necessitates advanced cooling systems. This not only increases operational costs but also impacts the sustainability of data centers.

GaN's material properties allow for faster switching speeds, higher conductivity, and better overall performance, leading to significant energy savings, therefore reduced heat generation. For instance, GaN technology has enabled a leap in power density for both AD/DC and DC/DC converters, which is crucial for supporting the high-power demands of modern servers and reducing the space occupied by power conversion units on server boards. This advancement in power density and GaN's efficiency directly contributes to less heat generation and a reduced need for extensive cooling systems.

Compact SMD packages, both bottom and top-side cooled (TSC), significantly enhance GaN device thermal management. These packages enable chip-scale GaN devices to maintain excellent thermal performance when mounted on standard PCBs.

The high power density requirements generally force the designers to adopt a two-dimensional component positioning inside the power supply unit, which makes thermal management even more challenging. The availability of TSC SMD packages provides the option of thermally decoupling the component from the PCB, thus enabling a better PCB routing, both on the top and bottom side of the board. The heatsink can then be easily placed on the top of the package. TSC is also beneficial in case liquid cooling is adopted in the data center, either by cold-plate approach or by full system immersion. These cooling concepts, a bit unusual in the legacy server applications, are actually becoming increasingly popular, especially due to the emerging use of AI computing and the large amount of power required by it.

These new advanced cooling concepts co-exist in server switched-mode power supplies (SMPS) alongside the more traditional forced air cooling. In this approach, the two-dimensional component positioning, needed for power density reasons, is also achieved by the daughter card design approach. Figure 6.71 shows an example of forced air cooling server PSU using the daughter card technique.

Figure 6.71 shows very clearly how the daughter cards (1, 2, and 3 in this example) allow to take advantage of the "second dimension" for the component assembling. The daughter cards are vertically assembled and positioned in order to experience the airflow produced by the fan, thus enabling optimal thermal management of the power devices assembled on it. For this kind of design approach, SMD packages with low height are essential. The PG-HSOF-8-3 or, more commonly, "TOLL" (TO-Leadless) package, shown on board 1 in Fig. 6.71, is a great example of it.

Moreover, the very compact and optimized layout on the daughter card of an interleaved totem pole PFC converter, e.g., (as in the specific case of 1 in Fig. 6.71), also provides additional benefits in terms of switching behavior of very fast devices

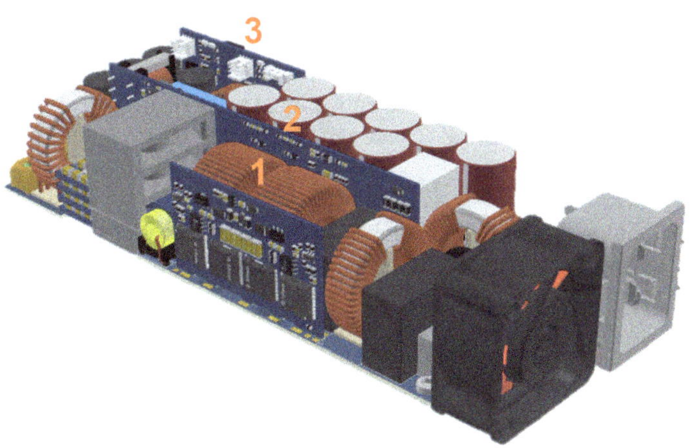

Fig. 6.71 Example of a modern air-forced cooled server PSU (Infineon demo board) adopting the daughter card design technique (e.g., 1, 2, and 3)

such as GaN. This also facilitates the converter's operation at a high switching frequency.

This modular design approach potentially fulfils a key requirement of high-volume applications such as SMPS for data centers: the availability of multiple sources for each component, especially semiconductors. This aspect significantly affects GaN devices, since different GaN technologies are currently available in the market, with different physical structure, gate driving circuit and, of course, package pin outs. Moreover, some of them are even available in packages with integrated gate drivers, but not all.

The daughter card technique provides, then, the option to qualify, in the same PSU designs, GaN switches, e.g., in the half-bridge configuration, with different driving circuits and pin-outs. In principle, it enables the optimal layout for each of them. The typical implementation of this concept consists in designing, for a given topology or switching cell, two more daughter cards, each of them optimized for a specific GaN vendor. All the cards must have the same pin-out, to fit with the one on the main PCB. The same heatsink can be placed, e.g., on the bottom side of the different boards, in order to guarantee the same overall dimension of the module, and, of course, for proper thermal management. The daughter card then becomes a sort of "module," potentially reusable in different PSU designs, by leveraging economies of scale and reducing the time to market. This, in the end, means "modular" SMPS design, fully enabled by GaN technology.

We can conclude that, by adopting GaN in the PSU, data centers can significantly mitigate the thermal management challenges, paving the way for a more efficient, reliable, and sustainable future in server technology and data center operations.

6.14.3 GaN's Impact on Data Center Infrastructure

Gallium nitride (GaN) technology significantly impacts data center infrastructure, especially server technology, by addressing critical areas such as server size, weight, energy consumption, operational costs, environmental impact, and future scalability.

Benefits for data center infrastructure

GaN technology offers unparalleled performance and efficiency, primarily because GaN transistors operate at faster switching speeds, providing higher operating efficiency and significantly reducing energy waste and heat generation.

In a typical data center environment, utilizing GaN transistors can lead to substantial yearly profit increase, reduction in CO_2 emissions, and decrease in operational expenses. This is especially relevant as the demands of generative AI increase, requiring more power density in data centers.

GaN's ability to close the "Power Usage Effectiveness" (PUE loophole by making power supply system more efficient is a notable advantage, as it addresses the uncertainty in the PUE metric related to power conversion within servers.

6.14.4 Challenges and Future Prospects

Building on the transformative role of GaN in data center infrastructure, it is crucial to address the specific challenges and prospects of GaN in alignment with its established benefits.

While GaN brings numerous advantages to data center infrastructure, its adoption faces challenges primarily related to early-stage perceptions of reliability and cost. However, strides in GaN technology have significantly mitigated these concerns.

Advances in manufacturing such as moving from smaller to larger wafer sizes have enhanced the reliability of GaN devices and driven down costs, making GaN increasingly competitive against traditional silicon-based power systems. Moreover, dedicated test/qualification results, performed by all the main vendors, demonstrate that GaN is ready to increase efficiency and density in a broad range of customer application/mission profiles, including modern data centers, while also assuring long-term reliability [31].

6.14.5 R&D for Enhanced Efficiency and Reliability

GaN technology advancements include the integration of multiple power switches on a single chip, enhancing power electronics‘ efficiency and reliability. The first full-function 650 V GaN power ICs introduced between 2016 and 2017 illustrate this progress. These developments underscore ongoing R&D efforts to fine-tune GaN's efficiency and reliability for server applications.

6.14.6 Future of GaN in Data Centers

GaN is pivotal in evolving server technology and data center infrastructure. The ongoing reduction in GaN device costs and continuous improvements in performance and reliability sets the stage for the broader adoption of GaN in data centers. As the technology matures, its impact on improving energy efficiency, reducing operational costs, and supporting the scalability of future data center needs will only grow, aligning with the increasing demands for sustainable and high-performance computing resources.

6.14.7 Key Takeaways

GaN technology significantly enhances server technology within data centers by providing key benefits, including increased power efficiency, reduced size and weight of power supply units, and significantly lowered energy consumption and heat generation. These advantages lead to substantial reductions in operational costs and CO_2 emissions, aligning with the growing demands for higher power density in data centers, particularly in the era of generative AI.

The broader impact of GaN adoption transcends server technology, marking a pivotal shift in the tech industry and promoting environmental sustainability. By enabling more energy-efficient operations, GaN contributes to reducing the carbon footprint of data centers, which are notoriously energy-intensive. This aligns with global efforts toward achieving net-zero emissions and underscores the critical role of innovative materials in shaping a sustainable technological future.

Moreover, GaN's influence stimulates progress in tech industry sectors, including telecommunications, automotive, and renewable energy, by offering a more efficient, reliable, and cost-effective alternative to traditional silicon-based components [32, 33]. The ongoing development and integration of GaN technology signify a promising path toward enhancing performance and sustainability across the tech landscape [29–33].

6.15 Extending the Benefits of GaN to Renewable Energy Solutions

Alex Zahabizadeh

The majority of gallium nitride's (GaN) implementation to date has been on the demand side of electronics. A large number of laptop and phone chargers have switched to GaN, e.g., power supply units (PSUs) leverage GaN for its ability to reduce inductor size and improve efficiency in power factor correction (PFC) circuits, helping PSU manufacturers meet industry standards such as 80 Plus Titanium.

Now comes the adoption of GaN on the supply side, with multiple solar inverter companies announcing GaN-based products in the last few years. There is, however, even more opportunity for GaN adoption in renewable energy solutions to improve the efficiency of energy generation and energy storage, and to help reduce overall size and cost.

6.15.1 Challenges in Battery Storage Solutions

Research company ResearchAndMarkets' 2022 report shows the battery energy storage systems' market is growing at a compound annual growth rate of 27.9%, from $4.4 billion in 2022 to $15.1 billion by 2027. You will find many such systems in the data center market, where uninterruptable power supplies (UPSs) increase the reliability of PSUs. A UPS system helps improve equipment safety and efficiency by protecting against unexpected power outages, preventing data loss at data centers and helping avoid shutdowns that can damage equipment or corrupt data. It reduces the cost of ownership and helps extend battery life in the equipment.

The same benefits of a UPS are applicable to the electric grid and power energy within residential and commercial energy storage systems (ESS). To reduce dependence on the grid and stabilize the availability of renewable energy resources, an ESS stores electricity during the day when energy sources such as solar are readily available, or when grid power is more available and cheaper. Then, during periods of high load when demand on the grid network increases (such as at night), the ESS returns power to the home at a cheaper rate. An ESS is also able to deliver power in the case of a power outage, e.g., during a hurricane or a blizzard.

In both UPSs and ESSs, round-trip efficiency is extremely important. There is a power conversion loss when charging the battery and again when discharging. This means a compounded 95% efficient AC/DC power stage, and the round-trip efficiency is approximately 90%. In an online UPS system that is converting and cleaning the AC grid power, the two-stage efficiency impact is a significant cost adder to the operational cost of a server farm. For example, if an ESS could send power to the grid for \$0.10 per kWh at off-peak periods and deliver power to homes during a peak period for \$0.15, the round-trip efficiency is losing 20% of the intended cost savings.

6.15.2 GaN Offers a Potential Solution in Solar Inverters

As demand for renewable energy grows, achieving higher efficiency in solar power conversion becomes crucial. GaN field-effect transistors (FETs) can effectively lower the levelized cost of electricity by helping enable solar panels to produce more energy from sunlight. This means that GaN can reduce the cost/kWh, making it better able to better compete with cheaper and traditional energy solutions. This benefit comes from the fact that GaN FETs differ from traditional silicon metal-oxide semiconductor field-effect transistors (MOSFETs) given their unique two-dimensional electron gas channel structure. Unlike silicon MOSFETs, GaN FETs lack body diodes and the associated reverse recovery loss. This makes them suitable for hard-switching topologies, such as the bridgeless totem pole PFC circuits commonly found in DC-to-AC inverters.

GaN FETs additionally play an important role in maximizing cell power output. Photovoltaic cells exhibit a complex relationship between temperature and total resistance, resulting in nonlinear output efficiency. Single-phase string inverters, commonly used in residential setups, employ maximum power point trackers (MPPTs) to optimize cell output. GaN FETs shine in this context, offering higher efficiency and smaller line filters. For example, the 400W GaN-based MPPT Charge Controller and Power Optimizer Reference Design from Texas Instruments (TI) demonstrates efficiencies of 98.3–98.5%, depending on the intermediate bus voltage. Previous designs for a solar power optimizer are two to three times larger, and only capable of efficiencies of 96–97%.

6.15.3 Challenges with GaN Are Already Solved for Battery Storage

While not all of the challenges of designing with GaN have been eliminated, the industry is coming to a point where the barriers to implementing GaN are significantly reduced. GaN was previously more difficult to design with given the more sensitive gate driver, which can easily be damaged. Many GaN devices are now available with fully integrated gate drivers that eliminate any risk of overdriving the GaN FET's gate and additionally bring switching performance improvements given the reduced parasitics.

As shown in Fig. 6.72, many devices such as the TI LMG3422R030 GaN FET include several built-in safety and reliability features that help monitor systems and ensure correct operation. Protection features include undervoltage lockout, which disables the GaN FET and gate driver operation at low supply voltages (since it is not possible to predict system behavior at those conditions), overcurrent protection for damage prevention, and short-circuit protection to ensure system reliability. The

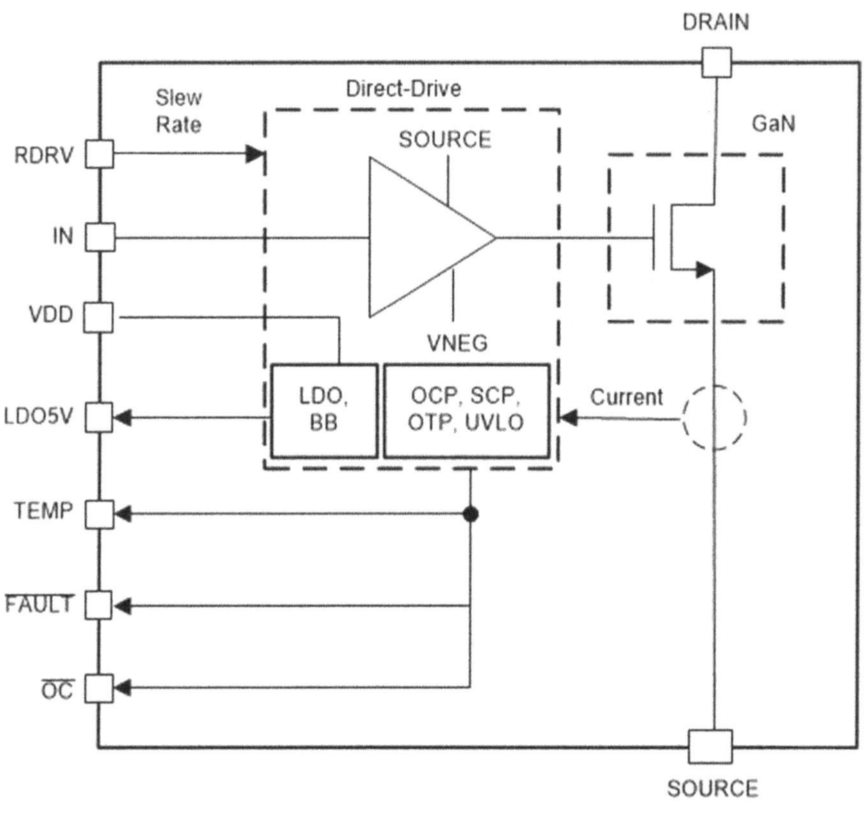

Fig. 6.72 Simplified block diagram of the LMG3422R030

Fig. 6.73 Fully assembled model of the 4 kW PFC reference design

LMG3422R030 has also built-in temperature reporting to a controller, which enables overtemperature protection that disables the power device if the device temperature goes above the overtemperature threshold.

Many reference designs demonstrate GaN in applications from 50 W to 5 kW that can fill the requirements of battery storage solutions. As shown in Fig. 6.73, the TI 4 kW Single-Phase Totem Pole PFC Reference Design with C2000™ and GaN features a 4 kW single-phase continuous conduction mode totem pole power converter capable of peak efficiency ≥99.1%, cost-effective current sensing, and have become an increasingly popular topology in high-reliability applications such as server PSUs, 5G telecommunications towers, and industrial power supplies.

6.15.4 Looking Forward

The world's need for energy will continue to increase, and technology will need to continue to significantly change and advance along with it, especially in the battery storage solution space. Innovations in both GaN and battery-monitoring technologies continue to improve the round-trip efficiency and long-term cost-effectiveness of solar and other renewable energy solutions to supplant the existing market of energy sources.

6.16 GaN ICs Power AI Server Roadmap to 100 kW+ Per Rack

Stephen Oliver

With the introduction of AI processors like NVIDIA's Grace Hopper and Blackwell, rack power has increased from around 30 kW to over 100 kW. It is critical to find a high power-density AC-DC solution that delivers 3–4× more power in the same space as legacy silicon chips, and exceeds the European Union's "Titanium Plus" efficiency standard. Leading-edge GaN power technology—including monolithically-integrated GaNSafe™ power ICs—is critical to develop 220 V_{AC} to 48/54 V_{DC} server power designs at 3200 W, 4500 W, and onwards to 10,000 W.

6.16.1 Data Centers Are Getting Big, Really Big

Spending on hyperscale data centers, with application-specific GPUs and high-density, AC-48-1 V power architectures has increased by 200% since 2016 [34]. In the same period, the introduction of AI processors like NVIDIA's Blackwell has seen data processing increase 1000×, now reaching 20,000 TFLOPs per GPU [35].

However, the corresponding processing-to-power ratio (TFLOPS/W) has not kept pace with this increase, and power-per-rack has increased to over 100 kW. AI is power-hungry. And it takes 2000 Blackwell GPUs 90 days and 4 MW to train GPT-MoE-1.9T. As a result, it is critical to find a high power-density AC-DC solution that delivers 3–4× more power in the same space as legacy silicon chips, and exceeds the European Union's "Titanium Plus" efficiency standard ("'Directive 2009/125/EC, 2019 Annex") (Fig. 6.74).

80 Plus test type	Icon	115 V internal non-redundant				230 V internal redundant				230 V EU internal non-redundant			
Percentage of rated load		10%	20%	50%	100%	10%	20%	50%	100%	10%	20%	50%	100%
80 Plus Gold			87%	90%	87%		88%	92%	88%		90%	92%	89%
80 Plus Platinum			90%	92%	89%		90%	94%	91%		92%	94%	90%
80 Plus Titanium		90%	92%	94%	90%	90%	94%	96%	91%	90%	94%	96%	94%

Fig. 6.74 80 Plus efficiency standard and European Union efficiency requirements (green highlight) "Directive 2009/125/EC, 2019 Annex"

6.16.2 "Silver-Box" Evolution

As power demand per rack has increased, the traditional n+1 redundant, 220 V_{AC} to 12 or 48 V_{DC} rectifier or "silver box" architecture has changed. Initially, designs evolved using existing silicon FETs but with more advanced topologies. For example, a diode full-bridge rectifier followed by a hard-switching single or interleaved PFC boost has migrated to the soft-switching "bridge-less" or "totem-pole" designs that integrate AC-DC rectification with the PFC function. Secondary rectification also upgraded from Schottky diodes to fully synchronous with low-voltage FETs, but frequencies remained low, and so passive energy storage elements—transformers, chokes, output caps, EMI filters, etc.—remained large, heavy, and expensive. As silicon-based silver boxes reached their power density limit—at around 4 W/cc (70 W/in^3)[35]—and as achieving the efficiency requirement uses larger, more expensive silicon FETs, new materials are required to meet the growing demands of AI processing.

6.16.3 GaN: The Upgrade Challenge from Cellphone Chargers to AI Server Power

Since the first gallium nitride (GaN) charger was introduced in March 2018 [36] using Navitas' GaNFast™ power ICs, this wide-bandgap technology now dominates new designs from 25–240 W (the upper power limit of the revised USB-C specification). Today all the top-ten mobile OEMs are in production with GaN power ICs. While over 125 million GaN power ICs have been shipped—with excellent quality performance and a 20-year warranty—the harsher environment, power, and reliability expectations of higher-power markets like AI server, solar, and EV had created a tough "glass ceiling" for GaN.

The baseline "consumer" GaNFast power IC has been strengthened and enhanced to take on the challenge. Beginning with the package, the upgrade from 5 × 6 or 6 × 8 mm QFN to proprietary 10 × 12 mm TOLL and 10 × 15 mm TOLT increases mechanical robustness, thermal performance and enables a larger—i.e., more powerful—GaN die. The original, monolithic "GaNFast" integration of GaN FET and GaN driver acts as the foundation for application-specific features and functions for high power. Now, application-optimized GaNSafe™ [37] power ICs are the most protected GaN power devices in the world, with key enabling features and performance vs GaN discrete FETs, with their exposed weak gates and complex lagging protection circuits. For data center applications, double-pulse and short-circuit testing are critical, and discrete GaN devices display erratic voltage spikes, and catastrophic damage [35].

6.16.4 GaN Enables AI Server Power

Leading-edge GaN power IC technology is critical to develop 220 V_{AC} to 48 V_{DC} server power designs at 3200 W, 4500 W, and onwards to 10,000 W.

The common redundant power supply (CRPS) specification [39] for AC inputs are defined in the "Hyperscale Open Compute" project by participants including Facebook (Meta), Intel, Google, Microsoft, and Dell. The specification dictates that the AC-DCs be modular for flexibility of application and easy replacement/maintenance and must fit in a standard 1U (40 mm) × 73.5 × 195 mm form factor. Power output from each AC-DC ranges up to several kW and the units are arranged in a parallel redundant configuration to supply higher power and maintain system availability if one or more units fail. The initial AC-DC stage in this example generates a regulated 12 V or 48 V nominal (54 V maximum) which is bussed around server blades, feeding non-isolated Point-of-Load (PoL) or isolated converters which provide the precise supply voltage for CPUs, GPUs, and other system ICs.

GaN power ICs enable high-density, high-efficiency CRPS designs with low conduction and switching losses allowing high switching frequency and small passive components, plus application-required protection features. To demonstrate, in 2023, Navitas designed a CRPS185-compliant AC-54 V server power supply rated at 3.2 kW output, achieving ~100 W/in^3 (8.4 W/cc), which exceeded the EU's titanium efficiency requirement and per customer feedback, did so at a lower cost ($/W) than lower-power silicon systems.

6.16.5 Navitas CRPS-compliant Power Supply Design Approach

An outline of the Navitas reference design, designated CRPS185, is shown in Fig. 6.75. The AC rectification and power factor correction stages are amalgamated into a "bridgeless, totem-pole interleaved converter." This is equivalent to the traditional boost converter in function, producing a regulated DC output at typically

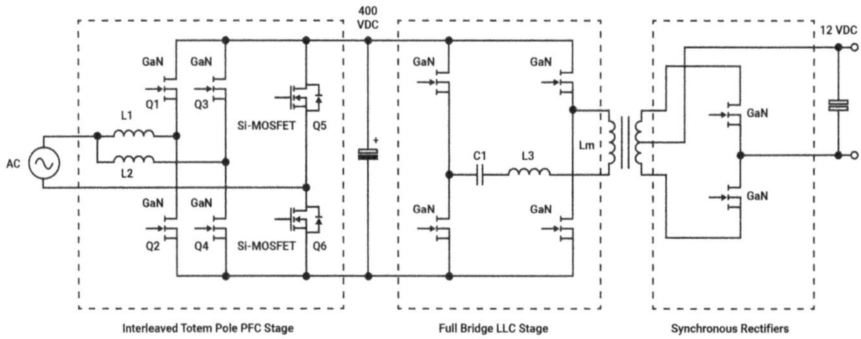

Fig. 6.75 The Navitas CRPS 3.2 kW reference design outline using GaNSafe power ICs

around 400 V_{DC}, while using pulse-width modulation to force input current to follow input voltage waveshape for a high power-factor. Q1 and Q2 form the switch and synchronous rectifier of a high frequency boost converter, with their functions swapping depending on the input AC polarity. Q5 and Q6 switch alternately at line frequency and provide a return path to the AC supply, so that a positive DC output is generated with both polarities of AC input. A second interleaved pair of boost switches Q3 and Q4 spread the current stress while reducing high frequency input ripple current, making EMI filtering easier.

Q5 and Q6 switching at 50/60 Hz have very low dynamic losses so can be silicon MOSFETs with no impact on efficiency, in some cost-sensitive designs they could be just silicon diodes. Q1–Q4 however are "hard" switched at high frequency, with potentially high dissipation from current/voltage overlap and any body-diode reverse recovery current. This is because the stage operates in "continuous conduction current mode" (CCM). The alternatives, discontinuous or boundary mode can be "soft switched" but are impractical at high power due to the higher peak currents causing stress and EMI. If silicon MOSFETs are used for Q1–Q4, losses are so high that the topology is impractical, but when GaN is used, efficiency dramatically improves from low conduction loss and absence of any reverse recovery in reverse or "third quadrant" conduction. GaN naturally reverse-conducts during forced commutation, but as this is through the channel, there is no body diode present. Low loss is maintained for high frequencies enabling specification of small magnetic components L1 and L2 as well as smaller EMI filter components (not shown).

The main, isolating DC-DC conversion stage is a "full-bridge LLC." LLC refers to the two inductors formed by the transformer magnetizing inductance Lm and L3, which is either a discrete inductor or sometimes deliberately introduced transformer leakage inductance. The "C" in LLC is C1 in the figure. The topology is a resonant one providing soft, low-loss switching, so its frequency can be pushed up to several hundred kHz allowing a small transformer to achieve high power density. The circuit operates with a full bridge of GaNSafe ICs, energizing the transformer at fixed PWM duty cycle and variable frequency with two "tank" resonances formed from the two inductors and capacitor. The switching frequency runs up and down the high frequency slope of the resonant curve and this varying impedance in turn increases or reduces the transferred voltage level to the transformer secondary. With feedback to the controller and its switching frequency, output voltage regulation is achieved.

In all positions, GaNSafe ICs contribute to high efficiency with their very low gate charge and consequently-low gate-drive power requirements along with low output capacitance C_{OSS} minimizing switching loss. Again, each device on the primary side sees no more than 400 V for a large margin away from their 650 V rating (750 V for transients).

As described above, silicon FETs are impractical for the bridge-less section. A study was done to focus on improvements of GaN power ICs over silicon FETs for the DC-DC portion only, as shown in Fig. 6.76. Silicon's lower efficiency eliminates the old technology from the high power densities required for the AI roadmap.

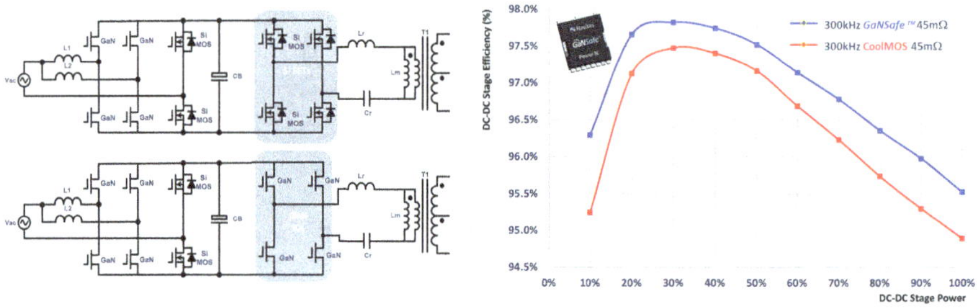

Fig. 6.76 Si to GaN IC comparison for 3.2 kW DC-DC stage

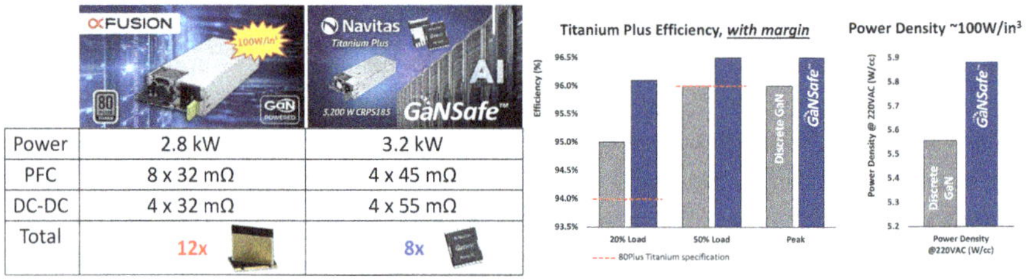

Fig. 6.77 CRPS185 comparison, discrete GaN vs GaN power IC

6.16.6 3.2 kW Results: Benefit of GaN Power ICs over GaN Discrete

However, the performance of GaNSafe power ICs (at 3,200 W) vs discrete GaN (at 2,800 W) can be evaluated using another reference design. As shown in Fig. 6.77, the GaNSafe IC design achieves almost 5.9 W/cc (100 W/in^3) density—14% more than the GaN discrete version, with margin over the Titanium efficiency specification, additional hold-up—and uses 1/3 fewer GaN components.

6.16.7 Roadmap to 4.5 kW, 10 kW0

As server power ships with GaNSafe ICs, we can also consider the longer-term roadmap for AI power. By considering another wide-bandgap material—silicon carbide (SiC)—we can revisit architectures to increase power density further. A new 4.5 kW CRPS185 system platform from Navitas' data center design center uses a hybrid GaN and SiC approach to push densities to 8.4 W/cc (138 W/in^3) and efficiencies over 97%, as shown in Fig. 6.78. Full details of the system design are available to qualifying customers including fully tested hardware, schematics, bill-of-materials, layout, simulation, and hardware test results. A new 8–10 kW

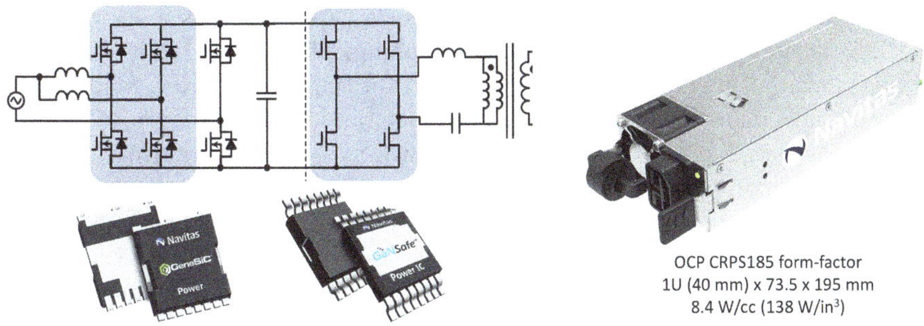

Fig. 6.78 Hybrid GaN + SiC approach (schematic, devices, form factor)

power platform is in development for the end of 2024 to support 2025 AI power requirements.

6.16.8 Conclusion

The aggressive power, and power-density demands of AI computing force the entry of new architectures and wide-bandgap semiconductor materials. GaN power ICs give a critical boost to both efficiency, and power density via high-frequency operation, to meet the CRPS standard for data center power supplies with some margin [38–41]. Devices such as the GaNSafe power ICs from Navitas save energy and associated circuitry costs, reduce system-level costs ($/W).

6.17 Foundry Services: Part 2, Wavetek

Maurizio Di Paolo Emilio

Compound semiconductors are composed of multiple elements, creating a mixture that inherently introduces defects due to thermodynamic considerations. Dealing with these defects and traps becomes a key element in the compound semiconductor industry. Managing and minimizing defects is crucial for achieving optimal device performance and reliability. This is in stark contrast to silicon, which is a single-element crystalline material that, while not perfect, allows for higher purity and better crystalline properties.

The challenge of growing compound semiconductor materials with the lowest defect density is a critical aspect of their manufacture. Achieving this involves skillful control over the growth process to mitigate defects and inconsistencies that can adversely affect the resulting devices. Additionally, the mismatch between different layers and materials within compound semiconductors can lead to issues like warpage and stress, further emphasizing the need for precise manufacturing techniques.

The consideration of material properties, such as spontaneous polarization in GaN, adds another layer of complexity. While imperfections are inherent in compound semiconductors, they can be harnessed for specific purposes. Maintaining material quality throughout the entire manufacturing process is essential for preserving the initial good properties of the grown material and ensuring that they translate into high-performance devices.

Lin's advice underscores the delicate nature of working with compound materials in semiconductor manufacturing. The emphasis on treating these materials gently and with meticulous control is crucial for achieving high-quality and high-performance devices.

Growing compound materials with precision and care is imperative to minimize the introduction of defects. While some level of defects is inevitable due to the nature of these materials, the goal is to keep them to a minimum to ensure optimal device functionality. This requires a deep understanding of the material properties, growth processes, and impact of defects on the final devices.

By prioritizing careful control and defect reduction during the growth and manufacturing processes, semiconductor manufacturers can enhance the overall quality, reliability and performance of compound semiconductor devices. This approach aligns with the complex and nuanced nature of working with compound materials and reflects a commitment to producing the best possible outcomes in semiconductor technology.

The key to successful compound semiconductor manufacturing lies in carefully managing defect density, layer composition, and doping. While simulation tools aid in predicting outcomes, each company's "recipe" for achieving optimal performance remains closely guarded. The analogy of a secret culinary recipe aptly illustrates the proprietary nature of these processes, with each manufacturer refining their methods based on years of experience and research.

Standardizing certain aspects, such as material stacking, can provide a foundation, yet the specifics of layer thicknesses and doping densities greatly influence device performance. The interplay between controlled processes and the need to maintain the best properties of the grown material is a delicate balance, showcasing the precision and expertise required in this industry.

Compound semiconductor manufacturing is a pursuit of excellence, with companies striving to minimize damage, maximize quality, and create devices that excel in today's demanding technological landscape. The competition and innovation within this field underscore the drive to push boundaries and deliver cutting-edge solutions.

Geopolitical

In the realm of compound semiconductor technologies, the creation of a thriving ecosystem is paramount for achieving critical mass and success. Such an ecosystem involves the synergy of various elements: from quality materials and advanced manufacturing processes to innovative design capabilities and effective packaging solutions.

This ecosystem concept underscores the intricate interplay between different components, each contributing to the growth and advancement of compound semiconductor technologies. Superior materials serve as the foundation for the journey, which manufacturing capabilities then harness. Design houses further contribute by creating applications that harness the unique properties of these materials.

"Notably, packaging plays a pivotal role in ensuring the final products meet performance expectations," Lin said.

The importance of international collaboration cannot be overstated, as it enables pooling of expertise, resources, and advancements. However, in the current geopolitical environment, countries like China are focused on gaining material sovereignty and developing self-sufficient ecosystems, which is resulting in a fragmentation of efforts.

The growth and success of compound semiconductor technologies, especially in emerging areas like WBG materials, rely on cohesive global efforts that transcend political boundaries. A thriving ecosystem, fostered by collaboration, investment, and strategic planning, is the foundation upon which the future of compound semiconductor innovation will be built.

The utilization of GaAs material and its relationship to gallium availability presents a significant aspect of this ecosystem. According to Lin, gallium, a critical component in GaAs growth, is a finite resource, and its usage is regulated by certain countries. As he pointed out, while some countries may have control over gallium quantity, the purity of gallium is overseen by different nations. This situation poses challenges in maintaining a consistent and pure supply of gallium for compound semiconductor growth processes.

Within this context, the market for GaA substrates is dominated by just three major suppliers, located in different countries. The significance of these substrates is amplified by their integral role in various electronic devices, especially in the booming smartphone market, which China dominates.

The need for a consistent and reliable supply chain becomes evident, as disruptions could potentially impact crucial industries and end users. This emphasizes the complexity of global supply chains and the need for strategic planning to ensure stable and robust material access.

Green energy

Companies across various sectors, including Wavetek, are setting energy-saving targets as part of their corporate responsibilities. This trend reflects a shared understanding that reducing energy consumption is not only economically beneficial but also essential for addressing environmental concerns and minimizing carbon footprints. These targets often guide companies in optimizing their energy usage, improving efficiency and minimizing waste.

Furthermore, the adoption of green energy is being embraced at both the corporate and governmental levels. According to Lin, Taiwan's efforts to transition toward renewable and green energy sources, such as solar power, align with global trends in mitigating climate change. While the integration of large-scale green energy solutions may have its challenges due to infrastructure limitations, there is a growing awareness of the need to incorporate sustainable practices wherever possible.

The commitment to energy-saving targets and the adoption of green energy practices are central themes in the corporate landscape today. Companies recognize their role in contributing to energy efficiency, environmental sustainability, and a greener future, both through internal initiatives and alignment with governmental efforts. While challenges remain, the collective commitment to these goals is a significant step toward creating a more sustainable and environmentally responsible business environment.

Reliability and packaging

Reliability is a critical factor for the successful adoption and integration of compound semiconductors like GaN and SiC into various applications.

As he noted, compound semiconductors are inherently more complex due to the presence of defects caused by lattice mismatch. However, advancements in material growth techniques and process control have played a pivotal role in mitigating these challenges. Over the past decade, significant progress has been made in growing high-quality GaN and SiC materials, addressing concerns related to reliability.

The initial skepticism surrounding the reliability of these materials has been gradually replaced by confidence, largely due to advancements in epitaxial growth techniques. Proprietary methods and recipes have been developed by various companies to grow GaN and SiC materials in a way that minimizes defects and optimizes performance. The improved understanding of strain management and lattice matching has contributed to the development of more reliable devices.

This evolution is reminiscent of similar challenges faced in the past, such as with pHEMTs and other technologies. Over time, these technologies matured, and their reliability improved as a result of refined growth techniques and manufacturing processes.

"With compound semiconductors like GaN and SiC, the emphasis on reliability is crucial to meet the demands of various industries, including automotive, telecommunications and power electronics," Lin said. "As these materials find their way into more applications, their reliability will continue to be a key factor determining their long-term success."

The inherent characteristics of WBG materials, such as high-voltage and high-power capabilities, necessitate specialized packaging solutions that differ significantly from those commonly used in the silicon industry.

Unlike silicon devices that often operate at lower-voltage levels, WBG devices require packaging that can handle much higher voltages without causing electrical breakdown or arcing. This demands a complete redesign of packaging structures and layouts. Furthermore, the different form factors and pin-to-pin spacings of WBG devices introduce challenges in ensuring proper insulation and preventing electrical sparking.

According to Lin, the absence of standardized packaging solutions in the WBG field reflects the current stage of its market development. The industry is currently in a rising phase of the "hype cycle," wherein companies are focusing on capturing higher margins and establishing their proprietary packaging designs. This reluctance to standardize packaging is driven by the competitive landscape and the pursuit of optimal profit margins.

References

1. P. Porambage, M. Liyanage, Evolution of mobile networks, in *Security and Privacy Vision in 6G: A Comprehensive Guide*, (IEEE, 2023), pp. 1–14. https://doi.org/10.1002/9781119875437.ch1
2. P. Launiainen, The golden age of wireless, in *A Brief History of Everything Wireless*, (Springer, Cham, 2018). https://doi.org/10.1007/978-3-319-78910-1_4
3. Infineon Technologies AG, Wireless Charging: A Guide to the Qi Standard. Neubiberg (Germany): Infineon Technologies AG; 2023. Document number: B189-I1417-V1-7600-EU-EC. Date: 07/2023. Available from: https://www.infineon.com/cms/en/product/gated-document/infineon-whitepaper-optiga-trust-charge-8ac78c8c8929aa4d0189966608227abe/
4. History of the Qi Specifications, Wireless Power Consortium. Available from: https://www.wirelesspowerconsortium.com/knowledge-base/specifications/history-of-the-qi-specifications/
5. T. Shi, P. Wiener, *High power Constant Current Class EF2 GaN Power Amplifier for AirFuel Magnetic Resonance Wireless Power Transfer Systems* (GaN Systems Inc., Ottawa, 2018) Source: https://gansystems.com/wp-content/uploads/2018/10/Class-EF2-PCIM_2018_Manuscript.pdf
6. Magnetic Resonance Charging Standard For Multi-Device Charging, AirFuel. Available from: https://airfuel.org/airfuel-resonant/
7. X. Lu, P. Wang, D. Niyato, D. Kim, Z. Han, Wireless Charging Technologies: Fundamentals, Standards, and Network Applications. IEEE Communications Surveys & Tutorials **1–1** (2015). https://doi.org/10.1109/COMST.2015.2499783
8. https://www.allaboutcircuits.com/news/gan-goes-green-three-ways-gan-is-good-for-the-planet/
9. https://gansystems.com/newsroom/2023-power-semiconductor-predictions/
10. https://www.infineon.com/dgdl/Infineon-Gallium_nitride_A_full_GaN_solution_for_high_power_density_chargers_and_adapters-Whitepaper-v01_00-EN.pdf?fileId=8ac78c8c7fb5929e017fb70ab8d5065f&da=t&_ga=2.2719670.723884135.1652167878-79998051.1617881807
11. https://seloreglobal.com/blogs/selore/gan-usb-c-chargers-powering-the-future-efficiently
12. https://www.researchnester.com/reports/gan-adapters-market/4249
13. https://www.eenewseurope.com/en/first-pfc-and-hybrid-flyback-combo-for-gan-based-usb-c-chargers/
14. https://www.infineon.com/cms/en/product/promopages/driving-gigantic-performance/
15. https://www.infineon.com/cms/en/product/power/gan-hemt-gallium-nitride-transistor/
16. https://www.persistencemarketresearch.com/market-research/gan-adapters-market.asp
17. https://www.transparencymarketresearch.com/ganbased-power-supply-adapter-market.html
18. https://www.gdgtme.com/features/gan-technology-for-charging-advancements-in-power-delivery/
19. https://ventiontech.com/blogs/technology-overview/advantages-of-using-gan-chargers
20. US Department of Energy, Wide bandgap semiconductors: Pursuing the promise. https://www.energy.gov/eere/amo/articles/wide-bandgap-semiconductors-pursuing-promise
21. G. Bueno Mariani, E. Alfawy, F. Wernegger and J. Gade, "Wide Band Gap Inverters and High-Frequency Effects on Motor Drives," PCIM Europe 2023; International Exhibition and Conference for Power Electronics, Intelligent Motion, Renewable Energy and Energy Management, Nuremberg, Germany, 2023, pp. 1-6, https://doi.org/10.30420/566091301
22. https://www.infineon.com/dgdl/Infineon-Wide_bandgap_switches_in_motor_drives_systems-Whitepaper-v01_01-EN.pdf?fileId=8ac78c8c8929aa4d0189cf9535897104&da=t
23. GaN Electronics Patent Landscape Analysis 2023, KnowMade, November 2023., https://www.knowmade.com/patent-analytics-services/patent-report/gan-electronics-patent-landscape-analysis-2023/

24. GaN electronics patent landscape 2023: A unified picture of the IP competition for power GaN and RF GaN technologies, December 2023., https://www.knowmade.com/technology-news/press-release/a-new-gan-electronics-ip-report-providing-a-deep-dive-into-the-power-gan-and-rf-gan-electronics-patents-worldwide/
25. Next-generation power devices: Status of the IP competition for vertical GaN power devices, KnowMade, November 2023., https://www.knowmade.com/technology-news/semiconductor-news/power-electronics-devices-news/the-ip-competition-for-vertical-gan-technology-has-just-started/
26. Eric Persson: Optimizing PCB layout for HV GaN power transistors. IEEE Power Electronics Magazine, Vol. 10, No.2 June 2023.
27. Juncheng (Lucas) Lu: Paralleling GaN E-HEMTs in 10 kW–100 kW Systems, 2017 IEEE Applied Power Electronics Conference and Exposition (APEC)
28. https://dieselnet.com/standards/cycles/wltp.php
29. Gallium nitride technology in server and telecom applications
30. New PCB Winding "Snake-Core" Matrix Transformer for Ultra-Compact Wide DC Input Voltage Range Hybrid B+DCM Resonant Server Power Supply
31. Reliability and qualification of high-voltage CoolGaN™ GIT HEMTs
32. Why GaN will be key to feeding power-hungry 5G networks
33. Towards Efficient, Reliable, and Cost-Effective Power Supply Units for 5G Telecom Rectifiers Applications Using Wide-Bandgap Semiconductors
34. Charter Equity Research, March 2024
35. Jensen Huang, NVIDIA GTC2024
36. Performance Evaluation of Bridgeless PFC Boost Rectifiers, Huber, Jang, and Jovanović, Delta Products Corporation, APEC 2007, https://ieeexplore.ieee.org/document/4195714
37. Navitas survey of industry CRPS185 silver boxes.
38. "Mu One", 45W USB-C, March 2018 using 2x GaNFast power ICs in active-clamp flyback, https://navitassemi.com/navitas-ganfast-enables-worlds-thinnest-travel-adapter/
39. GaNSafe Power ICs Create Highest Density and Efficiency in Data Center and EV Power Systems", SEMICON 2023, https://navitassemi.com/download/gansafe-power-ics-create-highest-density-and-efficiency-in-data-center-and-ev-power-systems/?wpdmdl=987500691&ind=987500692
40. "Electrify Our World" with Next-gen GaNFast and GeneSiC Power, Navitas, APEC 2024, https://navitassemi.com/download/electrify-our-world-with-next-gen-ganfast-and-genesic-power/?wpdmdl=987501733&ind=987501726
41. OCP CRPS specification "https://www.opencompute.org/documents/m-crps-r1-v1p0-rc4-pdf"

Nihit Bajaj After completing his masters in Electrical Engineering from the Arizona State University (USA) in 2009, Nihit Bajaj spent 12 years in several IC design leadership roles at Qualcomm, Appulse Power, and Silanna Semiconductor. Since 2021, Nihit has been at Infineon Technologies, initially serving as Director of Marketing for SMPS Applications, and currently as Director of Product Marketing in Infineon's new GaN business line.

Daniel Makus Application and Business Development Director for xEV power conversion including OBC and DCDC applications. Responsible for market research, roadmap development and G2M strategy in xEV applications with a focus on power conversion.

Rafael Garcia System Architect for onboard chargers and DCDC applications for electric vehicles. He focuses on the impact of the major trends in the selection of the different semiconductor technologies. Rafael also works on identifying the key drivers that will lead the next generation of products, thus accelerating the decarbonization of the application.

Igor Spinella is the founder, CEO, and CTO of Eggtronic, a leading global company in the fields of power electronics and wireless power. Igor was born in Modena, Italy and studied at the University of Modena and Reggio Emilia where he earned a bachelor's and master's degree in Mechatronics, Robotics, and Automation Engineering and a PhD in Mechanics and Automotive Design & Technology. Before founding Eggtronic, he spent a brief time as a university researcher and worked at some of Italy's famed luxury car manufacturers.

Paul Wiener leads strategic marketing efforts for Infineon's GaN business line. With 25 years of electronics industry experience in operations, sales and marketing, and business development, Paul has a proven track record leading successful new technology positioning and commercialization strategy and execution. Prior to joining Infineon, Paul was part of the leadership team at GaN Systems as Vice President of Strategic Marketing for 8 years. Prior to that, Paul led the power magnetics business unit at Eaton. He was Vice President of Sales for Fultec Semiconductor (now Bourns) and held several management roles at Genoa, BroadLogic, and Raychem. Paul has an MBA from Golden Gate University and a bachelor's degree in business from UC Berkeley.

Junyang Luo received M.Eng from Center for Power Electronics, Dept. of Electrical Engineering, National University of Singapore in 2001. Joined Infineon AP in 1999 as Application Engineer and following as AE manager to support business of Power Management & supply including AC/DC PWM ICs, PFC ICs, Power Discrete. Main responsibilities are to define and develop solutions for SMPS applications, support new product evaluation in SMPS platform, and define the new product requirement based on the application trend. In 2020, he was appointed as System Architect for Charger/Adapter to lead technical application activities focusing on Charger/Adapter, and consolidate and define the application roadmap to guide the new product development and business strategy.

Matthias Kasper received the M.Sc. and Dr.Sc. degrees in electrical engineering from the Swiss Federal Institute of Technology (ETH) Zurich, Switzerland, in 2011 and 2016, respectively. During his studies, he focused on power electronics, mechatronics, and energy systems. In his Ph.D. at the Power Electronic Systems Laboratory at ETH Zurich, he dealt with multi-cell converter systems for different applications, e.g., PV module integrated DC-DC converters, auxiliary supplies for high step-down ratios, and a highly efficient telecom rectifier module. Since January 2017, he is part of the Systems Innovations Lab of Infineon Technologies Austria AG, where he works on novel topologies, control schemes, and multi-objective optimization routines. He has authored and co-authored more than 20 scientific publications and holds 12 international patents.

Guilherme Bueno Mariani holds a Double Engineering degree in electrical engineering from the Grenoble Institute of Technology, Grenoble, France, and Universidade Estadual Paulista, Guaratingueta/SP, Brazil, which he received in 2012. In 2016, he obtained his Ph.D. in electrical engineering from the University of Grenoble Alpes, Grenoble, France.During the period from 2016 to 2022, Dr. Bueno Mariani worked as a research engineer focusing on motor drives at Mitsubishi Electric Research Center Europe in Rennes, France. Since 2022, he has been an Application Engineer with the PSS division at Infineon Technologies in Villach, Austria. His current research interests include solar energy and motor drives technologies.

Hubie Noto joined Silanna Semiconductor in November 2021 with over 36 years' experience in power management industry. Previously he held positions as Director of Marketing at Power Integration, Director of Marketing at iWatt/Dialog Semiconductor, Director of Marketing of ACDC and Lighting at Monolithic Power Systems, founder of Lumina International, and worked for Marvel for 6 years as General Manager of Power Management Division. Before joining Marvell, he held positions in such prestigious companies as Siemens, Motorola, and Micrel Semiconductor. He holds a bachelor's degree in electrical engineering from the University of Nebraska-Lincoln and has more than 26 patents.

Thierry Bouchet Doctor in Electrical Engineering in 2001 and worked for Ion Beam Services (IBS) and Atmel, two semiconductor companies. In 2011, Thierry created a fabless startup (ADIS) in the area of power electronics for niche market. After working as Strategic Marketing manager for energy and power electronics at CEA LETI, he launched a new startup: Wise Integration, the ultimate power supply miniaturization solution using GaN Integrated circuits.

Peter Vaughan is the Director of Automotive Business Development at Power Integrations and has over 30 years of power electronics experience. Most recently, Peter was Senior Director of DC power at ChargePoint developing power conversion solutions for ultra-high speed EV charging. He holds a BEng Hons Degree in Electrical and Electronic Engineering from Heriot-Watt University, UK.

Rémi Comyn holds a PhD in electrical engineering and physics from the University of Sherbrooke and Côte d'Azur University. For more than 5 years, Rémi has been working with KnowMade as a patent and technology analyst, unveiling business, IP and R&D strategies of players engaged in the development of compound semiconductors for power and RF applications. Previously, Rémi worked for several years as a research engineer, focusing on gallium nitride (GaN) technology.

Dieter Liesabeths is a senior management leader in the semiconductor industry with over 30 years of experience in sales, marketing, business development, new technology introduction, and new product definition. In his current role as SVP of Product at VisIC Technologies, he is building the new products for GaN-on-Silicon for the power inverter of battery electric vehicles (BEV) and transforming the industry to a sustainable future. Additionally, he owns the company DL Consulting where he serves as an industry expert in the wide-bandgap community.

Prior to consulting with VisIC Technologies he worked for 10 years at Wolfspeed GmbH as a Senior Sales & Marketing Director and built first the industrial and later the automotive department with an automotive application center. Dieter drove the industry's conversion from Silicon to wide-bandgap (SiC), by building an eco-system to support the fast-switching wide-bandgap devices and partnering with industry leaders for creating a multi-billion USD business.

Dieter holds an Dipl.-Ing. (TH) from the University of RWTH Aachen, Germany.

Dima Novo serves as the VP of R&D at VisIC Technologies, where he leads multi-disciplinary teams in the design, manufacturing, and testing of GaN power chips, devices, modules, and system-level technology demonstrators. These efforts are aimed at enabling both off-board and onboard EV charger systems and power inverters for the automotive industry. Dima brings a wealth of experience from his tenure in engineering and management roles within automotive start-ups and companies in the green energy and engineering sectors. He holds a B.Sc degree in Electrical Engineering from Ben Gurion University in Israel.

Denis Marcon received a M.S. degree from the University of Padova in 2006. In 2011, he received the degree of Doctor in Engineering (Ph.D.) from the Catholic University of Leuven and Imec with a thesis entitled "Reliability study of power gallium nitride-based transistors." Denis is a leading author or co-author of more than 50 journal papers or international conference contributions.

After his Ph.D. graduation, Denis has been leading projects aiming to develop GaN HEMTs for several applications (RF and power switching). Thereafter, he joined the business development team of Imec where he was directly responsible for the partnerships with Imec in the field of GaN power electronics as well as on dedicated development and manufacturing of Si-based devices, MEMS, sensors, and micro-systems.

Denis is the General Manager of Innoscience Europe (a subsidiary of Innoscience), and he is directly responsible for the Innoscience's GaN business and marketing in Europe.

Francesco Di Domenico is a Senior Principal Application Engineer at Infineon Technologies Austria AG. He is a System Architect in the Power & Sensor Systems (PSS) Division at Infineon; he is a specialist in high-power switched-mode power supplies (SMPS) for servers/data centers, telecom, EV charging, and general industrial applications. His competencies include server and telecom power architectures, and, more in general, the design of power converters, magnetic components, and control concepts, with special focus on resonant/soft switching topologies. Before joining Infineon, he was the R&D Lead Engineer in the professional SMPS industry in Italy.

Alex Zahabizadeh is a project manager in Texas Instrument's GaN business, and is involved with GaN product development and product execution. With TI, Alex has worked in several marketing and engineer roles supporting several TI product portfolios including TI's interface logic products and industrial GaN power devices. Alex joined TI in 2017 after completing a BS in Electrical Engineering with a concentration in semiconductors from The University of Texas at Dallas.

Stephen Oliver is VP Corporate Marketing & Investor Relations at Navitas Semiconductor, with over 30 years of experience in the power semiconductor and power supply industries worldwide.

Prior to Navitas, Stephen served as VP of the VI Chip product line at Vicor, and earlier at Motorola, Philips (NXP), and International Rectifier (Infineon). He holds a B.Eng (Hons) in Electrical & Electronic Engineering from Manchester University (UK) and an MBA in Global Marketing & Strategy from the UCLA Anderson School of Management. Stephen holds several patents in power semis, is a Chartered Engineer and has been VP, President, and Chairman of the Power Source Manufacturers' Association (PSMA).

Maurizio Di Paolo Emilio holds a Ph.D. in Physics and is a Telecommunications Engineer. He has worked in various international projects in the field of gravitational waves research, designing a thermal compensation system (TCS) and data acquisition and control systems, and on others about x-ray microbeams in collaboration with Columbia University, high voltage systems and space technologies for communications and motor control with ESA/INFN. TCS has been applied to the Virgo and LIGO experiments, which detected gravitational waves for the first time and earned the Nobel Prize in 2017. Since 2007, he has been a reviewer for scientific publications for academics such as Microelectronics Journal and IEEE journals. Moreover, he has collaborated with different electronic industry companies and several Italian and English blogs and magazines, such as Electronics World, Elektor, Mouser, Automazione Industriale, Electronic Design, All About Circuits, Fare Elettronica, Elettronica Oggi, and PCB Magazine, as a technical writer/editor, specializing in several topics of electronics and technology. From 2015 to 2018, he was the editor-in-chief of Firmware and Elettronica Open Source, which are technical blogs and magazines for the electronics industry. He participated in many conferences as a speaker of keynotes for different topics such as x-ray, space technologies, and power supplies. Maurizio enjoys writing and telling stories about Power Electronics, Wide-bandgap Semiconductors, Automotive, IoT, Embedded, Energy, and Quantum Computing. Maurizio has been an AspenCore content editor since 2019. He is currently editor-in-chief of Power Electronics News and EEWeb and a correspondent for EE Times. He is the host of PowerUP, a podcast about power electronics, and the promoter and organizer of the PowerUP Virtual Conference, a summit where each year great speakers talk about the power electronics design trends. Moreover, he has contributed to a number of technical and scientific articles, as well as a couple of Springer books on energy harvesting and data acquisition and control systems.

Chapter 7
Challenges and Future Trends

Patrick Le Fèvre, Geoff Haynes, Kennith Kin Leong, Vladimir Odnoblyudov, Cem Basceri, and Han Wui Then

Authored by industry experts including Patrick Le Fèvre, Geoff Haynes, Kennith Kin Leong, Vlad Odnoblyudov, Cem Basceri, and Han Wui Then, this chapter offers a multifaceted examination of GaN technology. The introductory passage by Patrick Le Fèvre sets the stage for discussing the transformative impact of WBG semiconductors, highlighting the excitement and challenges inherent in learning and implementing these cutting-edge technologies. Geoff Haynes explores the historical progress and future trajectory of GaN technology, tracing its evolution from early transistors to its potential resurgence in high-power applications.

Kennith Kin Leong discusses the groundbreaking development of GaN monolithic bidirectional switches (M-BDS) and their potential to revolutionize power electronics, shedding light on their advantages and implications for various applications. Vlad Odnoblyudov and Cem Basceri focus on advancing GaN power technology to new heights, detailing the scalability achieved through innovative manufacturing platforms and their impact on device performance and manufacturing costs. In the paragraph 7.6, Han Wui addresses critical challenges in GaN technology, emphasizing the importance of achieving cost-effectiveness while maintaining performance, reliability, and integration with other technologies like silicon CMOS.

P. Le Fèvre (✉)
POWERBOX (PRBX), Stockholm, Sweden
e-mail: patrick.le-fevre@prbx.com

G. Haynes
Consultant, Poole, UK

K. K. Leong
Infineon Technologies, Villach, Austria

V. Odnoblyudov · C. Basceri
QROMIS, Inc., Santa Clara, CA, USA

H. W. Then
Intel Foundry Technology Research, Intel Corporation, Hillsboro, OR, USA

© The Author(s), under exclusive license to Springer Nature Switzerland AG 2024

M. Di Paolo Emilio (ed.), *GaN Technology*,
https://doi.org/10.1007/978-3-031-63238-9_7

Overall, this chapter provides a comprehensive overview of the challenges and future trends in GaN technology, offering insights into its transformative potential and the efforts underway to overcome hurdles and drive its widespread adoption.

7.1 Learning a New Technology Is Fun But Always a Challenge

Patrick Le Fèvre

For a power designer any new technology that makes it possible to improve performance while simultaneously making products smaller and more energy efficient is a very exciting concept—we are talking holy grail territory. Over the last century, the world of power electronics has witnessed many inventions and innovations and without going back to the Thyratron, the latest major one was the move from analog to digital control. However, we are now witnessing a new, huge stride forward in technology, the implementation of wide-bandgap (WBG) semiconductors. Gallium nitride and silicon carbide have been used in radio power amplifiers and high voltage diodes for years, but it was only a few years ago that they have become part of the power switching chain in the form of transistors. Adopting a new technology is full of challenges that somewhat surprisingly are not always technical. Learning is an important part of the road to success but market adoption and building a new ecosystem are far more complicated than it may at first seem. Let us take a snapshot of where GaN currently stands and what are the remaining challenges.

For all new technologies, Time-to-Market is inevitably a long process, and from original research, patenting, technology introduction, and market adoption, this could be more than 10 years. We are all aware of the camel-back curve (Fig. 7.1) and if many of us belong to the Technology Enthusiasts category, the success of a new technology will come from the pragmatists and conservatives. Introduced in 2005, digital control in power supplies is still considered by skeptics to be a curiosity. And undoubtedly, in normal circumstances, it would have been the same for the adoption of WBG, but ironically an unexpected event by the name of COVID may have contributed to the speed of the learning and implementation processes.

For many technology analysts, the cornerstone of WBG took place at the Applied Power Electronics Conference (APEC) 2018 when "challengers" demonstrated the commercial potential of WBG technology. It is not possible to name all of them but among the leaders promoting GaN, I would say that the efficient power conversion (EPC) idea to implement GaN in Light detection and ranging (LIDAR) was really interesting, especially with that technology becoming preponderant in the new generation of motor vehicles. It was also the year in which USB adapter manufacturers started to consider implementing WBG. Navitas is another example of an innovative company that in the early days pushed GaN integration to a higher level by packaging drivers and switches on the same substrate.

The general enthusiasm for WBG was real, and we estimate that more than 60% of power supply companies paid attention to the technology, with a decent

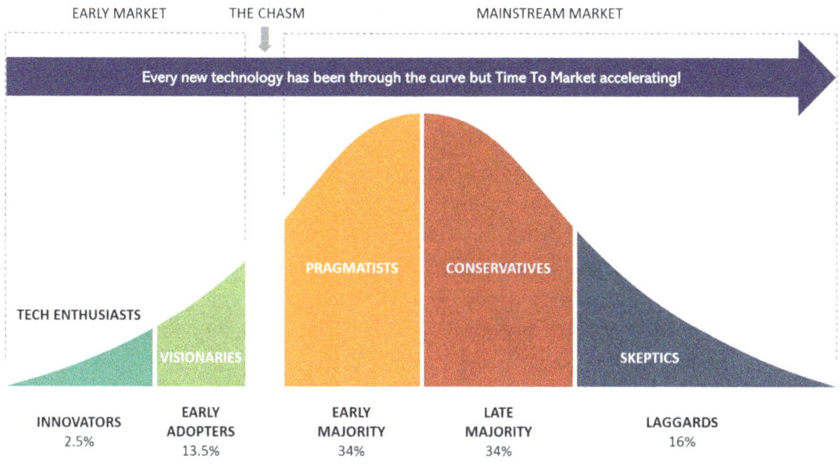

Fig. 7.1 Experienced power designers have crossed that technological chasm many times and GaN adoption follows the same pattern. (Source: PRBX)

proportion beginning evaluation and prototyping when the pandemic struck. In a matter of weeks, travel restrictions, curfews, work-at-home, and other rules changed our way of working. When that happened, the power electronics component manufacturers quickly adjusted to this new situation, increasing the number of online webinars, adapting R&D resources unable to work at the office to operate in the areas of technical support, and contributing to online content. In fact, some semiconductor manufacturers shipped ten times more evaluation kits than usual, and the number of WBG new product announcements was truly impressive.

There is no doubt that the pandemic created many issues and dramas but it also created a unique opportunity for companies to give time to their power designers to learn about WBG and even to run online simulations, thus shortening time for prototyping and adoption when back in the office.

It is hard to put an exact figure on how much the conditions brought about by the pandemic have contributed to accelerate the power designers' learning of WBG, but some estimate that within 2 years, power designers learned as much as they would have in 8 years under normal conditions. Moving slowly out of the pandemic, this situation is confirmed by the number of new products being launched, with the big winner in the low power segment being nomad power solutions, e.g., USB chargers.

7.1.1 Early Adopters Opened the Way to Volume

When first presented, WBG power semiconductor utilization was limited by the number of drivers available, making it difficult for power designers to consider the technology. Also, new technologies are always questioned regarding reliability and sustainability. Market adoption depends on how simple it is for power designers used to conventional MOSFETs to use WBG, and semiconductor manufacturers' speed in developing "ready-to-use" solutions that include driver, protection, monitoring, and many other functionalities into a single chip. This not only simplifies implementation but also reduces the overall size of the power stage, and combined with higher switching frequencies make it possible to reduce the size of magnetics, thus increasing power density while reducing the overall volume and mass of the power supply.

Among the many products we could name are three examples that highly integrate the power-train and control, contributing to packing more power into a smaller space. As end-users we all expect USB chargers to deliver more power, to charge faster, to be lighter, and industrial power supplies to follow the same pattern.

Already in 2020, this wish became a reality and one example of the benefit of using WBG GaN to achieve that is a 110 W Mini fast charger that is over 12 times smaller than the 96 W charger supplied with the Apple MacBook Pro 16 launched by OPPO. This has been made possible by combining the Navitas GaNFast power ICs with a planar-transformer, an optimized topology and a higher switching frequency. At the same time, efficient power conversion released a GaN IC integrating everything to make it simple for power designers to implement in to their new designs. In the path of the consumers segment, power electronics designers implemented similar technology (GaN + Planar) for industrial power supplies, making it possible to offer very compact power solutions that are easy to integrate into small space environments (Fig. 7.2).

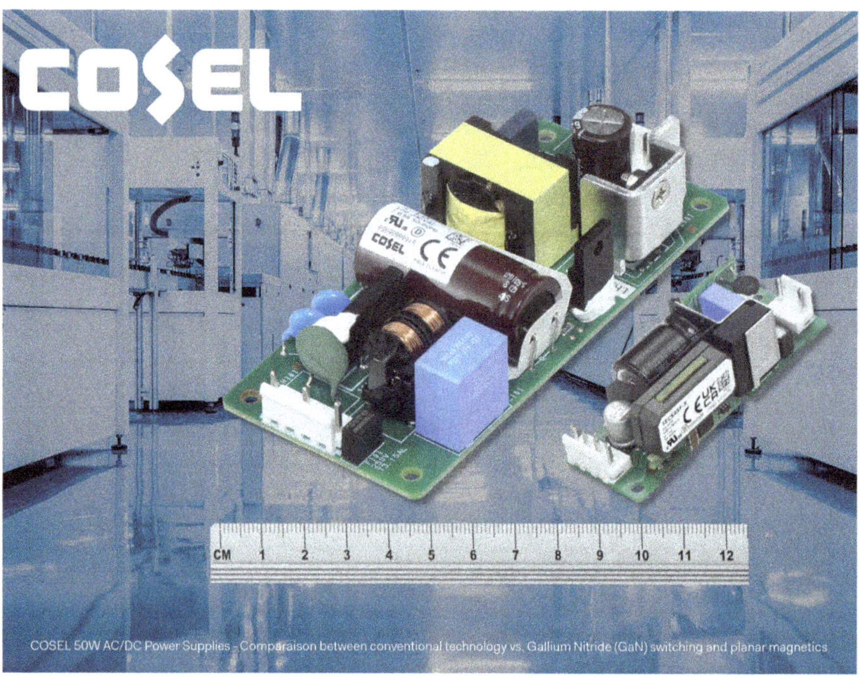

Fig. 7.2 COSEL industrial power supplies adopting GaN and integrated magnetics make it easy to integrate in small space environment. (Source PRBX/COSEL)

Those examples illustrate how WBG GaN manufacturers rapidly moved from "complex" to "simple" to implement the technology, contributing to generate volume and market adoption.

7.1.2 High Power GaN Setting-up Foundation for Future

As we have seen, driven by the consumer segment, power designers soon realized the benefits offered by GaN to offer more power in smaller packaging. Power designers had to face a number of challenges to develop high switching frequency using GaN technology in very compact packaging. It was not easy but somewhat simpler than the challenges faced by high-power applications such as multikilowatt power supplies for data centers or Electric Vehicles (EV).

Switching faster at higher frequencies and achieving higher efficiency is what we are all aiming for, and of course not all power supplies will require megahertz switching, but the range of applications benefiting from WBG is significant. By combining digital control, WBG, and a higher switching frequency, it is possible to reduce the size of the front end power supply to house the battery backup in the same volume. Power supplies and WBG semiconductor manufacturers have already presented a number of case studies promoting the benefits of that combination, and there are already products on the market developed for such applications. For example, the concept of decentralized energy storage in industrial applications is growing and while there is no doubt that will contribute to the implementation of high switching topologies using GaN, selecting the right supplier of GaN and associated driver is a concern for many power electronics designers. As presented by the market analysts Yole in a recent study (Fig. 7.3), despite GaN having been on the market for several years the market remains fragmented with each GaN manufacturer offering different combinations of products and services addressing specific segments.

To get the best out of GaN, power designers must work in close cooperation with semiconductor manufacturers and embrace one-stop solutions (GaN transistor, driver, protection, etc.) tightened to a single source, albeit raising concerns about the risks of using products from a new supplier with limited history and financial background.

Crossing the chasm is always a risk but also an opportunity to win new business ahead of the competition. Here again it is impossible to name all the suppliers, but it is interesting to note that GaN Systems and Transphorm were selected by many power supply companies to power the next generation of telecom and renewable energy applications, with both companies being acquired by market leaders, respectively, by Infineon and Renesas, securing long term business for the combined portfolios.

As we move forward in the adoption curve, as it has been for all previous technology evolutions, the market will consolidate and we will continue to see acquisitions and extended partnerships to strengthen market positioning and technology leadership. That is without mentioning the critical importance of protecting Intellectual Property Rights (IPR) that in emerging technologies is a "must do" process.

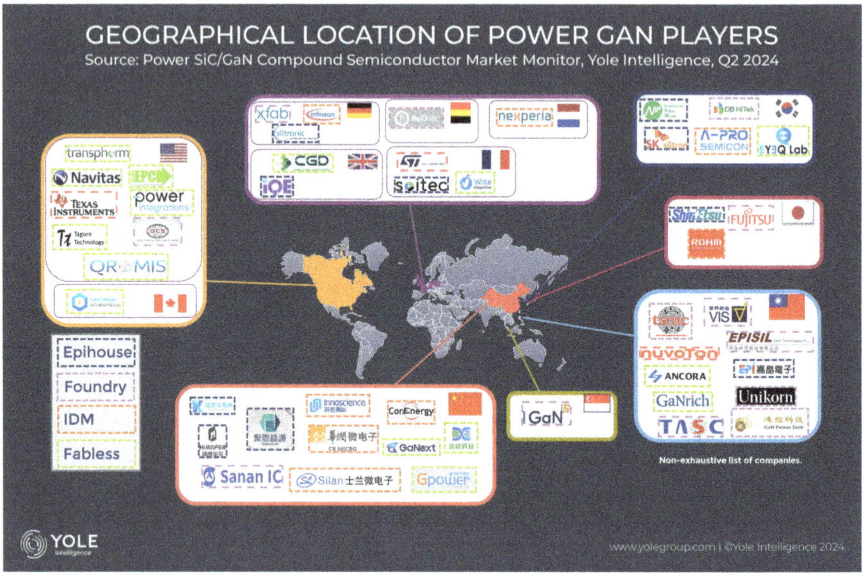

Fig. 7.3 Overview of the Power GaN players ecosystem in 2023. (Source: Yole Group, image courtesy of Cem Basceri)

7.1.3 Conclusion

Many of the challenges faced by power designers when GaN was presented at APEC 2018 have been overcome, but at the same time that GaN reached this new level of maturity, conventional MOSFETs also improved in performance. Out of any marketing buzz, power designers will always perform a technology assessment to select the most appropriate power solution and both technologies will coexist for a long period.

The main challenge might not be technical and in many applications, such as USB adapters, GaN is already perceived as a real benefit by end-users, but in a number of industrial applications and transportation such as railways it is still considered as being "too new!" The remaining challenge will be to build trust and confidence with the conservatives, which I have no doubt will happen.

7.2 Innovations in GaN

Geoff Haynes

A historical view of the progress of compound semiconductor technology and applications seen from my personal perspective working at GaN Systems and subsequently as Industrial Champion for the UK EPSRC Power Electronics Research Centre helps to explore likely future trends and innovations for GaN transistors, technologies, and applications.

7.2.1 *Early GaN Transistors*

GaN Systems' first practical GaN transistor was developed on a mature RFMD, RF GaN on SiC process as early as 2012. It could be switched in tens of nanoseconds through more than 1500 V at 25 A and was eminently scalable. The obvious advantage was switching speed. It had a "quasi-vertical" structure that combined many small, normally on, HEMT transistors sharing a common gate. All the resulting 'island' transistors are connected in parallel through two copper layers: one that consolidates the source connections at the substrate level using through-silicon vias, and another that unifies the drain connections on the wafer's top surface. Could this perhaps re-emerge now as an important element of the future for high-power, high voltage, GaN transistors?

7.2.2 *SiC Takes the High-Power Ground*

Unfortunately, at that time, the market believed it needed normally off transistors and the benefit of a silicon substrate to allow low-cost manufacture of GaN power transistors in legacy CMOS wafer fabs. That set things back perhaps 3 years until normally off GaN on silicon transistors dominated the 20 finalists in the Google Little Box Challenge. Luckily, during that time GaN Systems were privileged to co-develop an optimum transistor structure working closely with TSMC in the development of their production GaN on silicon process. Sadly, the high operating voltage and better thermal relief of the SiC substrate had to be sacrificed to the cause. Acceptance of the technology in production systems was then delayed by further 2 years by the necessity to qualify the production process and prove long-term reliability of the novel element of the GaN transistor, its HEMT structure.

In the meantime, the SiC transistors' advantages of improved specific on resistance, switching high voltages at high ambient temperatures in high-power converters using conventional circuits, packages, and system architectures focused the power industry's design attention. The EV market's attraction to 800 V batteries for reduced operating and charging current have recently created a massive pull for the technology, accelerating manufacturing volume, cost, and maturity of supply.

In the last 2 years, GaN on silicon transistors have finally started to be adopted in high volume in production of low power applications. First as the obvious solution to extremely power dense, high-frequency resonant switched power supplies for laptops. Now, they are being increasingly adopted in novel designs for PFC, LLC, and synchronous converter elements of larger power supplies switching at a few hundred kHz, where power density and efficiency are critical. Applications in DC-DC converters for electric vehicles now promise rapid growth for higher power GaN transistors used in resonant switching circuits operating at much higher frequencies.

7.2.3 What Comes Next for the GaN Transistor?

What happened to the early promise of our high-voltage quasi-vertical "Island" transistors? Could the chains of the silicon substrate now be released? It is the conducting silicon substrate that constrains the GaN transistors operating voltage to a few hundred Volts, not the breakdown of its fantastically fast switching, HEMT channel. It is the crystal interface mismatch between the GaN and its silicon substrate that requires a thick substrate to avoid physical strain in the wafer during manufacture. It is the poor thermal conductivity of that thick silicon substrate that makes cooling the transistor such a challenge. With a thinner, more thermally efficient, isolating substrate, the GaN transistor comes alive. Early exploration of alternative substrates or even lift-off processes for Si transistors formed on a silicon wafer are clearly now due to be re-visited along with further optimizations of the interface between the GaN transistors on a silicon substrate. The massive demand for SiC wafers for the EV market will have driven down the cost of SiC as a GaN substrate and GaN on SiC processes themselves have matured in RF applications that are also growing rapidly in communications markets. Perhaps its advantages as a substrate for GaN transistors will re-emerge as the increasing market pull for higher, faster switching transistors.

7.2.4 High Speed Switching Brings Its Own Multi-Physics Design Challenges

GaN Systems recognized that matching the transistor design to the transistor's packaging more effectively in terms of thermal dissipation, current carrying capability, and low inductance switching path is critical to switching at higher speeds. The GaN Systems embedded "GaNpx" single transistor package developments went a long way in providing solutions to customers willing to adopt design and manufacturing change.

This optimized embedded packaging has enabled dramatic power density gains in DC-to-DC applications where soft (resonant) switching allows the transistor to be switched when no current is flowing.

In 2019, at PCIM, TTPi, a spin out from Nottingham University, demonstrated an amazingly power dense, 50 kW, DC-to-DC converter with an open loop bandwidth of more than 100 kHz. It used a novel electrical and physical design that reduces the EMI generated and contains that within the converter package.

What are the lessons that this history teaches?

The key advantages of the GaN transistor are as follows:

- It switches very fast.
  - Fast switching demands much lower inductance across the entire switching loop but the switching transition in hard switched systems still has to be limited to avoid unwanted resonance at the switching edge.

Fig. 7.4 An isolated 50 A half bridge embedded module printed in panel form

- It has a high positive thermal resistance.
  - Reducing its operating temperature by half can halve its on-resistance.
- It has a low specific on-resistance.
  - The die is small, contributing to the improved switching speed but making it harder to cool.
- It has no body diode or output saturation voltage.
  - These enable novel circuit design.

The development of embedded power transistor packaging is allowing several switching transistors to be co-located on a thermally efficient insulating substrate and interconnected using thick copper layers within an embedded pre-package component for use in higher power converters. The printed nature of the manufacturing process allows a number of these embedded power components to be assembled on a large format board and separated after manufacture (Fig. 7.4).

Extending these forms of packaging to encompass the entire switching element of a fast converter will accelerate the adoption of the advantages of fast switching GaN transistor converters. By including input and output filters and digital control in a screened embedded package, a complete switch building block component can be developed that can be simply adopted in the design of many different converter types. TTPi are proposing a family of isolated high power bidirectional converters constructed in this way. Modular blocks can be configured in series or parallel for voltages up to 1200 V and powers from 1.5 kW to 50 kW (Fig. 7.5).

7.2.5 Promise of Next Generation Very Fast Hard Switching of GaN Transistors

QPT, a company that is just now emerging from stealth mode, is developing hard switching converters using existing transistors enabled to switch in just 1 ns. Reducing the switching time by as much as a hundred times can have a dramatic

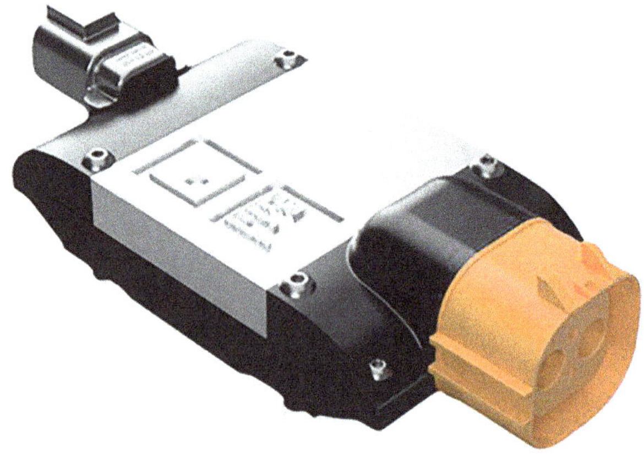

Fig. 7.5 Bidirectional isolated power-dense DC-to-DC converter. https://ppmpower.co.uk/products/power-supplies-main-content/bi-directional-dcdc-converters/ultramod-dc-dc-converters/. PPM power

effect on the power density and efficiency of a converter. QPT needed to solve three major challenges to achieve this:

1. Create a very fast isolated drive signal.

   The QPT driver is designed to achieve an output slew rate of greater than 600 V/ns to support a 1 nS switching transient for hard switching of a 400 V motor. Standard GaN drivers do not reach anything like this speed. To achieve it, the gate of the GaN transistor must switch in hundreds of pico-seconds. Very low propagation delay with minimal dispersion over time, voltage, and temperature to accurately control the deadtime is also required to avoid presenting uncontrolled impedance to the output filter that would still result in excessive under and overshoot of the switched waveform.
2. Design a low inductance packaging capable of efficiently removing the heat directly from the GaN die.

   QPT have developed new passive cooling solutions that conduct heat away from the GaN dies without creating parasitic coupling between them, despite their much smaller surface area compared to their silicon equivalent. QPT's solution enables the transistors full power handling capability. There is no point having a GaN device designed to switch 8 kW and then getting it hot so that it is only able to switch at a quarter of that power because of the increase in its channel temperature. Just adding copper around the die is not sufficient to conduct the thermal energy away. That will load the die with parasitic capacitance coupling, seriously degrading its switching characteristic. Operating the transistor at much lower temperatures also improves its operating lifetime.
3. Manufacturing entirely new types of filters and transformers capable of operating in the gigahertz domain represented by the much higher harmonic frequencies created by faster switching transitions.

   A typical LC filter used in a conventional power converter will have a self-resonant frequency of no more than 10 MHz. As a result, it will resonate badly when pulsed at high speed. This wasted switching energy will pass straight through such an LC filter. QPT has developed complex microwave filters with exceptionally high self-resonant frequencies capable of handling the required system power. Not only must the LC filter block these very high frequencies, but it must also be matched to the complex impedance of the GaN half-bridge output. There is no conventional way to terminate the output over a multiple gigahertz bandwidth without using a purely resistive load, which would totally degrade the efficiency of the drive, to the point where there would be no advantage in using GaN. QPT has resolved this by using an entirely new termination structure that allows hard switching with almost no penalty in terms of switching loss. Similar, wide band filters are also employed to block conducted EMI on the DC bus and phase input connections.

7.2.6 Compelling Application that Will Benefit from Fast Hard Switching

The one application that immediately benefits from such fast hard switching converter is the PWM-driven variable-frequency drive. Almost half of the world's power is dissipated in electrical motors. These drives typically employ three half-bridge switches that hard-switch a DC bus with a PWM signal at a low frequency that is ideally chosen to be just above the audible limit of the human ear. These three-phase drive signals are typically crudely filtered by the motor itself. This works well if the drive is mounted close to the motor but, when mounted remotely, requires a physically very large and expensive filter or costly screened cables to avoid radiated EMI created by the harmonics of the switching frequency. By using a higher PWM frequency, a sinusoidal drive signal can be created.

In these applications, the QPT GaN fast hard switched drive can reduce power loss by as much as 80%—say, from 25% to just 5% overall. The large, heavy, expensive output filter, if fitted, just disappears, saving a further 1–2% of power loss. Screened cables are no longer required for remotely located motors. Early prototypes are currently being designed for 7.5 kW industrial motor applications, but the technology is expected to scale to 100 kW and more.

7.2.7 How Does All This Development Drive Future Trends in the Design and Manufacture of GaN Power Transistors

The compelling potential reduction of wasted power from global application of these very fast hard switching drives in industrial systems and their extension to higher power automotive and aerospace applications can be expected to create a massive market pull for the technology. Just the circumstances to drive a seed change in the design and manufacture of the GaN transistor itself. Scaling the power in these systems will drive the need for higher voltage operation and accelerate the requirement for higher voltage GaN switches. This will in turn drive the need to adopt more appropriate substrate materials for high power GaN switches. For me, it heralds new novel HEMT GaN transistor designs on alternative substrates. Perhaps even a return to the ideas we originally pioneered at GaN Systems. I doubt that simply resurrecting the MOSFET transistor in a vertical GaN process is a compelling answer. It just does not support the fast hard switching that this major disruption in power system design requires.

A promising GaN process development and new transistor design, originally pioneered in Japan, is attracting my attention. The PSJ GaN transistor employs a hole gas to balance the excessive charge gradients in the electron gas of the GaN HEMT channel. Further development of the technology into an exciting 3 kV, bidirectional switch is being explored at Sheffield University in the UK. It certainly fits well with the requirements of very fast, high voltage, hard switching applications and is well suited for application in high voltage circuit breakers for control of industrial power systems and for semiconductor transformer replacements for deployment in the last mile of the national grid.

7.3 Emerging Trends and Innovations—Introducing the GaN Monolithic Bidirectional Switch and AC Application Outlook

Kennith Kin Leong

In 2007, Morita et al. presented the first true GaN monolithic bidirectional switch (M-BDS) in a landmark paper [1]. Since then, there has been a quiet race to bring such a device to the market due to its potential ability to enable new topologies and applications at an attractive cost position. Traditionally, the most common method to create an effective bidirectional switch (BDS) is to connect two uni-directional switches (UDS) in anti-series, either in the common-drain or collector configuration, but more commonly the common-source or emitter configuration, also known as back-to-back configurations (Fig. 7.6). The disadvantages of this approach are the doubling of the drift region (two times resistance) to sustain the rated voltage in both directions, and the required paralleling of the two additional UDS to bring the total effective resistance back to the original value.

Effectively, this configuration is four times the cost compared to the same UDS with the same $R_{DS(on)}$, and also increases the bill of material (BOM) and space. There are other methods and configurations to create a BDS, but they all have either a higher conduction or a higher cost penalty. The cost advantage of the GaN M-BDS comes from the double-gate structure, which uses the same drift region to sustain the voltage in both directions in a quasi-common drain configuration but without a

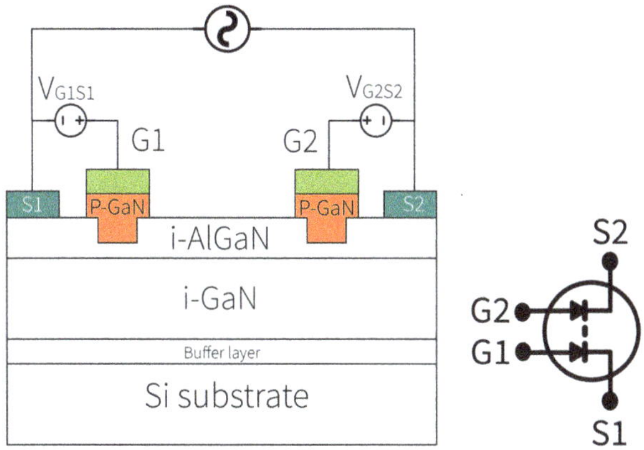

Fig. 7.6 Left, illustration of the GaN M-BDS using GIT technology and right, the basic device symbol

physical drain, with independent gate control. The resulting device is a monolithic lateral structure that is significantly smaller than the traditional back-to-back counterpart that can be easily fit into a single package.

There are some disadvantages in the quasi-common drain configuration:

- Each gate requires its own auxiliary supply in reference to its own source node, which increases the BOM when compared to the common source configuration,
- With the significant decrease in die area compared to the back-to-back BDS, thermal handling capability is also reduced.

7.3.1 Modes of Operation

The M-BDS has four electrical terminals (G1, G2, S1, and S2) and is capable of independent gate control that results in a power device with four modes of operation. The traditional on and off modes and the two diode modes, where, depending on the gate conditions, the device can behave as a diode in either direction (Fig. 7.7). Since the M-BDS can block voltages in both directions, care must be taken in the off mode to avoid any unclamped inductive switching, which could induce overvoltage conditions. The M-BDS can transition from the off mode to the diode mode where the device can continue to sustain a voltage in one direction, or allow the current to conduct in the opposite direction. Furthermore, for soft-switched operations, transitions from diode mode to ON/ON mode (and back) are recommended.

With the additional gate, four modes of operation, and even more transitional possibilities, the driving of this device becomes much more complex. For example, in some applications, both the voltage and current polarity have to be dynamically monitored to enable the correct control. This complexity is reflected in the descriptive parameters where some traditional values like gate threshold voltage, gate charge, etc., can also be mode-dependent.

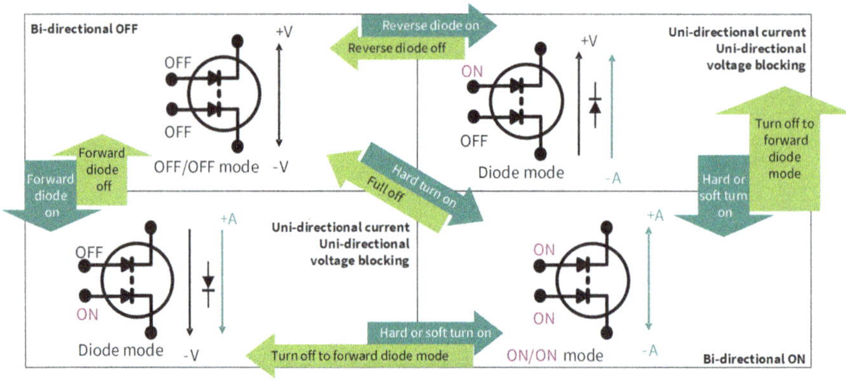

Fig. 7.7 M-BDS modes of operations

7.3.2 *GaN GIT M-BDS*

In the gate injection transistor (GIT) GaN technology, an UDS injects holes into the channel via the gate, and conductivity modulation takes place. This is further enhanced by the hybrid drain under hard switching turn-on. Both these characteristics lead to better $R_{DS(on)}$*A figure of merit when compared with other technologies. This behavior is further enhanced in M-BDS since both gates are injecting holes into the channel during on-state, and depending on the hole injection level, the dynamic $R_{S\text{-}S(on)}$ ratio could even be lower than one when compared to the static $R_{S\text{-}S(on)}$ condition under the same temperature.

7.3.3 *Dynamic Substrate Control*

Conventionally, the substrate of a UDS is electrically connected to the source node to keep it always at the lowest potential (Fig. 7.8, left). This avoids any back-gate effects that could lead to lowering the charge concentration in the channel. In the M-BDS situation (Fig. 7.8, right), the substrate has to be disconnected from either of the sources, leading to a five-terminal device where the subnode (Sub) must be used only for thermal coupling and any electrical connection should be avoided. This is because fixing the substrate node to either of the sources will create an asymmetry during AC operations and connecting to both would create a complete short. On the other hand, it may also be unwise to fully float the substrate since the potential at the subnode could drift and become detrimental to the device.

One method to eliminate this problem is to dynamically connect the substrate to the lowest source potential; this can be done in a number of ways. However, the best method would be to perform this completely on a die and without any control or additional BOM from the end user.

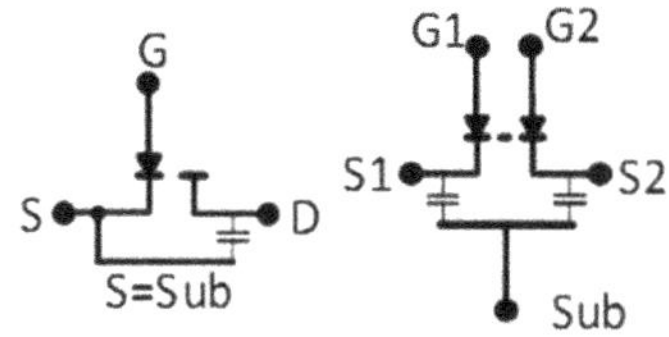

Fig. 7.8 Substrate connections for UDS (left) and M-BDS (right)

7.3.4 GaN M-BDS Types

Up to this point, the M-BDS under discussion has been the normally-off (NOFF-BDS) type. This is the focus since it is used in most applications. However, there are, in fact, two more fundamental types of M-BDS. They are normally-on type (NON-BDS) and hybrid type (H-BDS) as shown in Fig. 7.9).

The NON-BDS has both gates being normally-on and requires a negative gate voltage to turn off. One use-case for this device is in current source converters (CSC) where one of the phase-legs can be NON-BDS to ensure a conduction path even under auxiliary power loss conditions. This enables a type of fault tolerance in CSC, which is difficult to implement with NOFF-BDS. Another use-case for NON-BDS is to turn it into a NOFF-BDS by connecting low voltage switches on each gate and each source in a cascode manner.

The H-BDS is a new type of asymmetrical BDS where one gate is the normally-off type and one gate is normally-on type. This device on its own has limited usage but when combined with other elements can enable new characteristics and functionality. Two examples are shown below:

Self-reverse blocking M-BDS (based on H-BDS, Fig. 7.10, left)—first proposed in [2], where the normally-on gate (G2) is connected to the anode of the low voltage Schottky diode and used as a virtual S2 for the H-BDS, with the cathode of the Schottky diode connected to the real S2. This configuration leads to a device that can block voltages in both directions but can actively control the current only in one direction; analogous to the inverse of a power MOSFET, where voltages can be blocked only in one direction, but the current can be conducted in both directions (but actively controlled only in one direction). There are a few circuit topologies that could make use of such a device, but potentially, this configuration could make the biggest impact in CSC where it can reduce the total number of gate drives by half (one driver per BDS) and therefore, simplify the control significantly. Furthermore, the required gate-supply configuration is similar to a half-bridge, which means bootstrapping could be possible. The only disadvantage is the slight increase in the conduction loss from the series low-voltage Schottky diode.

Voltage doubling X-stacked H-BDS (Fig. 7.10, right)—two H-BDS can be connected in series in an X-stacked manner, where the normally-on gate of each H-BDS is coupled to the source furthest away from it via a capacitor. In this manner, a NOFF-BDS is created with two times the voltage blocking capability but with the same number of gates to be driven (two instead of four). Furthermore, blocking of

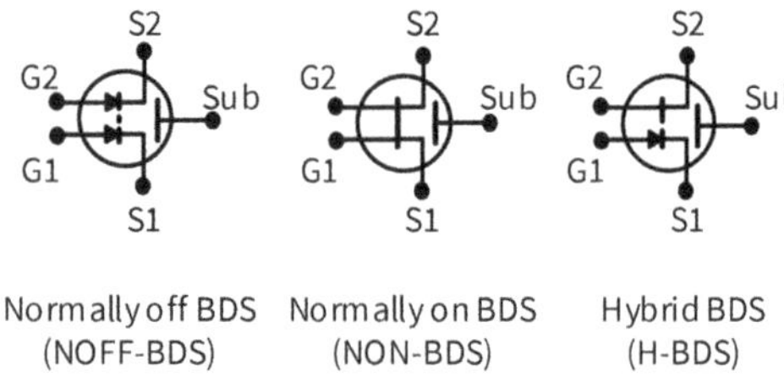

Fig. 7.9 Fundamental types of M-BDS

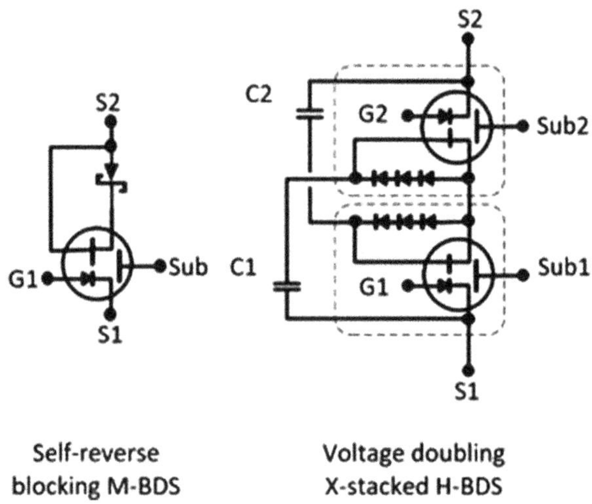

Fig. 7.10 Examples of H-BDS configurations to create new functionality

an even higher voltage is possible by connecting a NON-BDS in series between the two H-BDS, as the normally-on gate connections of the NON-BDS follow the same logic as the voltage doubling X-stacked concept shown in Fig. 7.10.

7.3.5 GaN M-BDS AC Application Outlook

There are three main target applications for the M-BDS:

- Replacement of back-to-back switches in existing applications,
- Single-stage AC power conversion,
- Current source converters.

Replacement of back-to-back switches—In many known topologies like Vienna rectifier, T-type inverters, HERIC, and so on, the use of two UDS to create a back-to-back BDS is a common approach. The main disadvantage of using two UDS is the increased die area and the cost penalty. With the advent of the M-BDS, the easy target is to replace those existing back-to-back BDS with the lower-cost

M-BDS. Furthermore, by replacing the Si MOSFET back-to-back BDS with GaN M-BDS, higher switching frequency can be easily reached and the overall system power density can be improved by reducing the sizes of the passive components. As a side benefit, in most of these existing topologies, unclamped inductive switching is not a concern since there is always an alternative diode path to discharge the current.

Single-stage AC power conversion—It is relatively well known in academia that the BDS can be utilized in the direct single-stage topology for AC/DC, DC/AC, and AC/AC conversion. The key system benefits of these approaches are numerous, designs without DC-links, higher efficiency, higher power density, etc. The main resistance from the industry to adopt these approaches is the higher cost (two times UDS) which becomes significant in topologies where six or more BDS are commonly required. With the availability of M-BDS, the industry can reexamine these single stage approaches and perhaps create more novel topologies.

Nevertheless, the adoption of these approaches is not straightforward. For one, it requires a different design mind-set from the current status quo, and for another, new problems arise from such a paradigm shift in topology concepts. To give an example, a DC-link-less design would significantly increase reliability and power density while reducing the cost at the same time; on the other hand, surge handling becomes significantly more challenging. With such a device that has no current path under zero-power conditions, overvoltage conditions must be very well handled.

Current source converters—Similar to the single-stage approach, CSC is well-known in both academic and industrial research. However, it suffers the same cost penalty in solutions using back-to-back BDS and therefore, has not been adapted in the industry despite having many advantages over traditional voltage source converters (VSC), such as no dv/dt limitations, improvements in motor designs, and reliability. It does have some additional disadvantages beyond the use of BDS, such as a two-stage requirement for DC voltage source for the motor, complexity in driving and commutation schemes, limitations in fault tolerance when using traditional UDS back-to-back designs, etc. Some of these problems could be solved with the additional types of M-BDS, such as NON-BDS and H-BDS, but only time will tell how these advances will be employed in the industry.

7.3.6 Conclusion

With the development of the true GaN M-BDS coming closer to the open market, the benefits over the traditional back-to-back BDS will be seen in the industry. New topologies and new capabilities are being explored, all with overall system cost savings, which is generally uncommon when new technology is introduced, prompting a wide interest in these devices. However, such paradigm shifts also introduce new challenges, and each of these challenges must be thoroughly investigated before wide adoption.

7.4 Taking GaN Power to the Next Level of 100 V to 2000 V and Beyond Scalability with 200 mm and 300 mm QST Manufacturing Platform

Vladimir Odnoblyudov and Cem Basceri

The CMOS fab-friendly and SEMI standard thickness 200 mm QST® (QROMIS Substrate Technology) substrates, which are similar to Silicon-on-insulator (SOI) substrates with respect to manufacturing processing and cost, enable fabrication of long awaited commercial high-performance GaN power devices with breakdown voltages ranging from 100 V to 2000 V and beyond. This is achieved by utilizing a substrate core with the coefficient of thermal expansion (CTE) matched to GaN over a wide temperature range as well as with its high thermal conductivity and high mechanical strength, breakage-free fully integrated QST® substrates.

In this section, introduction to CTE-matched, CMOS fab-friendly QST® substrate technology and high-quality, low dislocation density, high-breakdown voltage (up to 2400 V), crack-free, and low cost GaN epitaxial layers are presented. The results from a comparative thermal performance study between GaN-on-Si and GaN-on-QST®, revealing that the peak channel temperature of GaN-on-QST® HEMT being 32.5 °C lower than that of GaN-on-Si HEMT, are also included. Similar study for RF HEMT device operation, which is included in this section, show that the steady state operation temperature of GaN-on-QST® device is 8 °C lower than that of GaN-on-Si device. Commercial production of 200 mm SEMI standard thickness QST®-based 650 V E-mode GaN HEMT devices (40 and 150 mohm) on a high-volume CMOS foundry manufacturing platform and also, the results from the development work in 200 mm CMOS fabs for 1200 V E-mode HEMTs, 650 V wafer-level monolithically integrated circuits (ICs) and semi-vertical / vertical power devices, including crystal film bonding (CFB®) technology, on QST® substrates are also presented.

In recent years, GaN, with a bandgap of 3.4 eV, has become a forefront wide-bandgap semiconductor material for next-generation, energy-efficient power electronics. With its higher breakdown strength, faster switching speed, and lower on-resistance, GaN power devices can convert power far more efficiently than Si-based devices, and thus enable smaller, faster, lighter, and low-cost power conversion systems as well as significantly reduced CO_2 footprint. For these reasons, and also for achieving full sustainability, the demand for GaN has grown considerably in recent years as companies work to develop next-generation, energy-efficient power electronics. While its performance is commercially demonstrated, widespread GaN adoption requires a scalable and high-yielding manufacturable technology platform enabling economies of scale and a full spectrum of low-cost products such as lateral and vertical power switches extending from 100 V to 2000 V and beyond, wafer-level monolithic ICs, and rectifiers. To further scale up both the substrate size and GaN stack thickness and also to achieve wafer-level monolithic ICs of the current mainstream 150 mm, 650 V GaN-on-Si manufacturing platform, even with non-SEMI standard substrate thickness (1 mm or thicker), is extremely challenging, due to the significant mismatch in coefficient of thermal expansion (CTE) between GaN and Si, and the IC device cross-talk challenges arising from the conducting Si substrates.

A disruptive commercial solution enabling high volume, wafer breakage-free, high mechanical strength, low cost, and scalable GaN device manufacturing is the CMOS fab-friendly and SEMI standard thickness engineered substrate called "QST®" (QROMIS Substrate Technology) with a polycrystalline, high thermal conductivity AlN ceramic core material having a thermal expansion that very closely matches the thermal expansion of the GaN/AlGaN epitaxial layers over a wide temperature and prevents excessive stress or GaN cracking or wafer breakage during the cooling stage of GaN epitaxy growth (Figs. 7.11 and 7.12) [3–5]. High thermal conductivity (170–230 W/mK) AlN ceramic core material is covered by several encapsulation layers, on top of which is a SiO_2 bonding layer and a single crystalline Si (111) layer which serves as the nucleation layer for the epitaxial GaN growth

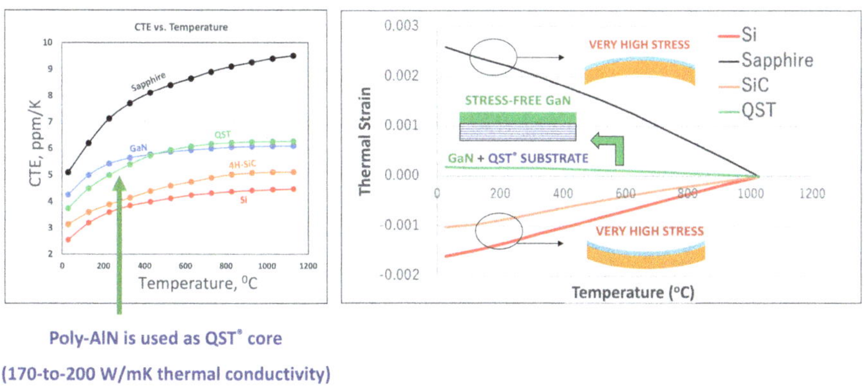

Fig. 7.11 Polycrystalline, high thermal conductivity AlN ceramic core material having a thermal expansion that very closely matches the thermal expansion of the GaN/AlGaN epitaxial layers over a wide temperature range

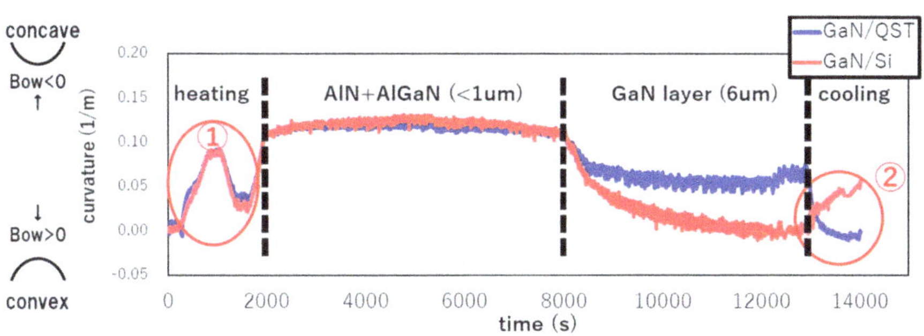

Fig. 7.12 QST® prevents excessive stress, GaN cracking,or wafer breakage during and after GaN epitaxy growth

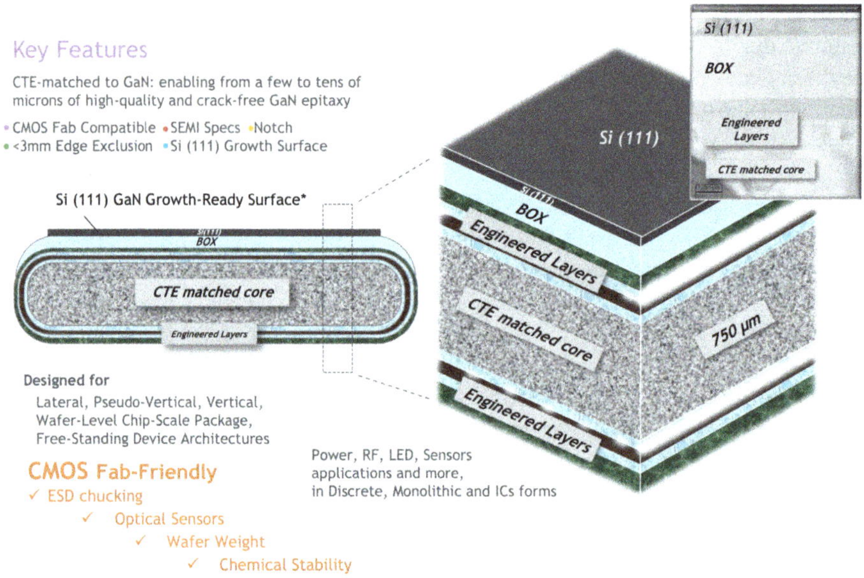

Fig. 7.13 Cross-section of a commercial QST® substrate and its key features. Note: Si (111) GaN growth-ready surface can be changed to single crystal GaN, SiC, or other GaN growth-ready surfaces

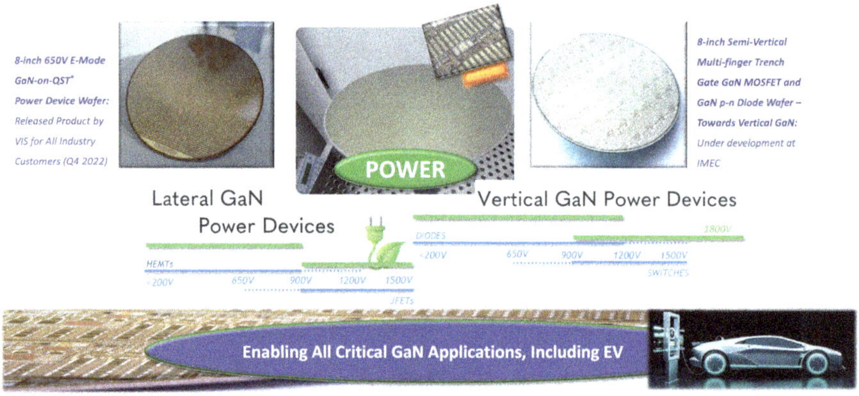

Fig. 7.14 GaN power devices enabled by QST® substrate

(Fig. 7.13). Si (111) GaN growth-ready surface can be changed to single crystal GaN, SiC, or other GaN growth-ready surfaces.

Commercial QST® substrates, available from QROMIS Inc. and Shin-Etsu Chemical Co. Ltd., enable wafer breakage-free high-volume manufacturing of 200 mm GaN device wafers, covering all GaN applications including 100–2000 V and beyond high performance discrete and wafer-level monolithic IC power devices, RF devices, and display microLEDs (Fig. 7.14). 300 mm QST® substrates are currently being developed for initial shipments in 2024.

While QST® platform enables higher performance and application scale for GaN devices, it also brings a critical economies of scale advantage for device manufacturers and foundries since all GaN devices (power, RF, microLED, and others) can be manufactured on the same CMOS-fab friendly, SEMI-spec platform for efficiency and further cost reduction (Fig. 7.15).

Another important feature that QST® solution brings is that the QST® substrates are assembled and manufactured in traditional CMOS fabs (similar to mainstream Silicon-on-insulator (SOI) substrates with respect to manufacturing processing and cost) by utilizing energy efficient mainstream semiconductor process tools with >98% yield and less than 7 days of process cycle time (Fig. 7.16) while e.g., SiC substrates are manufactured in isolated, non-CMOS compatible fabs with crystal

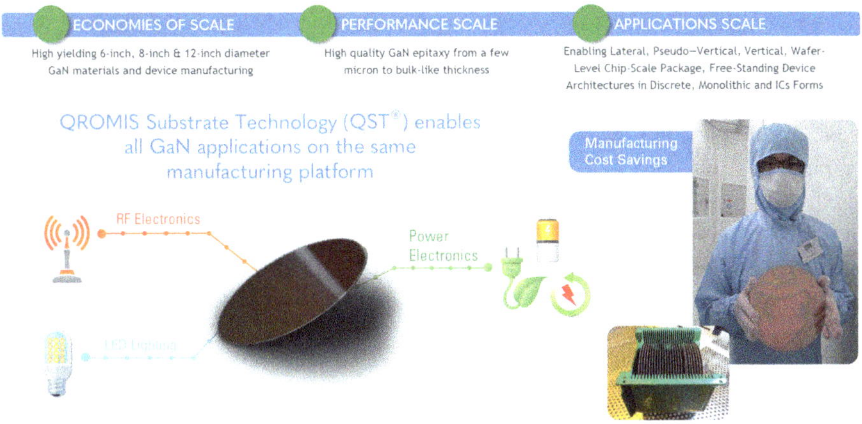

Fig. 7.15 All GaN devices (power, RF, microLED, and others) can be manufactured on the same CMOS-fab friendly, SEMI-spec platform for efficiency and further cost reduction

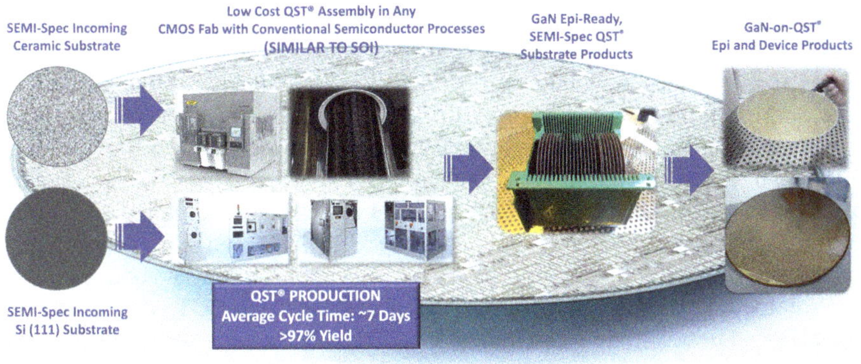

Fig. 7.16 QST® substrates are assembled and manufactured in traditional CMOS fabs (similar to mainstream Silicon-on-insulator (SOI) substrates with respect to manufacturing processing and cost) by utilizing energy efficient mainstream semiconductor process tools with >98% yield and less than 7 days of process cycle time

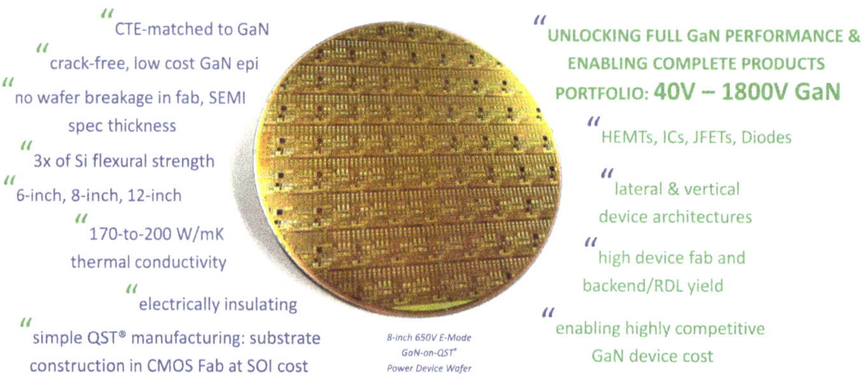

Fig. 7.17 Summary of the key features of QST® for GaN power applications

growth processes at extreme temperatures >2200 °C for several days which result in excessive amount of power consumption, water cooling, and consumables such as graphite and felt. This is a critical component of the QST® innovation which was carefully designed as per the projected cost and sustainability requirements in the industry.

Finally, the key features of QST substrates for GaN power applications are summarized in Fig. 7.17.

7.4.1 GaN Power Epitaxy on QST®: 100–2000 V and Beyond Voltage Scalability, Dislocation and Epitaxy Cost Reduction

As the QST® substrates are thermally matched to GaN, thick, high-quality GaN device layers with low defect density, high crystal quality and low wafer bow can be achieved at SEMI standard thicknesses (Figs. 7.18 and 7.19) [6]. Moreover, QST® substrates open perspectives for very thick GaN buffers, including realization of free-standing and very low dislocation density GaN substrates by >100 μm thick fast-growth epitaxial layers. These unique features enable long awaited scalable commercial GaN power switches and rectifiers suitable for high voltage and high current applications presently dominated by Si IGBTs and SiC power FETs and diodes.

With the CTE matched structure, GaN epitaxy layers on QST® do not require complex and costly strain management layers which are essential for GaN-on-Si and other heteroepitaxy platforms. As such, GaN-on-QST® platform enables very significant GaN epitaxy cost reduction, process simplification, and increased reactor uptime via reduced process times, while yielding thicker GaN layers. Figure 7.20 depicts such drastic cost and performance improvement from a comparative study

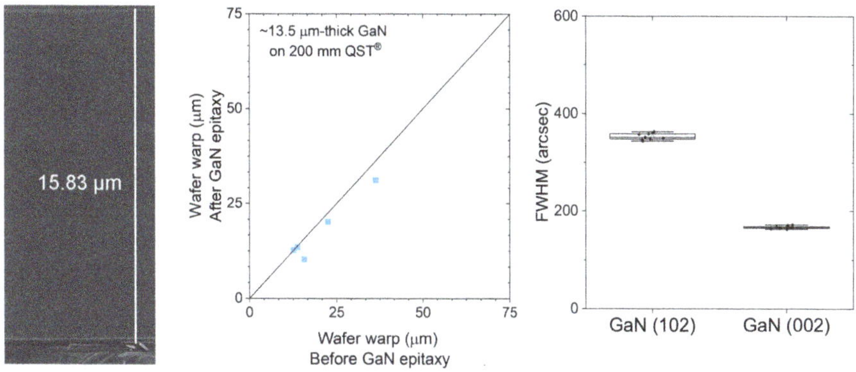

Fig. 7.18 (**a**) SEM cross-section of 16 µm GaN epitaxy on 200 mm QST® substrate. (**b**) Wafer warp before and after GaN epitaxy, and (**c**) crystal quality of 13.5 µm GaN stack on QST® using a 10 µm thick drift layer for semi-vertical/vertical devices

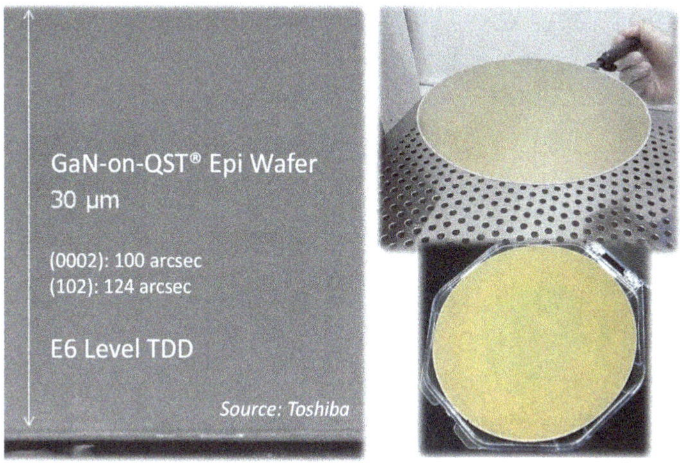

Fig. 7.19 SEM cross-section of 30 µm thick, low dislocation density (E6 level threading dislocation density, cm^{-2}) GaN epitaxy on QST® substrate

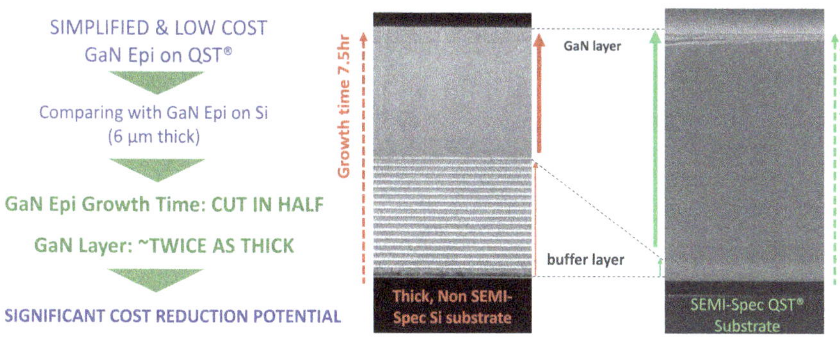

Fig. 7.20 GaN epitaxy cost and performance improvement on QST®

led by Shin-Etsu Chemical: GaN-on-QST® vs GaN-on-Si. This critical feature of QST® platform, in addition to its voltage and wafer diameter scalability, enables accelerated GaN adoption.

Use of the CTE-matching technology also results in crack-free edges of 200 mm GaN-on-QST® epitaxy wafers for wide variety of thicknesses and applications, which significantly improves device yield by enabling small edge exclusion of ~5 mm and eliminating potential particle and flake generation areas. Micrographs of four different GaN-on-QST® epitaxy wafer edges, studied by various techniques such as SEM, CAMTEK, optical microscope, and XRT, are shown in Fig. 7.21. The thicknesses of the studied epitaxy layers vary from 5 μm to 14.3 μm and every technique confirms crack-free wafer edges.

Substantial GaN thickness budget of tens of microns of crack-free epitaxy enables seamless scalability of insulating buffers and breakdown voltages on QST® substrates. As an example, a series of HEMT buffer vertical leakage curves is presented in Fig. 7.22. It can be seen that scaling the buffer thickness from 2.5 μm to 15 μm enables breakdown voltages in the range of 200–1800 V (*breakdown voltage is defined as a voltage @ 1 μA/mm² leakage current density*). As shown by the

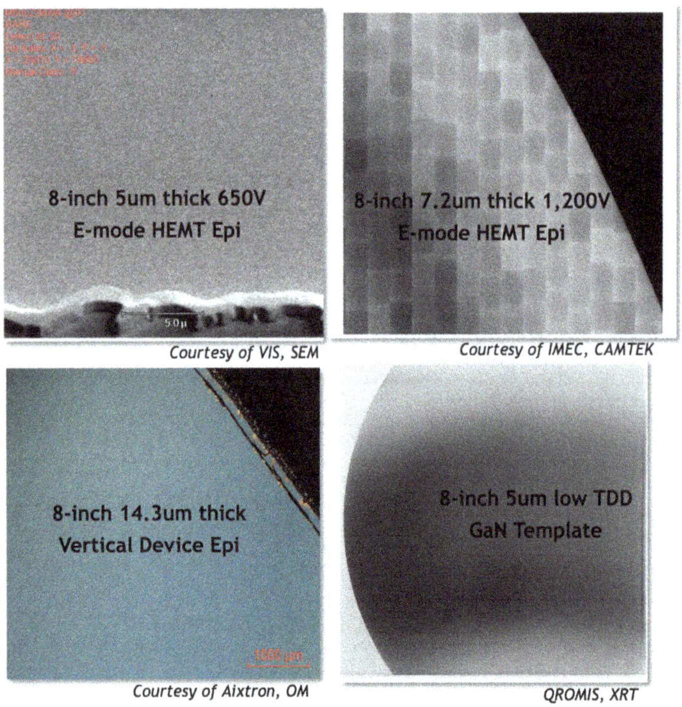

Fig. 7.21 Crack-free edges of various 200 mm GaN-on-QST® epi wafers studied by different techniques

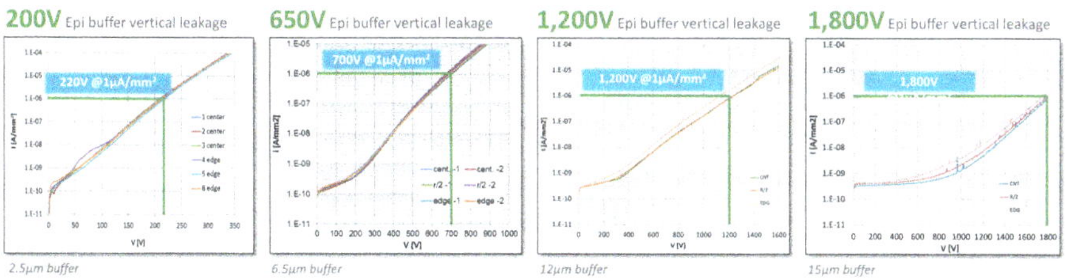

Fig. 7.22 Breakdown voltage scalability for lateral HEMT epi wafers with different buffer thicknesses

multiple leakage curves taken from the center to the edge of each wafer, the leakage curves are quite uniform within each wafer.

One of the unique features of GaN epitaxy on QST® substrates is the ability to control threading dislocation density in a wide range of between low 10^6 cm^{-2} and mid 10^9 cm^{-2} levels which in turn enable the GaN material quality from "GaN-on-GaN" type to "GaN-on-Si" type. Such dislocation density control can be achieved by using several approaches: (1) special dislocation reduction techniques, such as SiN micro-masking, superlattices or graded buffer layers; (2) using the substantial GaN thickness budget available on QST® at high growth rates for dislocation reduction as shown in Fig. 7.20; and (3) introducing other single crystal growth interfaces besides Si (111), such as single crystal GaN or SiC, which are lattice-matched or closely lattice-matched to GaN epitaxy layers.

Such approaches could be used separately or in combination with each other. While it is well-known that lower dislocation density ($<10^7$ cm^{-2}) is desired to enable low leakage and high performance vertical GaN power devices, the same can also be applied to lateral HEMT devices. It has been demonstrated that lowering the dislocation density enables GaN HEMT devices with: (1) better thermals; (2) lower R(on); (3) dispersion-free operation; (4) lowest leakage; and (5) improved reliability [7].

As such, QST® substrate can enable low dislocation, all-GaN, and fast growth rate HEMT epitaxy layers on 200 mm platform which in turn may enable "GaN-on-GaN" type device performance for lateral device architectures (Fig. 7.23).

In summary, the key features of GaN epitaxy layers on QST® substrates are as follows:

- Thickness scaling from a few to tens of microns,
- 100–2000 V and beyond power voltage scaling
- Low stress and crack-free,
- Low dislocation densities, down to low 10^6 cm^{-2} levels,
- Low cost by fast growth and thin buffers,
- Enabler for high quality lateral GaN power HEMTs and vertical GaN power devices,
- Enabler for high quality GaN RF and GaN LED/micro-LED devices.

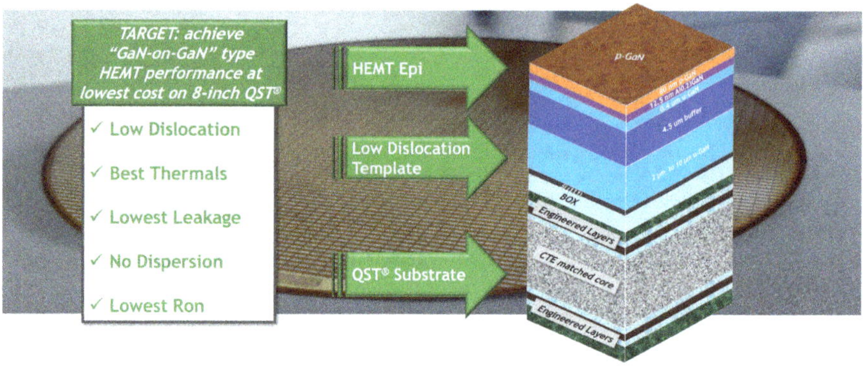

Fig. 7.23 Enablement of low dislocation, low cost, and high performance GaN-on-QST® HEMTs

7.4.2 *Cooler GaN Devices on QST: Improved Thermal Conductivity*

Thermal resistance plays a significant part in determining the performance of a transistor. Several main factors determine overall thermal performance of the device: (1) thermal conductivity of the GaN epitaxy layers, including the buffer layers; (2) thermal resistance of the interface(s) between substrate and GaN epitaxy; and (3) thermal conductivity of the substrate.

Thermal conductivity of GaN epitaxy is determined by the thickness and the quality of the epitaxy layers. GaN thermal conductivity increases from ~100 W/mK to ~230 W/mK when dislocation density decreases from 10^9 cm^{-2} to low 10^7 cm^{-2} (Fig. 7.24). Typical GaN-on-Si (orange) and measured GaN-on-QST® (blue and purple) data points are presented. It should be noted that GaN-on-QST® data points were independently obtained from a systematic study [8]. As can be seen, all three data points follow very well the GaN thermal conductivity vs dislocation density dependence and 10 μm thick GaN-on-QST® epitaxy layers have higher thermal conductivity than 5 μm thick GaN-on-QST® epitaxy layers. As detailed in previous sections, CTE-matching of QST® substrate to GaN enables flexibility for dislocation density and GaN epitaxy thickness. Furthermore, polycrystalline AlN ceramics, used as QST® core material, has thermal conductivity between 170 and 230 W/mK, compared to 130–150 W/mK for silicon substrate, depending on doping and orientation.

Comparative thermal performance studies of GaN-on-Si (*with 3 μm thick buffer*) and GaN-on-QST® (*with 6 μm thick buffer*) were conducted by using corresponding HEMT devices which were processed side-by-side through the same fabrication flow (Fig. 7.24) [4]. The channel temperature measurement results revealed that peak channel temperature of GaN-on-QST® HEMT was 32.5 °C lower than that of GaN-on-Si HEMT. As expected, the high thermal conductivity of the thick GaN layer on QST® and the QST® core material of polycrystalline AlN yield a lower total

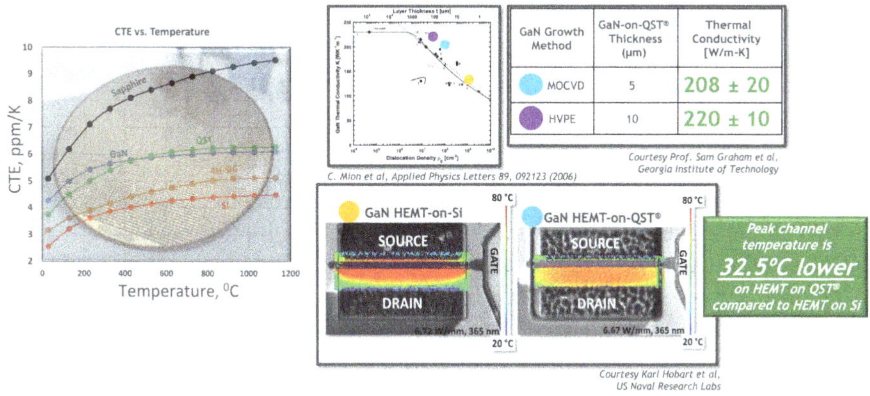

Fig. 7.24 Dependence of GaN thermal conductivity on dislocation density (*upper middle*), thermal conductivity of GaN-on-QST® with different GaN layer thickness (*upper right*) and peak channel temperature of GaN-on-QST® HEMT vs GaN-on-Si HEMT

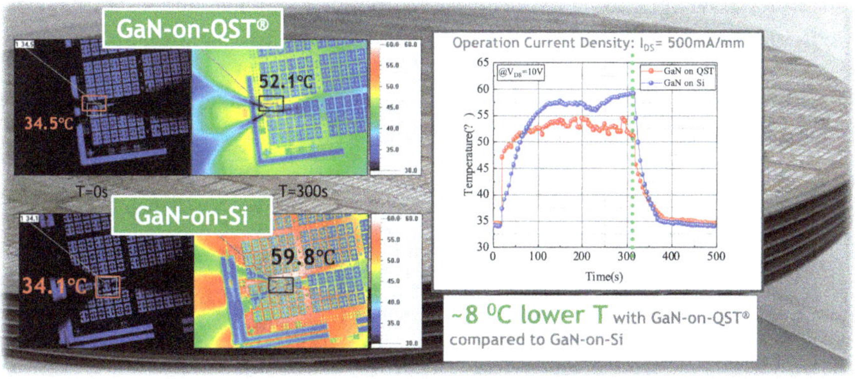

Fig. 7.25 Time dependence of operation temperature for GaN-on-QST® vs GaN-on-Si HEMT devices

thermal resistance for GaN-on-QST® as compared to GaN-on-Si. With these results, it is also demonstrated that lateral diffusion of heat in GaN epitaxy layers distributes the heat over a larger effective area than the gate area and reduces the impact of thermal resistances of the interfacial and engineered wrapping layers for QST®.

Similar results were independently demonstrated for RF HEMT device operation where GaN-on-QST® and GaN-on-Si RF HEMTs were processed side-by-side and in this case with similar buffer thicknesses, thus similar dislocation densities in the GaN epitaxy layers [8]. As can be seen in Fig. 7.25, the steady state operation temperature of GaN-on-QST® device is 8 °C lower than that of GaN-on-Si device, despite the presence of additional several microns of thick engineered layers in QST® substrate structure. The sole impact of QST® substrate on improved thermal characteristics, with its high thermal conductivity polycrystalline ceramic core material, can be concluded from these systematic studies.

7.4.3 Production of 200 mm 650 V E-Mode Gan-on-QST® HEMT Commercial Devices on a CMOS Foundry Manufacturing Platform and >650 V Timelines

The first generation (Gen1) 650 V E-mode GaN HEMT devices (40 and 150 mohm) on 200 mm SEMI standard thickness QST® substrates entered mass production in second half of 2022 after passing the full JEDEC reliability tests including 1000 h HTRB and HTGB on packaged devices, led by Vanguard International Semiconductor (VIS). This achievement marked the establishment of the first 200 mm GaN device foundry services for all industry customers. Figure 7.26 shows the status of GaN-on-QST® manufacturing at VIS and enablement of one of the first applications: fully processed 200 mm Gen1 650 V, E-mode GaN device wafer, options of two- and three-level metal layer device architectures and an example of the first commercial fast chargers (100 W) implementing Gen1 GaN-on-QST® devices.

Manufacturing volume of Gen1 GaN-on-QST® E-mode HEMT devices is ramping for initial 650 V applications which will be followed by Gen2 devices in second half of 2024, designed for industrial and automotive applications. Gen2 devices, which are currently in design/pilot shuttle stage, will be qualified at full rating 650 V hard switching. 1200 V GaN-on-QST® development is also underway (see the next section) and scheduled to be released by end of 2025. One of the target applications is 800 V EV charging.

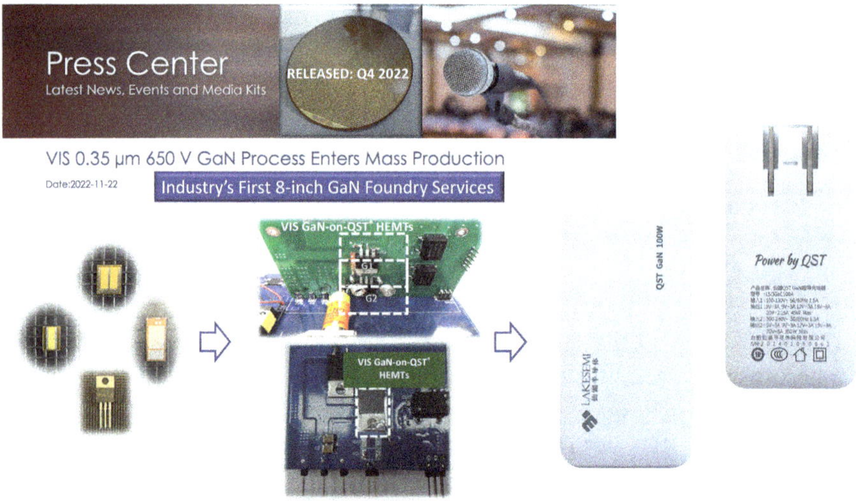

Fig. 7.26 200 mm Gen1 650 V GaN-on-QST® E-mode HEMT device wafer image manufactured by VIS, bare chip and packaged variations, and 100 W fast charger adopting VIS' Gen1 650 V GaN-on-QST® device

7.4.4 *Development Status: 1200 V and Beyond 200 Mm GaN-on-QST Power Devices*

For important applications beyond 650 V such as electric cars and renewable energy, it has become difficult to further increase the GaN buffer thickness for GaN-on-Si based solutions, because of the mismatch in CTE between GaN/AlGaN epitaxial layers and silicon substrates. One can envisage to use thicker Si substrates for GaN-on-Si platform for 900 V and 1200 V applications, but experience has shown that for these higher voltage ranges, the mechanical strength is a major concern in high volume manufacturing which manifests itself as wafer breakage/cracking in GaN epitaxy layers and/or wafer shape issues, as well as compatibility issues in wafer handling in some processing tools. Carefully engineered QST® substrates are now enabling 1200 V buffers and beyond on a SEMI standard thickness 200 mm substrate, scalable to 300 mm [10].

In this work, led by imec, p-GaN gate E-mode HEMT devices with a gate-to-drain spacing of 24 μm were fabricated on 200 mm QST®, with an 8 μm thick (Al) GaN epitaxial buffer stack grown by using the commercial, high volume capable AIXTRON G5+ C Planetary Reactor® (Fig. 7.27). The off-state drain and gate leakage of the devices is shown to stay below 5×10^{-7} A/mm until ~1800 V, which makes the buffer and fabricated devices well suited for 1200 V applications. For the selected power transistor design, a low $R_{DS\text{-}on}$ of 18.5 Ω·mm was extracted resulting in a specific on resistance ($R_{ON,sp}$) of 4.2 mΩ·cm². The transfer characteristics are depicted in Fig. 7.28. The fabricated transistors have a threshold voltage of 1.8 V, extracted by linear extrapolation.

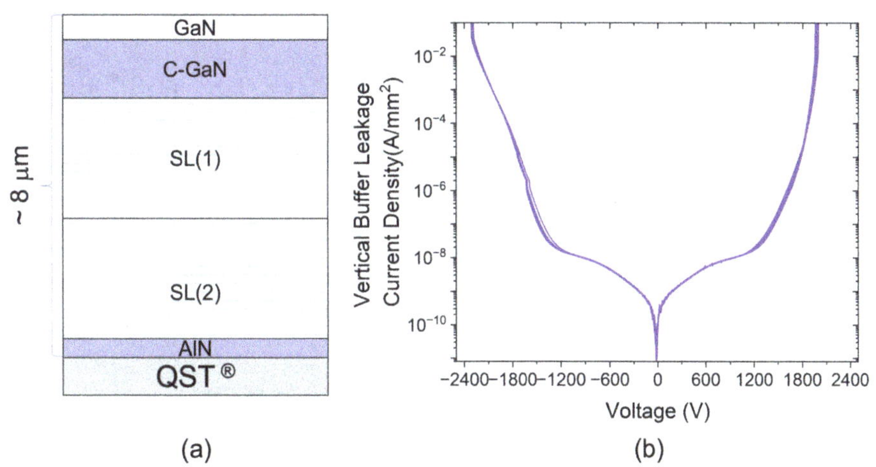

Fig. 7.27 Demonstration of (Al)GaN buffer epitaxial growth in commercial AIXTRON G5+ Planetary C reactor, qualified for 1200 V applications on 200 mm QST® substrates with an 8 μm thick stack

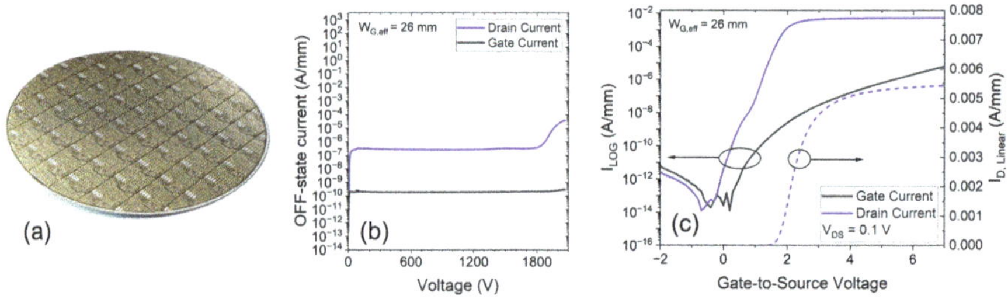

Fig. 7.28 (**a**) 1200 V p-GaN gate power HEMTs fabricated on 200 mm GaN-on-QST® substrates, (**b**) typical off-state leakage ($V_{GS} = 0$ V), and (**c**) transfer characteristics

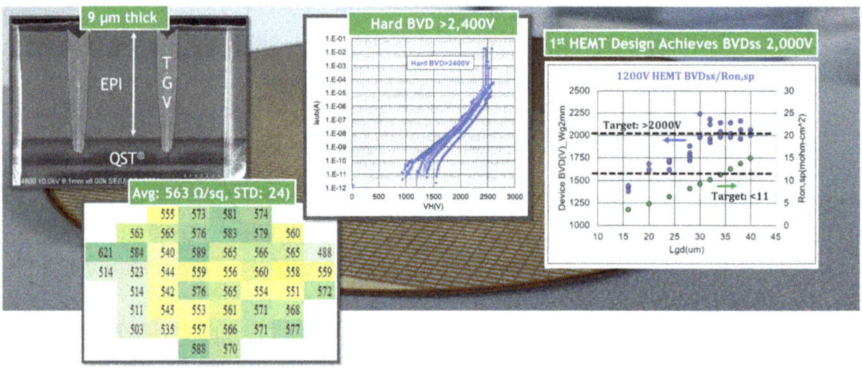

Fig. 7.29 1200 V rated E-mode GaN HEMT epitaxy cross-section on 200 mm QST® with "Through GaN Vias" (TGV), and the corresponding leakage curves and the sheet resistance uniformity

Additionally, Fig. 7.29 shows a snapshot of the 1200 V GaN-on-QST® HEMTs development results from a commercial, high volume 200 mm CMOS foundry manufacturing platform at VIS, with hard breakdown exceeding 2400 V.

7.4.5 Development Status: Wafer-Level Monolithically Integrated 200 mm Gan-on-QST® Power Integrated Circuits

Both in the industry and academia, GaN power ICs have been attracting a lot of interest because of the great advantages of achieving higher switching speed, better power conversion efficiency, and reduced form factor with custom functionalities such as integrated drivers, controllers, and sensors [9]. The holy grail of the GaN power ICs is the realization of a monolithic integration of low-side and high-side transistors, with other custom functionalities, on the same wafer which results in system on chip GaN ICs enabled by 200 mm QST®-based platform (Fig. 7.30).

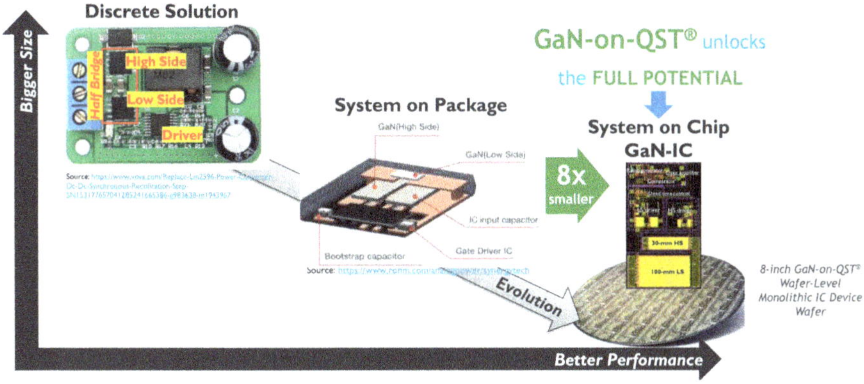

Fig. 7.30 Evolution of GaN power technology solutions

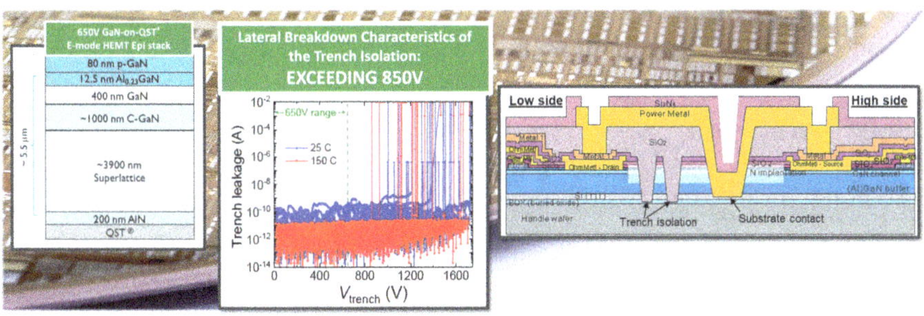

Fig. 7.31 650 V monolithic GaN-on-QST® epitaxy structure and monolithic half-bridge IC device cross section with trench isolation and 650 V rated trench leakage

In this work, led by imec, 650 V GaN buffer and p-GaN gate power HEMTs, trench isolation, and wafer level monolithic GaN power ICs on 200 mm QST® substrates are demonstrated as a showcase to verify the feasibility of this concept [11–13]. Fully functional 200 mm GaN power ICs of a HEMT with integrated driver, half-bridge, and half-bridge with integrated drivers have been processed (Fig. 7.31, 7.32, and 7.33). Figure 7.31 shows 650 V monolithic GaN-on-QST® epitaxy structure and monolithic half-bridge IC device cross section with trench isolation and 650 V rated trench leakage, as reported in [11]. The monolithically integrated half-bridge comprises of a low side HEMT and a high side HEMT in which the source terminals of these two devices are connected to their respective substrate for suppressing the back-gating effect.

Figure 7.32 depicts the schematic ICs of the power HEMT with integrated driver, which can significantly reduce the parasitic inductance in the gate loop. The driver consists of a push-pull driver and two resistor-transistor logic (RTL) inverters, because of lacking p-MOSFET. The driver and the power HEMT are electrically isolated by oxide filled trenches around the devices, which are connected to the buried oxide (BOX). Figure 7.33 shows a schematic and circuit of a GaN-on-QST® monolithically integrated circuit of a half bridge, in which contacts to the Si (111)

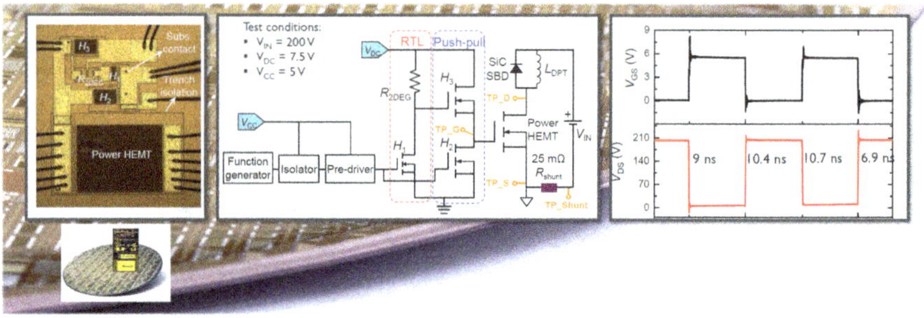

Fig. 7.32 Wafer level monolithic GaN-on-QST® ICs: power HEMT with integrated driver and its switching performance

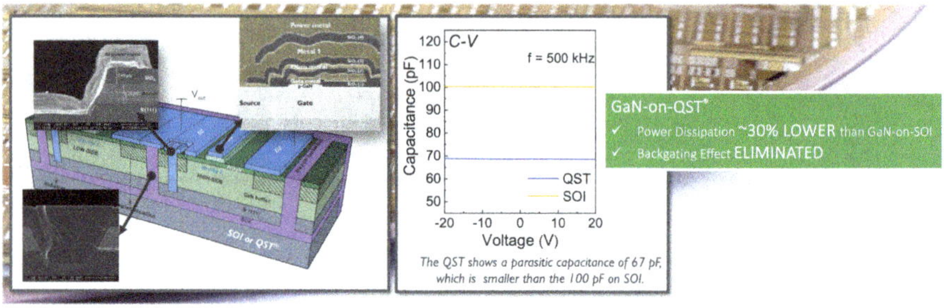

Fig. 7.33 Wafer level monolithic GaN-on-QST® ICs: eliminating back-gating effect and ~30% parasitic capacitance reduction in comparison to GaN-on-SOI

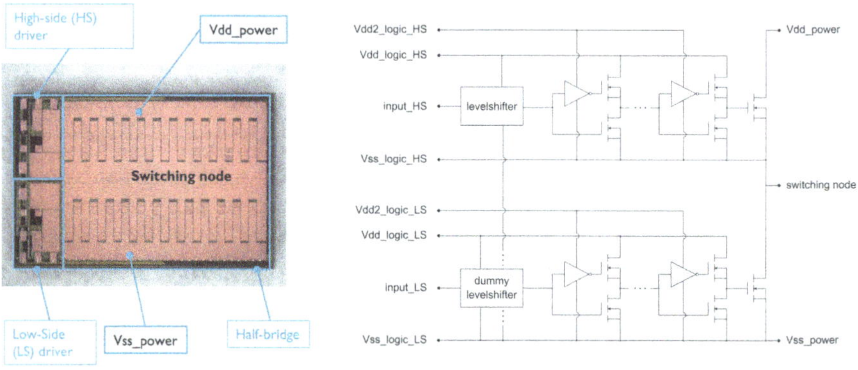

Fig. 7.34 Example of 650 V GaN-on-QST® monolithic IC, depicting a symmetrical 140 mm half-bridge with integrated driver and level shifters

layer are formed to avoid back-gating effects and more importantly ~30% parasitic capacitance reduction compared to GaN-on-SOI is achieved. Figure 7.34 demonstrates a symmetrical 140 mm half-bridge with integrated driver and level shifters. This design is beneficial for suppressing oscillation, reducing device area, and enabling ever higher operating switching frequencies.

7.4.6 Development Status: Semi-Vertical and Vertical 200 mm GaN-on-QST Power Devices

Vertical device architectures have several advantages compared to lateral devices, when scaling to high voltage operation [14–16]. First, the fields are vertical over the bulk of the device, rather than peaking at the surface, improving reliability and stability (dispersion) of the devices. There is a potential for avalanche robustness, which has not been demonstrated for lateral devices. And finally, the area of lateral devices becomes prohibitively large when scaling to large voltage due to the increased gate-drain distance, whereas for vertical devices, the thickness of the epilayer (drift region) determines the breakdown voltage. Most of the vertical GaN devices to date are grown on bulk GaN substrates which are small in substrate size and expensive. Si wafers are available in large substrate sizes, but the total GaN stack thickness that can be grown is very limited [17]. The CTE matched QST® substrates offer a breakthrough in the growth of thick GaN layers. As demonstrated in an earlier work, GaN Schottky barrier diodes on QST® exhibited a forward *I-V* characteristics comparable to that on GaN-on-GaN native substrates [18].

In this work, led by imec, and in reference [3] epitaxial growth by MOCVD in a commercial AIXTRON G5+ C Planetary Reactor® with >11 μm GaN stack on 200 mm QST® substrates using a 7.5 μm thick drift layer for fabrication of semi-vertical/vertical devices is demonstrated. p/n- diodes fabricated on 200 mm QST® substrates reaching >900 V in reverse bias for stacks with 7.5 μm thick drift layer and active net donor concentration (N_D) of about ~1 × 10^{16} cm^3 are shown in Fig. 7.35. As shown in Fig. 7.36, multi-finger semi-vertical trench gate MOSFET devices ($W_{G,eff}$ = 60 mm) were also fabricated, as a test vehicle to demonstrate the device performance in on-state (2.5 nm Al_2O_3/100 nm SiO_2 is used as gate dielectric). A low on-resistance of 6.2 mΩ•cm^2 with 2.3 V threshold voltage is achieved.

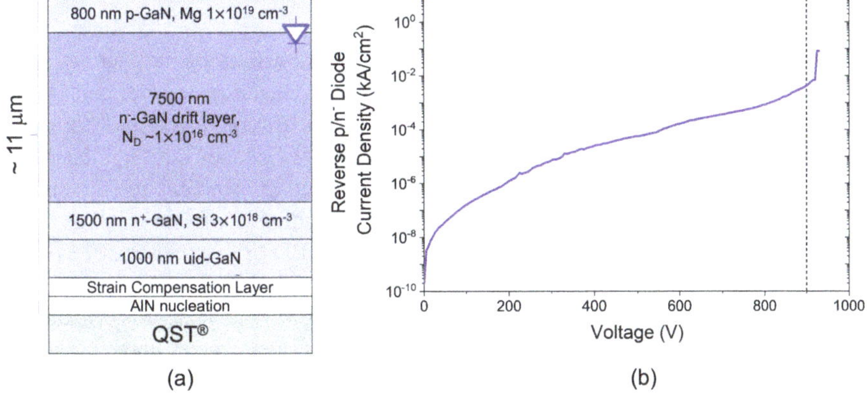

Fig. 7.35 (**a**) Schematic representation of stack for vertical GaN and (**b**) p/n- diodes fabricated on 200 mm QST® substrates reaching 900 V in reverse bias for stacks with 7.5 μm thick drift layer

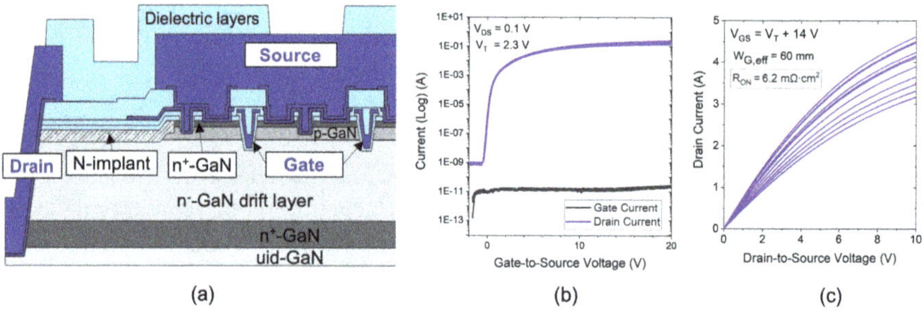

Fig. 7.36 Multi-finger semi-vertical trench gate MOSFET devices (**a**) ($W_{G,eff}$ = 60 mm) (2.5 nm AlON and 100 nm SiO_2 as gate dielectric with a 2.3 V threshold voltage (**b**), and a low on-resistance of 6.2 mΩ•cm² (**c**)

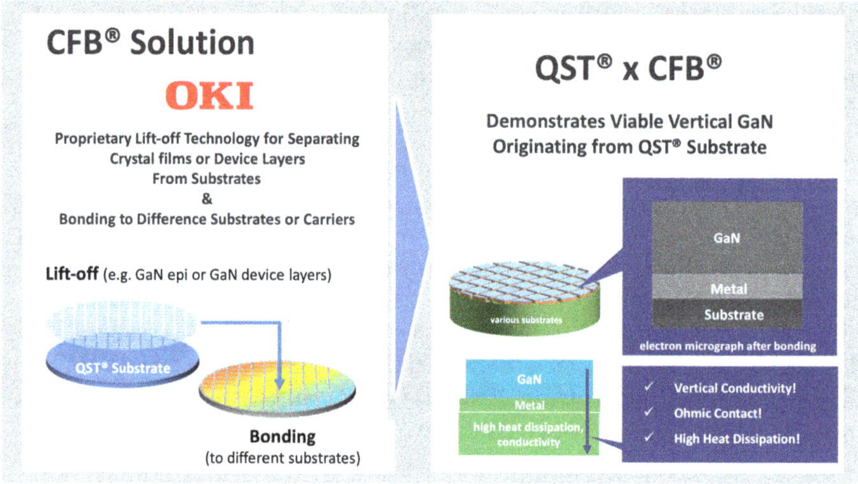

Fig. 7.37 CFB® lift-off process for GaN-on-QST®

Recently, as shown in Fig. 7.37, results from a demonstration study led by OKI Electric Industry (OKI), revealed that GaN epitaxy or GaN device layers on QST® substrates can be successfully separated and bonded to different substrates for realizing vertical GaN devices and high performance lateral GaN devices (e.g., HEMTs on high thermal conductivity substrates). This important achievement utilized OKI's proprietary lift-off technology called Crystal Film Bonding (CFB®) which has been used by the company for microLEDs on IC chip for LED printers since 2012 (over 100 billion microLEDs processed by CFB®). CTE-matching feature of GaN-on-QST® enables this successful achievement since having very low stress is critical for successful separation of GaN layers. Also, having different layers under Si (111) surface of QST® (engineered layers), enables variety of selective lift-off process options. Overall, this is a process module option, available from OKI, for manufacturing vertical GaN devices on QST®.

7.4.7 Recap

Commercial QST® substrates, which are similar to Silicon-on-insulator (SOI) substrates with respect to manufacturing processing and cost, are available from QROMIS and Shin-Etsu Chemical for enabling wafer breakage-free high-volume manufacturing of 200 mm low cost GaN device wafers, covering all GaN applications including 100–2000 V and beyond high performance discrete and wafer-level monolithic IC power devices, RF devices, and display microLEDs. 300 mm QST® substrates are currently being developed for initial shipments in 2024. GaN-on-QST® platform enables very significant GaN epitaxy cost reduction and process simplification via reduced process times while yielding thicker GaN layers. The results from a comparative thermal performance study between GaN-on-Si and GaN-on-QST® reveal that the peak channel temperature of GaN-on-QST® HEMT being 32.5 °C lower than that of GaN-on-Si HEMT. Similar studies for RF HEMT device operation show that the steady state operation temperature of GaN-on-QST® device is 8 °C lower than that of GaN-on-Si device. Manufacturing volume of Gen1 GaN-on-QST® E-mode HEMT devices (40 and 150 mohm) is ramping for initial 650 V applications (being adopted in fast chargers), followed by Gen2 devices in second half of 2024, designed for industrial and automotive applications such as charging piles. Gen2 devices, which are currently in design/pilot shuttle stage, will be qualified at full rating 650 V with hard switching. Commercial 1200 V GaN-on-QST® development is also underway and scheduled to be released by end of 2025. The demonstration results from 1200 V HEMTs, wafer-level monolithically integrated ICs, and semi-vertical/vertical devices along with CFB process technology on QST® platform open the path for realization of long awaited and scalable GaN power devices.

Acknowledgments
The authors are sincerely grateful to the members of Shin-Etsu Chemical Co. Ltd., Vanguard International Semiconductor Corp., imec, AIXTRON and OKI Electric Industry Co. Ltd. for their contributions and support. Authors would also like to thank ASM for the depositions of AlON as interface gate dielectric in the semi-vertical device fabrication. The development projects at imec received funding from the ECSEL Joint Undertaking (JU) under grant agreement No. 101007281. The JU receives support from the European Union's Horizon 2020 research and innovation program and Austria, Germany, Slovenia, Netherlands, Belgium, Slovakia, France, Italy, and Turkey. The work at imec also received funding from the Chips Joint Undertaking supporting the ALL2GaN Project under grant Agreement No 101111890 and its members including the top-up funding by Austria, Belgium, Czech Republic, Denmark, Germany, Greece, Netherlands, Norway, Slovakia, Spain, Sweden, and Switzerland.

7.5 Towards a Cost Effective GaN Technology

Han Wui Then

Researchers in industry and academia are working to address the critical challenges in GaN technology. These include: (a) cost, (b) performance, (c) reliability, and (d) integration. Cost, performance, and reliability are key to driving the adoption of GaN in the market. As it will become clear later in the chapter, GaN technology alone will not be able to address all the applications, hence integration with other technologies, especially silicon CMOS, becomes increasingly important.

Many factors go into determining the cost competitiveness of GaN, including production volume, cost of manufacturing, as well as the premium that one would attach to the unique high-power characteristics and capabilities of GaN. A significant proportion of the cost of a GaN technology is born by the cost of epitaxial growth of GaN, and wafer substrate cost. The idea here is to drive towards lower cost per unit area by using more cost-effective substrates, and larger wafer sizes (measured by diameter).

Today, there does not exist a cost-effective method to produce large bulk native GaN substrates. State-of-the-art native GaN substrates are typically 50 mm in diameter, grown by vapor phase epitaxy techniques such as halide VPE (HVPE), or halide-free VPE (HFVPE) [19]. Absence of large bulk GaN substrates, GaN crystals are typically epitaxially grown on foreign substrates such as SiC, sapphire, and silicon. Today, SiC wafers are available up to 200 mm in diameter, but GaN-on-SiC wafers in production are of 75–100 mm diameter in size, with plans announced to manufacture GaN on 150 mm SiC wafers. GaN-on-sapphire wafers are more commonly found in the LED market with 100–150 mm diameter in production, while sapphire wafers up to 300 mm diameter are available in small quantities. GaN-on-silicon (111) wafers are pervasive today due to the low cost of silicon wafers and its use driven by growth in the mass consumer market for GaN fast-chargers. Most GaN-on-silicon production are on 150 mm, but with the increasing adoption of GaN, the industry is marching towards 200 mm silicon substrates for GaN transistor technology. In scaling up, industry is anticipating gains in production volume, quality, and cost advantage over today's 100 mm and 150 mm wafers. The largest GaN-on-silicon wafer under development measures 300 mm in diameter, first demonstrated by Intel [20]. These 300 mm GaN-on-silicon wafers are processed in the most advanced CMOS fab, and excellent wafer uniformity has been demonstrated. From today's point of view, GaN-on-silicon appears to be the front-runner in the race towards a cost-effective GaN technology.

It is not straightforward to scale a GaN solution to larger diameter wafers on a foreign substrate such as silicon. Because of the mismatch in crystal properties, for e.g., differences in lattice constant and thermal expansion coefficients, between GaN and silicon, this leads to issues in crystal defectivity, wafer stress/bow control and cracks, and even wafer breakages [21]. These issues become worse for larger wafer sizes. Careful engineering of stress compensating layers in the buffer of the

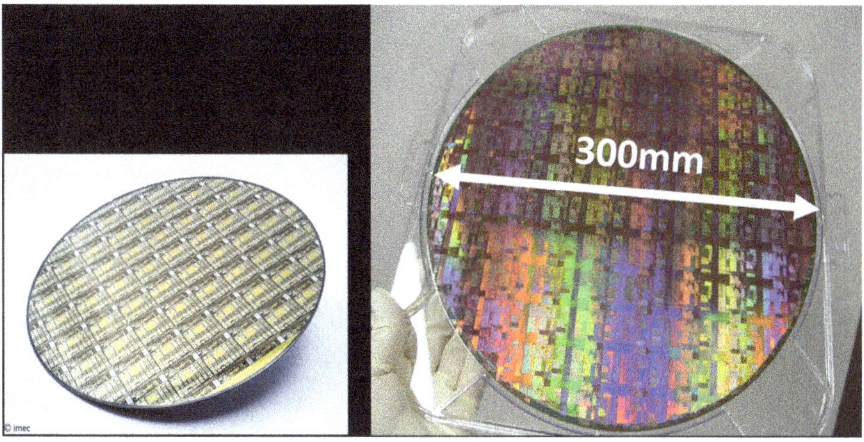

Fig. 7.38 (Left) 200 mm GaN-on-silicon wafer fabricated by IMEC [25], (right) 300 mm GaN-on silicon wafer fabricated by Intel [26]

GaN crystals are needed to achieve good epitaxial quality of GaN on large-diameter wafers [22–24].

Figure 7.38 shows an example of the 200 mm GaN-on-silicon wafer from IMEC, and 300 mm GaN-on-silicon wafer from Intel. Significant progress has also been made in demonstrating high-performing GaN devices on these large substrates. This will be discussed in the next chapter.

7.5.1 High Performance and Scaling of GaN Transistors

The scaling of GaN transistors is motivated by the push to achieve higher performance. It follows from physics that transistor scaling, i.e., shrinking of the channel length would lead to higher cut-off frequencies, f_T/f_{MAX}, and better transistor figure-of-merits (FOM), which translates into higher gains and better efficiencies for circuits. Scaling was initially driven by research efforts in the RF application space on GaN-on-SiC platforms [27] to achieve higher cut-off frequencies, f_T/f_{MAX}. This is followed subsequently, by research effort in power electronics on GaN-on-silicon platforms [20, 28]. The effort by the HRL team led to the demonstration of a record 20 nm channel length GaN Schottky HEMT on SiC wafer that achieved f_T of 310 GHz and f_{Max} of 364 GHz [27].

The various achievements of scaled GaN transistors with their accomplished f_T/f_{MAX} are compiled in Fig. 7.39 [29]. The shortest channel length in GaN transistors so far, have been demonstrated by HRL (20 nm) and Intel (30 nm). HRL's 20 nm GaN HEMT on SiC wafer achieved f_T/f_{MAX} ~ 331/517 GHz [30], and Intel's 30 nm GaN MOSHEMT on silicon wafers achieved f_T/f_{MAX} ~ 300/400 GHz [28] and 130/680 [31].

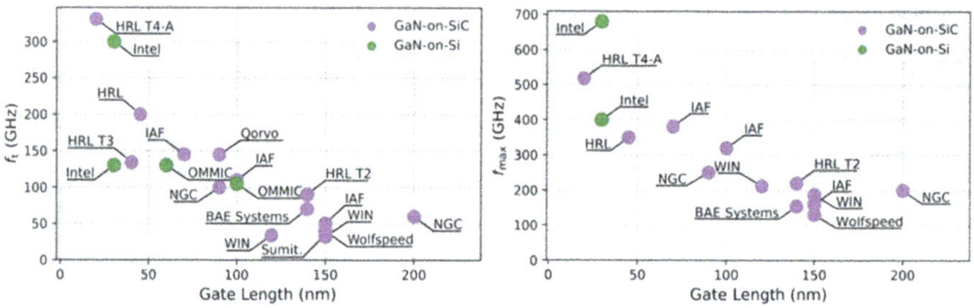

Fig. 7.39 Benchmarking of current-gain cutoff frequency, f_T and maximum oscillation frequency, f_{MAX} vs gate length from [29]

Besides f_T/f_{MAX}, the scaling of GaN transistors also leads to better figure-of-merits (R_{DSON} x Q_{GG}) for power applications. Shorter channel length results in smaller ON-resistance, R_{DSON}, and smaller gate charge, Q_{GG}. GaN transistor technology combines these excellent figure-of-merits (R_{DSON} Q_{GG}) with its ability to sustain high drain breakdown field, a characteristic attributed to its large bandgap. Because of this, in principle, at any given voltage of operation, a properly scaled GaN transistor would ultimately match or surpass a transistor made from a semiconductor of a smaller bandgap such as silicon, silicon germanium, III-V, or SiC, in an application in power electronics (as measured by Baliga's FOM) and RF (as measured by Johnson's FOM).

However, scaling of GaN transistor has its share of challenges and difficulties. In this paragraph, we will start by examining the scaling challenges in Schottky GaN HEMT. Schottky GaN HEMT is the state-of-the-art transistor architecture used in RF applications. It employs a Schottky gate that consists of a metal gate formed on a semiconductor barrier such as AlGaN/AlN or AlInN/AlN, over the GaN channel (Fig. 7.40).

Most Schottky GaN HEMTs are depletion-mode (normally-ON). A well-designed stack of $Al_{0.5}Ga_{0.5}N/AlN$, e.g., [27] could result in enhancement-mode (E-mode) operation (normally-OFF with threshold voltage greater than zero). However, classic scaling law [32] requires that the semiconductor Schottky barrier be made thinner as the channel length is scaled shorter to control short-channel effects. However, thinning the barrier also results in reduction of 2DEG charge density in the access regions between gate and source-drain, and increases the access resistance. This would lead to degraded transistor performance. Moreover, thinning the barrier results in such high gate leakage currents as the result of (quantum) tunneling of electrons through the barrier, that a scaled thin-barrier Schottky GaN HEMT may no longer be used.

The depletion-mode Schottky GaN HEMT is not used for power electronics due to its normally-ON nature. The E-mode pGaN HEMT (Fig. 7.41) is the state-of-the-art transistor architecture used for power electronics. It employs a p-doped GaN layer in a p-GaN/AlGaN/i-GaN junction at the gate of the transistor. P-type doping in GaN is accomplished by doping GaN with Mg-impurity [33, 34]. Not to be

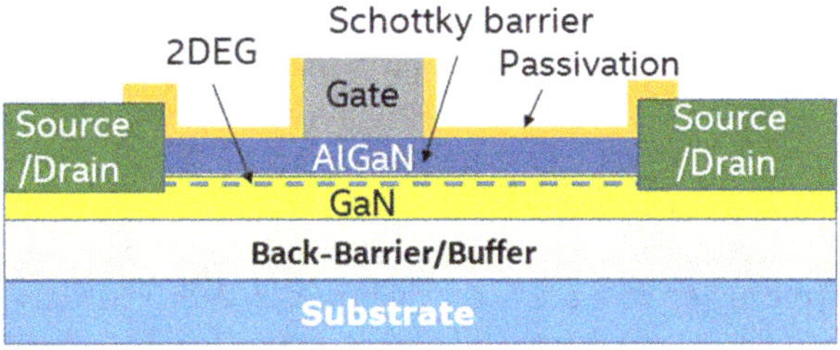

Fig. 7.40 Schematic of an AlGaN/GaN HEMT

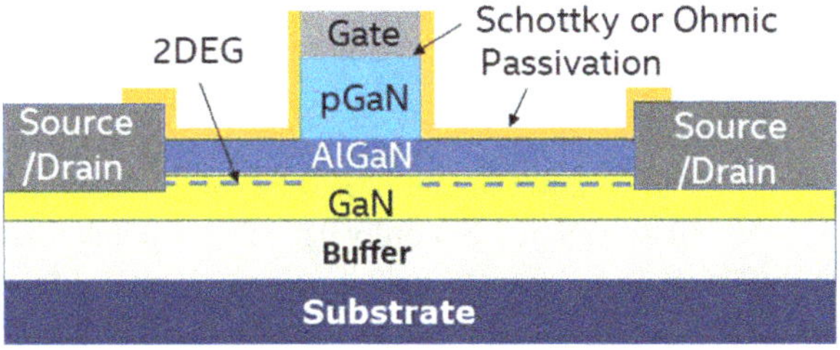

Fig. 7.41 Schematic of an e-mode pGaN HEMT

mistaken for the p-channel GaN transistor [35, 36], a pGaN HEMT is a n-channel device. In a pGaN HEMT, the pGaN layer is not utilized as a transistor channel. Instead, it is used to deplete the 2D electron gas (2DEG) in the GaN channel to achieve the condition for positive threshold voltage, or enhancement-mode (E-mode) operation. However, because of the limited active p-dopant concentration in GaN, it must be made sufficiently thick (typically 50~70 nm) to deplete the high density 2DEG. A thin p-GaN layer would not be able to deplete the high density 2DEG in the channel, hence resulting in the transistor operating in the depletion-mode. Assuming that one could maintain E-mode operation by thinning the underlying AlGaN (polarization charge-inducing layer) while at the same time, thinning the p-GaN layer, one would then have to contend with a high gate leakage current due to carriers (electrons) tunneling through the thin p-GaN/AlGaN barrier. Thinning the AlGaN layer could pose another problem, i.e., the 2DEG in the access region outside the gate would become less conductive, hence incurring a penalty in the transistor ON-resistance. These limitations make it very challenging to scale an E-mode pGaN HEMT.

7.5.2 Reliability

Much progress has been made in understanding the reliability of GaN transistors [37]. However, the bulk of the literature has been devoted to the reliability of GaN transistors with relatively long channel lengths. Traps have become the focal point of reliability studies in GaN transistors [38–43]. These traps cause degradation such as dynamic shifts and memory effects in ON-resistance and threshold voltage when the GaN transistor is subjected to stress voltages applied to the gate (gate lag effects) or the drain (drain lag or "current collapse" effects). While much of these insights are still relevant for scaled short-channel devices, many aspects of reliability need to be reexamined due to several reasons.

High electric fields can be attained in scaled devices, even though these devices typically operate at lower voltages <48 V. Scaled GaN transistors also employ new epitaxial structures such as carbon-doped GaN buffer, and AlGaN back-barrier for carrier confinement. High carbon-doping in GaN buffer, while CMOS-compatible, aggravates current collapse. The use of AlGaN back barriers, in general, help to counter (suppress) current collapse and hot-electron trapping, which leads to improved device reliability [43].

Moreover, to enable scaling, the introduction of new transistor technologies is accompanied by introduction of new processes, e.g., new epitaxial materials such as AlScN and InAlN for Schottky HEMT, N-polar GaN epitaxy for N-polar GaN HEMT, and atomic layer deposition (ALD) and atomic layer etch (ALE) for GaN MOSHEMT. Much work is needed to study and understand the reliability characteristics of these GaN transistors developed using these new materials and processes.

We will cover several key emerging trends driving innovations and research in GaN transistor technology. First, we will describe the research towards demonstrating high performance GaN transistors on large 200–300 mm GaN-on-silicon wafers. In device research, emerging trends include (a) wider-bandgap AlN, AlScN HEMT, (b) N-polar GaN HEMT, and (c) GaN MOSHEMT. A common theme is the need to scale GaN transistors to achieve higher performance. This follows from fundamental device physics that a GaN transistor with shorter (scaled) channel length stores less charge and can operate at higher speeds, hence providing higher (amplification) gains and better switching figure-of-merits (higher density and efficiency). The scaling of the transistor feature sizes must also be accompanied by better contact technologies such as epitaxial regrown N+ source-drain contacts and adoption of advanced backend technology such as copper interconnect. We will then discuss progress in reliability research for scaled GaN transistors. The prospects of integration of GaN with other transistor technologies, especially silicon CMOS, will be addressed in the next paragraphs.

7.5.3 Towards 300 mm GaN Wafers

Figure 7.42 shows data from the 300 mm GaN-on-silicon wafer by Intel [20]. It shows high uniformity in electrical Hall characteristics, and good wafer bow, which is a key criterion for running wafers through the fab. These 300 mm GaN wafers meet the stringent requirements to realize transistor fabrication in the most advanced CMOS fab. Scaling to large-diameter substrates such as 300 mm silicon also enable the adoption of advanced 300 mm process technologies such as high-k gate

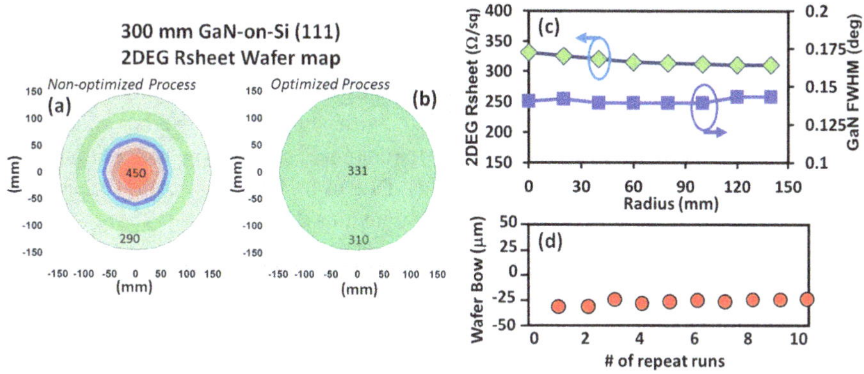

Fig. 7.42 Distribution of the 2DEG sheet resistance on a 300 mm GaN-on-silicon (111) wafer grown by (**a**) non-optimized process, (**b**) optimized process, (**c**) uniform distribution of 2DEG sheet resistance, and crystal quality (FWHM of GaN crystal) achieved across the 300 mm high-resistivity Si (111) wafer, and (**d**) the consistent fab-compatible wafer bow achieved across multiple growths [20]

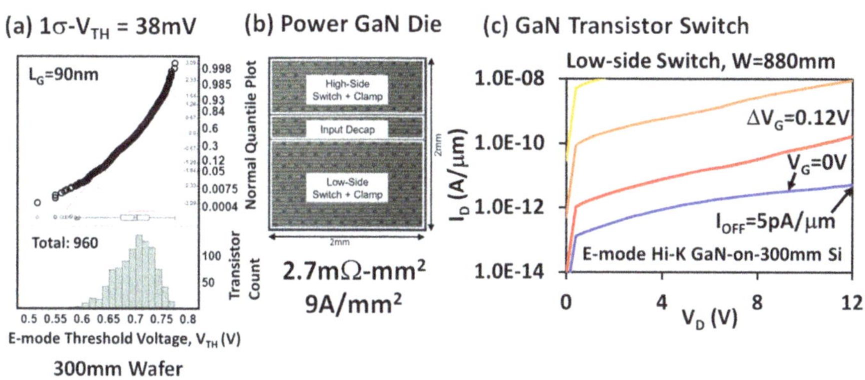

Fig. 7.43 (**a**) V_{TH} distribution across a 300 mm GaN wafer showing highly uniform 300 mm process with small 1 σ-standard deviation in V_{TH}, and tight V_{TH} spread; (**b**) Power GaN die photo with measured R_{DSON} and capable of high current density; (**c**) Low leakage demonstrated for a large 880 mm wide GaN switch transistor fabricated using the 300 mm GaN-on-Si process [28]

dielectric, DUV lithography, copper interconnect, and 3D integration [28]. The work achieves a milestone with the demonstration of a highly uniform threshold voltage variation of 38 mV (1 σ) on 300 mm silicon (Fig. 7.43) [28]. Large-scale GaN switch with a transistor width of 880 mm, ON-resistance of 2.7 mΩ/mm^2, and capable of high current density ~9A/mm^2, was demonstrated (Fig. 7.29) [28]. Low leakage ~5pA/µm is measured for an 880 mm wide low-side switch GaN transistor on 300 mm silicon at $V_D = 12$ V, $V_G = 0$ V [28]. These demonstrations are excellent indicators for the status and capability of the 300 mm GaN-on-Si process.

7.5.4 *Scaling of GaN Transistors*

One promising approach is to employ wider bandgap barriers such as AlN [44] and AlScN [45] for the GaN Schottky HEMT. A wider bandgap barrier such as AlN, allows it to be made thinner to support shorter channels, and yet, be able to maintain sufficiently low leakage (reduced tunneling). Due to its high polarization charge, it can also maintain a high 2DEG sheet charge densities in the access regions. An AlN GaN HEMT [46], e.g., could achieve sheet carrier charge density >3 × 10^{13} cm^{-2} with an AlN barrier thickness of only 3 nm. An aggressively scaled 20 nm channel length AlN (3.5 nm) GaN HEMT by HRL [47] achieved close to 3.5 A/mm of maximum drain current, and f_T/f_{MAX} of 454/444 GHz. Besides AlN, AlScN is another candidate. In addition to a high polarization charge, it also has a much higher dielectric constant (ε_r ~ 21), which is the highest amongst known III-nitride materials [48]. A high polarization charge combined with a high dielectric constant could be very useful in controlling detrimental short channel effects. Recent work achieved a short-channel 90 nm AlScN HEMT with maximum current density of higher than 4 A/mm [48].

N-polar GaN HEMT [49–51] offers another potential solution to scaling. In an N-polar GaN HEMT, the wurtzite GaN crystal is epitaxially grown in the (000 $\bar{1}$) direction instead of the usual Ga-polarity in the (0001) direction [52]. The epitaxial growth is accomplished by plasma-enhanced MBE on SiC and sapphire wafers. Instead of a top barrier, a bottom AlGaN barrier acts as a polarization charge-inducing layer that forms the 2DEG and confines the electrons in the GaN channel. As a result, the GaN 2DEG is above the AlGaN barrier, hence is situated closer to the surface. Because there is no AlGaN top barrier between the gate and channel, excellent contacts can be achieved by regrown source and drain from the (top) surface of the N-polar GaN crystal. With a thin GaN channel and small distance between gate and channel, the N-polar GaN transistor could offer excellent short channel control and enable short-channel lengths for higher performance. The results measured from recent N-polar GaN HEMTs are promising. A 50 nm gate length N-polar GaN demonstrated peak f_T of 176 GHz and peak f_{MAX} of 307 GHz, and a 77 nm gate length device achieved 94 GHz large signal performance with a linear gain of 10.5 dB, 50.2% peak power-added efficiency (PAE), and power density of 2.8 W/mm [53].

AlN HEMT, AlScN HEMT and N-polar GaN HEMT are depletion-mode transistors (normally-ON) because there exists a layer of polarization charge-inducing layer in the gate region of the transistors. In many applications, it is advantageous to have an enhancement-mode (normally-OFF) transistor. For example, an enhancement-mode (E-mode) GaN transistor is required for power electronics. Moreover, a negative power supply is not needed, hence simplifying, and improving circuit design. In fact, an enhancement-mode GaN MOSHEMT offers another potential solution to scaling—its structure is shown in Fig. 7.44.

In a GaN MOSHEMT, the gate region employs a metal-oxide-semiconductor or MOS structure instead of the p-GaN or Schottky barrier. The MOSHEMT structure retains the extended drain structure that maintains the high breakdown field. The

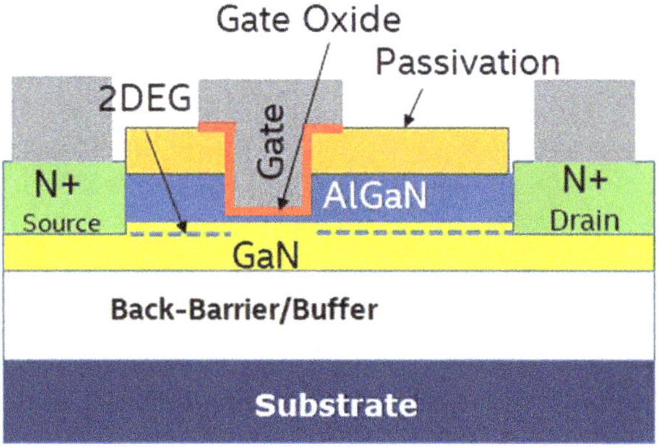

Fig. 7.44 Schematic of an E-mode GaN MOSHEMT [20]

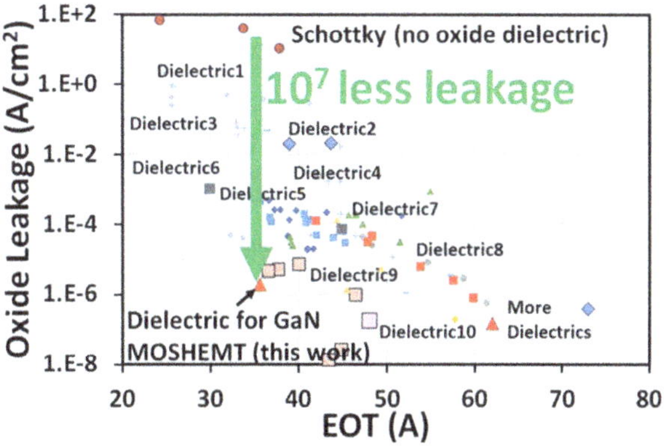

Fig. 7.45 The leakages and equivalent oxide thicknesses (EOT) of various ALD high-k gate dielectrics developed and investigated for the GaN MOSHEMT in [55]

gate oxide or gate dielectric provides an insulating layer that suppresses the gate leakage. The gate oxide therefore, allows the (AlGaN) barrier to be thinned or even completely removed. A GaN MOSHEMT provides a solution to scaling of electrical oxide thickness (EOT) to angstrom-scale [28] to achieve good short-channel control. An example of such an effort is embodied in the work done by the Intel team in [28]. The team wielded insights from Moore's law [54] and Dennard MOSFET scaling [32], which led to the use of high-k gate dielectric technology. With high-K dielectric, the gate leakage can be reduced by up to 7 orders of magnitude (Fig. 7.45). With atomic-layer-etch (ALE) and advanced deep-ultraviolet (DUV) lithography in a 300 mm fab, the Intel team demonstrated a 30 nm gate enhancement-mode high-k gate dielectric GaN MOSHEMT on 300 mm silicon. Figure 7.46 shows the scaling of high-K gate dielectric on GaN from T_{OXE} = 21.5 Å to 14.8 Å. For comparison, a typical equivalent T_{OXE} for an E-mode pGaN HEMT could be more than ~10× thicker. The achieved T_{OXE} of 14.8 Å is the thinnest gate stack demonstrated on a GaN transistor. In Fig. 7.47, the I_D-V_G characteristics of the L_G = 30 nm E-mode high-k GaN transistor is compared with a Schottky GaN HEMT of equivalent channel length. It illustrates the better controlled short channel effect with improved DIBL/SS, and higher transconductance G_M achieved for the high-k GaN MOSHEMT

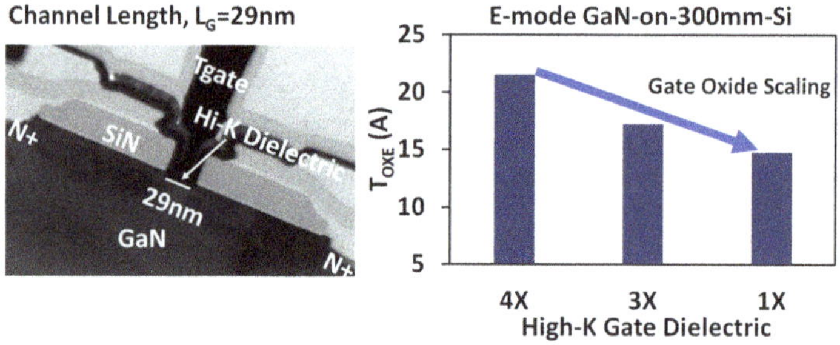

Fig. 7.46 (Left) TEM micrograph of a short channel length of 29 nm E-mode high-k GaN MOSHEMT in [28]; (right) high-K gate dielectric scaling from $T_{OXE} = 21.5$ Å to 14.8 Å, measured by C-V. These are record thin gate stacks demonstrated on a GaN transistor [28]

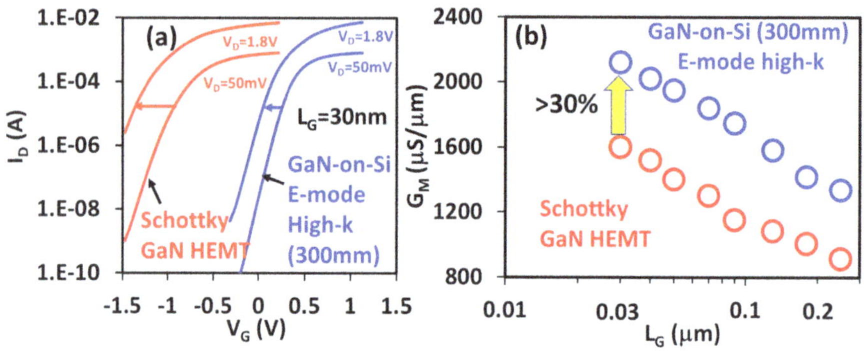

Fig. 7.47 (**a**) Comparison of I_D-V_G of a L_G = 30 nm E-mode high-K GaN MOSHEMT vs a L_G = 30 nm Schottky gate GaN HEMT shows better controlled short channel effect with improved DIBL/SS and (**b**) higher transconductance, G_M for the E-mode high-k GaN MOSHEMT [28]

transistor due to the scaling of T_{OXE} with high-k gate dielectric. The 30 nm high-k GaN MOSHEMT transistor achieves cut-off frequencies, f_T = 300 GHz and f_{Max} = 400 GHz, a record for GaN-on-Si transistors (Fig. 7.48).

Field-plate technology is frequently used to extend the voltage capability range of GaN transistors. By simultaneously scaling the channel length and the field-plate length to submicron dimensions of <100 nm, the penalty (increase in parasitic capacitances) incurred by the incorporation of field-plates can be minimized. An example of this work is presented in [31] (Fig. 7.49).

A measurement of the breakdown of (a) L_G = 30 nm and (b) L_G = 90 nm GaN transistor with source field-plates in Fig. 7.50 shows capability to withstand drain voltage (BV_{DS}) of 80 V (a record demonstration of short channel GaN transistors) and 100 V respectively, with excellent ON-resistance (R_{ON}). A high breakdown field strength ~1.7 MV/cm can be accomplished for the field-plated GaN MOSHEMT transistors [31]. A source field-plate (Fig. 7.49) with a length (e.g., 90 nm) that is comparable to the channel length (e.g., 30 nm) can act as a cascoded gate to boost the gain of the GaN transistor. This source field-plate also shields the first gate

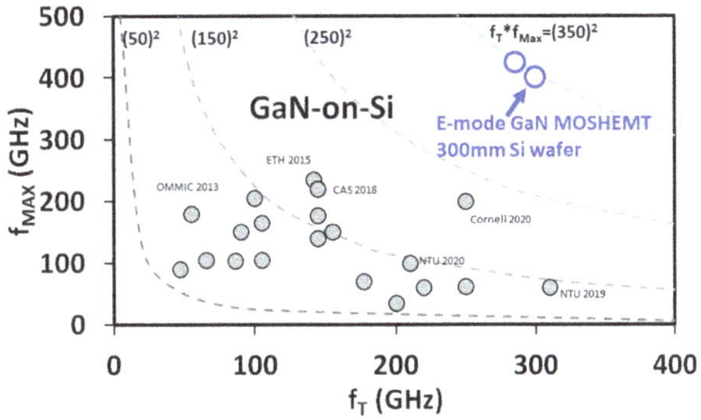

Fig. 7.48 Benchmarking of the f_T/f_{Max} achieved by GaN MOSHEMT in literature for GaN-on-Si [28]

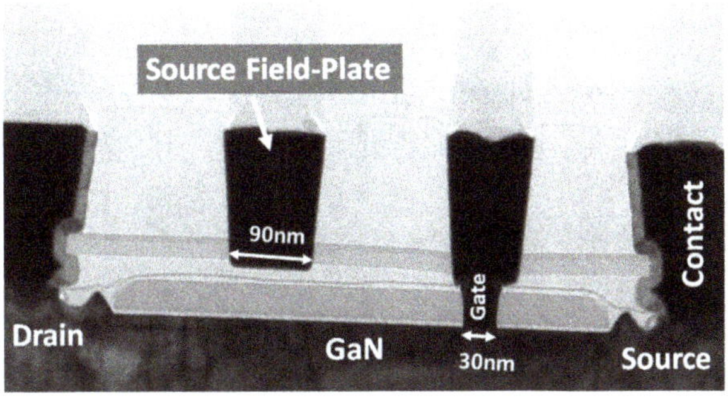

Fig. 7.49 Shows the source field-plated GaN MOSHEMT in [31]. Note that the submicron spacings and feature sizes of the source and gate field-plate are achievable only with advanced lithographic capabilities

(30 nm) from the drain, allowing significantly smaller values of gate-to-drain capacitance, C_{GD}. As a result, a short-channel $L_G = 30$ nm GaN MOSHEMT with a 90 nm source field-plate, achieved a record cut-off frequency, $f_{Max} = 680$ GHz ($f_T = 130$ GHz).

With field-plates, the GaN-on-Si (111) MOSHEMT technology (Fig. 7.50) achieves power FOM (= $R_{ON} \times Q_{GG}$), 10–17× better than Si LDMOS and pGaN HEMT at 12–15 V; and ~20× better than state-of-the-art E-mode pGaN HEMT and ~30× better than Si LDMOS at 40 V [31].

Transistor contacts and backend interconnects are also critical enablers for scaling. Contact and interconnect routing resistances need to be minimized to ensure that the intrinsic transistor gains from scaling is not diminished by extrinsic resistances. The best transistor contact technology for GaN transistor is provided by epitaxially regrown N+ source-drain [27, 56, 57]. State-of-the-art regrown source-drain contacts could reduce the contribution of contact resistances to <~0.1 ohm-mm, i.e., less than a third of the total ON-resistance of the best GaN transistors. Today, GaN transistors that are commercially available in the industry, either employ aluminum or gold interconnects in the backend. Both aluminum and gold are worse thermal and electrical conductors than copper. Therefore, employing copper interconnects for GaN transistors offers advantages in both electrical conductivity and heat dissipation. Copper interconnect for scaled GaN transistors is first

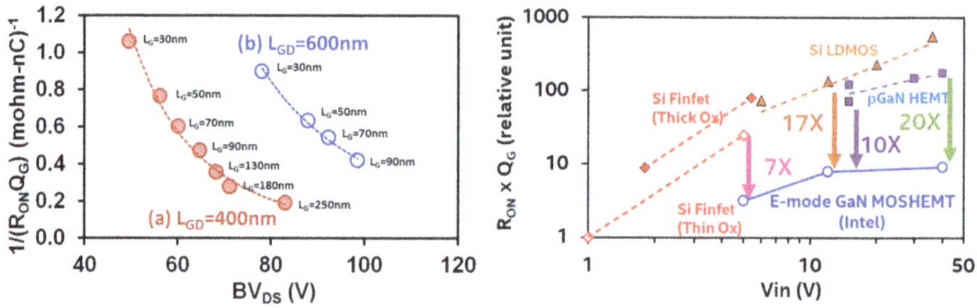

Fig. 7.50 (Left) shows the Figure-of-Merits (FoM) of a GaN MOSHEMT employing field-plate technology, extending its breakdown voltage capability up to 100 V. For the first time, the FoM of GaN transistor achieved 1 (mohm-nC)$^{-1}$. (Right) Figure-of-merit (R_{ON} x Q_G) vs application voltage, Vin for various transistor technologies employed for power electronics [31]

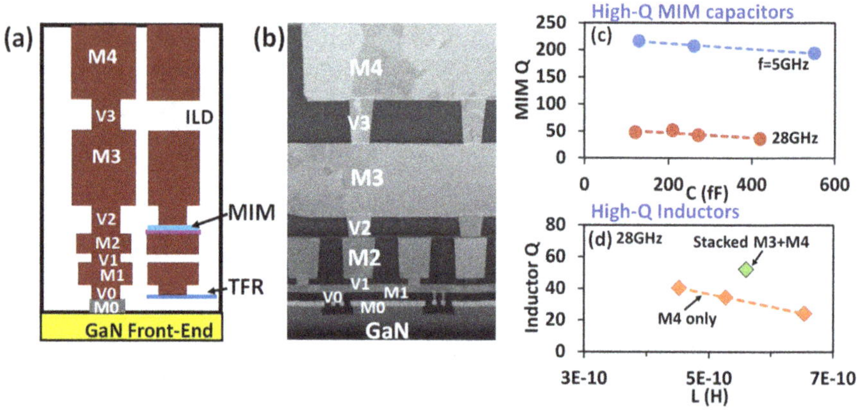

Fig. 7.51 (**a**) Schematic of the four-layer Cu backend interconnect stack, including metal-insulator-metal (MIM) capacitor and thin-film resistor (TFR), (**b**) SEM micrograph of the fabricated Cu backend interconnect over the E-mode high-k GaN MOSHEMTs on 300 mm silicon wafer, (**c**) fabricated MIM capacitors showing Q = 50 (28GHz), and (**d**) inductors showing Q = 52 (28 GHz) [20, 28].

demonstrated by Intel [20, 28, 31]. Besides enabling low resistance paths to the GaN transistors, it also enables high-Q metal-insulator-metal (MIM) and inductor (Fig. 7.51). The scaled GaN MOSHEMT with regrown N+ contact and copper backend technologies enabled the demonstration of fully integrated mm-Wave power amplifier and low-noise-amplifier [58, 59] that showed higher power density (> ~3X), high efficiency, and smaller solution area (> ~ 3X shrink), when compared to state-of-the-art silicon/SOI, SiGe, III-V, and GaN technologies (Table 7.1).

Looking ahead, research into new architectures including GaN finFET [28, 64], and superlattice castellated FETs (SLCFET) [65] are explored for further scaling GaN and to achieve even higher current densities. The scaling of GaN transistor combines high performance of GaN with high voltage capability; enhances the

Table 7.1 Shows that a mm-wave power amplifier implemented in a 90 nm GaN MOSHEMT process [59] can have better FOM, higher power density, and smaller solution area than other technologies [60–63]

Technology	90 nm GaN MOSHEMT 300 mm Si#	0.15 μm GaAs pHEMT	0.13 μm SiGe	45 nm RF SOI	65 nm CMOS
Topology	Two-stage differential	Two-stage differential	Differential cascode	Mixed-signal Doherty	Four-way combining
V_{DD} (V)	4	5	3.2	2	2.2
f_c (GHz)	30	27	28	27	28
3-dB bandwidth (GHz)	5.5	5$	10.5	13.5	3.5
Linear G_p (dB)	26.6	21.2	19.4	19.1	18
P_{sat} (dBm)	25.3	24.4	22.7	23.3	23.2
PAE_{max} (%)	38.7	43.4	38.1	40.1	35.5
Modulation	5G OFDM 64-QAM	5G OFDM 64-QAM	64-QAM	64-QAM	64-QAM
	11.7 dB PAPR&	N/A	8.3 dB PAPR	6.5 dB PAPR	8.3 dB PAPR
P_{out_avg}@EVM/ACPR	16.2	21.3*	16.2	15.9	16.1
EVM (%)	5.2&	3*	5.5	5.4	4.9
ACPR (dBc)	−31.8&	−20$*	−32	−29.6	−28.3
Core area (mm^2)	0.097	2	0.162	0.52	0.2
FOM+ (dB)	97.3	90.6	86.9	87.1	85.6
Saturated Power (W/mm^2)	3.49	0.14	1.15	0.41	1.04

#Multiple LG available; $Estimated; &No DPD or CFR; *Equalized results;
+FOM(dB) = Psat(dBm) + Gp(dB) + 10log(PAEmax[%]) + 20log(fc[GHz])

figure-of-merits (FOM) of GaN; and opens the door to applications, especially in integrated power electronics and RF mm-wave, that demand ultra-fast switching, high efficiency, and high-power density.

7.5.5 Progress in Reliability of Scaled GaN Transistors

Of all the III-nitrides, AlN has the largest lattice mismatch compared to GaN. In general, the larger the bandgap, the larger the lattice mismatch is between GaN and the barrier material. Such a mismatch needs to be carefully accounted for when designing the transistor stack. A study by researchers at IEMN [66] found, through RF reliability stress testing, that a 3 nm AlN HEMT showed vastly superior reliability over a 4 nm AlN HEMT. The 4 nm AlN HEMT degraded even at room temperature at a much lower V_{DS} of 12 V, whereas the 3 nm AlN HEMT showed no such degradation up to higher V_{DS} of 30 V for 24 hours. This is because a 4 nm thick AlN exceeds the critical thickness for the epitaxial growth of AlN on GaN. Exceeding the critical thickness results in excessive strain in the AlN film, which leads to a degraded reliability for the 4 nm AlN HEMT transistor.

Another important aspect of these new barrier materials is its surface passivation. For example, pulsed measurements of a AlScN HEMT with different surface passivation schemes [67] found that surface passivation of AlScN by SiN resulted in vastly different performance and reliability characteristics compared to passivation by GaN or AlGaN. SiN passivation reduced the dc/RF dispersion of the AlScN HEMT, and as a result, RF gain and output power were significantly improved. However, AlScN HEMT with SiN passivation suffered from a reduced breakdown voltage compared to devices passivated by AlGaN. This was attributed to the low dielectric constant of SiN which resulted in a higher electric field at the drain edge of the gate. Moreover, SiN passivation on AlScN also caused a large positive threshold voltage shift. It is not clear if the 2DEG is depleted by surface effects, or by trapped charges in the SiN film. More work is needed to explore the various passivation schemes for the surface of AlScN, and to understand their effects on the dynamical behavior of AlScN HEMT.

N-polar GaN HEMTs are shown to be capable of low dispersion, high breakdown voltage, and low parasitic resistances. It is important to note that unlike GaN-polar HEMT, in an N-polar GaN HEMT, the bottom AlGaN barrier is "buried," i.e., there is no exposed (Ga-polar) AlGaN surface. Therefore, in an N-polar GaN HEMT, the challenge is in passivating the N-polar GaN surface above the 2DEG. Various techniques including insertion of an AlGaN cap (which also acts as an etch-stop layer), and SiN gate insulator over the N-polar GaN surface, have been explored. These enabled the N-polar GaN HEMTs to push the boundaries with record performances in the W-band. For example, an N-polar GaN HEMT achieved an impressive 1 W of power at 94 GHz [68]. In [69], the dynamic (trapping) performance of 80–100 nm channel length N-polar GaN HEMT with SiN gate insulator was investigated with pulsed and transient measurements. It was shown that a higher Al concentration in the cap layer between the SiN and GaN channel can substantially suppress current collapse. Trapping of carriers (injected from the gate) is observed with large threshold voltage shifts resulting from gate stress. Such instability in threshold voltage could pose reliability issues in a practical application. Despite the encouraging results demonstrated by N-polar GaN HEMTs, more studies are needed to better understand the nature of these traps, and to learn how to passivate them.

High-k gate dielectric GaN MOSHEMT offers a unique approach to passivation. While the extended drain and source region is passivated with a suitable film such as SiN, a gate dielectric with a different (typically higher) dielectric constant can be used for passivation in the gate region. In fact, one of the most difficult challenges encountered in a GaN MOSHEMT is caused by inadequate gate dielectric passivation, which leads to unstable transistor characteristics. The instability is typically referred to in literature as positive bias-temperature-instability (PBTI). Here when under stress from an applied gate voltage (V_G), the threshold voltage (V_T) of the GaN MOSHEMT in the ON-state, drifts higher (shifts positive) over time, due to trapping of carriers at the gate metal-oxide-semiconductor (MOS) stack. As a result of that positive shift in V_T, the transistor gate bias overdrive (V_G-V_T) diminishes over time, causing an increase in the channel resistance (R_{ch}). This contributes to an overall increase in ON-resistance, R_{DSON}, hence, degrading the transistor performance over time. Most research in PBTI of GaN addressed MOSHEMT or MISHEMT transistors with long channel lengths, thick gate oxide T_{OXE} ~ 200–00 Å, and low-k gate dielectrics including SiO_2, SiN, and Al_2O_3 [70, 71]. In [72], the PBTI of GaN MOSHEMT with short channel length of 90 nm, and ultra-thin high-k gate dielectric with T_{OXE} ~ 20 Å is studied. Such thin dielectrics are deposited by atomic layer deposition (ALD). The PBTI of these scaled thin hi-k GaN MOSHEMTs are found to saturate over time, and the degradation eventually stops [72]. The observed PBTI evolution could not be modeled by a conventional PBTI model [73] that assumes a simple power law with a constant time-slope. Instead, the experimental observation could be explained by a complex power law with time-dependent trap dynamics derived from the Reaction-Diffusion (RD) model [74, 75]. In a scaled short channel length GaN MOSHEMT with ultra-thin gate oxide, the impact of PBTI on the ON-resistance, R_{DSON} is small, because the affected channel resistance, R_{ch} component, comprises only a fraction of the total R_{DSON}. Hence, despite the PBTI degradation, the FOM for GaN MOSHEMT remains attractive when compared against Si LDMOS and E-mode pGaN HEMT [72]. The work by Intel shows promise that PBTI in GaN MOSHEMT could be resolved. However, work is still needed to understand the physics behind the RD model, and to further improve the reliability of the MOS stack in a GaN MOSHEMT.

7.5.6 Prospects for Integration with Other Technologies

With the advent of massive compute and AI in data centers, wafer-scale computers, supercomputers, smart cars, etc., and transition to 5G/6G networks, we expect to see an exponential increase in demand for compute power, accompanied by an explosion in the number of connected devices. This will require semiconductor solutions to deliver ever higher performance, efficiency, and density. Solutions will also need to be packed with more functionalities in smaller packages. As it is becoming clear that there is not one transistor technology that can meet all the demands, there will be a need for heterogeneous integration of two or more dissimilar technologies.

While silicon CMOS has become the workhorse of logic/digital, switches, and power electronics, it is becoming clear that to meet the demands of RF applications, especially at high frequency and high power, III-V technologies such as GaAs HBT and InP HEMT, SiGe HBT, and GaN HEMT will be needed. On the other hand, depletion-mode III-V/GaN HEMTs due to its normally-ON nature, and GaAs HBT due to its input being current-driven rather than voltage-driven, could not be readily used for power electronics. To this end, an enhancement-mode GaN MOSHEMT transistor, is well suited for *both* power electronics and RF, with high Baliga's and

Johnson's figure-of-merits. The scaling of GaN MOSHEMT further enhances its prospect for meeting the requirements for future power electronics and high frequency RF applications.

However, unlike silicon, GaN is today primarily a n-channel only technology. P-channel GaN remains challenging due to its low active p-doping and low hole mobility. Without a complementary p-channel, many CMOS functionalities could not be readily implemented on the same GaN die, and must be sourced from a separate companion die.

Today, it is common to find entire modules made from multiple discretes and separate chips of dissimilar technologies soldered on a package or on a board. Progress is being made today in advanced packaging with die-stacking techniques such as die-to-die and die-to-wafer stacking [76–80]. These techniques enable package integration of dissimilar component technologies to achieve higher component count, and smaller packages. Here industry is making steady progress toward commercial high-volume manufacturing. One major advantage of employing advanced packaging for integration is the ability to assemble technologies that are vastly different without the need to re-develop the processes. For example, a III-V die can be packaged with a CMOS die from an advanced leading process node as well as from an older depreciated process node. However, packaging approaches involving die-to-die and die-to-wafer stacking may become increasingly challenging as die sizes and feature sizes shrink. Besides, there are factors that need to be considered including thermals, metal density, warpage control, wafer dicing quality, surface treatment conditions, bond tool accuracy, and particle control [81].

Many research groups are exploring next generation technologies for monolithic *on die* integration of GaN with Si CMOS. This includes techniques such as heterogeneous materials growth/deposition [82], wafer-to-wafer bonding [83], and layer transfer [20, 28, 84]. Monolithic on die integration comes closest to being able to place and "squeeze" two dissimilar semiconductor technologies together to achieve the highest possible solution density and performance. Figure 7.52 shows a survey

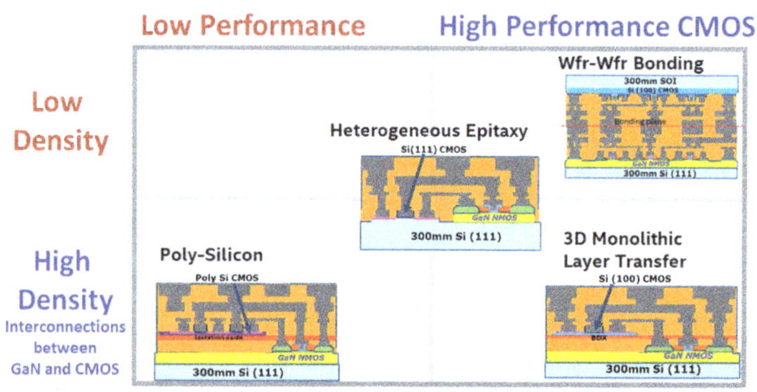

Fig. 7.52 Different approaches to GaN and Si CMOS integration including: (**a**) poly-silicon CMOS, (**b**) heterogeneous epitaxy of GaN and Si (111) CMOS, (**c**) wafer-to-wafer bonding [83], and (**d**) 3D monolithic Si (100) layer transfer using bonding techniques [20, 28, 84]

of the current landscape for different approaches to heterogenous integration of GaN with silicon CMOS [84]. In one example, polycrystalline silicon is readily deposited on GaN wafers and processed to form polysilicon CMOS. However, polysilicon CMOS is inhomogeneous and does not demonstrate sufficiently high carrier mobilities to deliver the required performance.

Selective area heterogenous epitaxy of GaN is explored in [82, 85]. In this approach, GaN is selectively grown by epitaxy in areas where the crystalline Si (111) surface is exposed. Si transistors are fabricated on the Si (111) surface area that is protected during the GaN epitaxy. Because of the lower carrier mobilities in the (111) plane, the drive currents of the Si (111) transistors are lower than that of the standard Si (100) transistors. Moreover, it is very difficult to achieve truly selective area growth by epitaxy. For example, non-uniform polycrystalline AlN/GaN is deposited over the oxide mask [84]. Another challenge in selective area epitaxial growth is to achieve truly *uniform* thickness of the grown GaN/AlN epi everywhere on the wafer. There is a significant difference in the thickness (or height) of the GaN "islands" due to "loading" effects, as a result of the varying dimensions of the exposed Si (111) surface area [84]. Such non-uniformity imposes strict design rules on the process and layout.

Wafer-to-wafer bonding is a promising path to integrate silicon CMOS with GaN. This is accomplished by bonding a silicon CMOS wafer to a GaN wafer. Each wafer is fabricated using a different process flow. An embodiment of such an approach was demonstrated by Raytheon [83]. In such an approach, the wafer substrate on the top is grinded away (sacrificed) and through-wafer-vias (TWV) are fabricated to access the devices on both wafers. The additional routing resistance, and capacitance added by the through-wafer-vias must be carefully considered and accounted for in design. Because the die size on the Si CMOS wafer is constrained to match that on the GaN wafer, the partitioning of the content between the two wafers must be thought through carefully, to ensure maximal utilization of die area on both wafers.

Monolithic integration of GaN with silicon CMOS using a 3D layer transfer approach has made significant strides in recent years [20, 28, 84, 86]. This is a technique that employs wafer bonding in the front-end processing of the GaN transistors. It is accomplished by first bonding a blank silicon donor wafer, to a GaN wafer. Upon separation, the silicon donor wafer is cleaved along a weakened crystal plane, and thereby, a thin layer of silicon channel material is left on the GaN wafer. The process results in the transfer of a thin layer of unpatterned silicon channel material onto a separate GaN wafer. Therefore, unlike wafer-to-wafer bonding, there is no need for alignment of patterns or structures on the donor silicon wafer to those on the GaN wafer. The transferred silicon channel is subsequently patterned and processed into silicon CMOS transistors on the GaN wafer.

There is only one backend in a monolithic integration flow by layer transfer. Both the GaN and silicon CMOS transistors share the same backend copper interconnect. In contrast, a wafer-to-wafer bonding approach requires two backend interconnect stacks: one on the silicon CMOS wafer, another on the GaN wafer, and a TWV, hence, it more than doubles the required interconnect resources. Using a sequentially monolithic process flow, the Intel team demonstrated working GaN MOSHEMT and Si PMOS transistors as well as functional hybrid GaN-Si CMOS inverters on a 300 mm GaN-on-silicon wafer [28]. The GaN transistors were completed before the processing of the silicon transistors. By doing so, the gate stack of the GaN transistors was exposed to the high temperature dopant activation and annealing steps in silicon transistor processing. Because the GaN MOSHEMTs employed a high-k gate dielectric, the high temperature exposure caused the high-k dielectric to densify, accompanied by an increase (degradation) in the gate leakage of the GaN MOSHEMT [84]. To preserve the quality of the high-k gate dielectric, it is therefore important that the high temperature steps for the fabrication of the Si CMOS transistors are completed before the gate dielectrics of the GaN MOSHEMT and Si CMOS transistors are deposited. This is resolved in an improved process flow for 3D monolithic integration of GaN and silicon CMOS transistors, Fig. 7.53 [86]. The Intel team successfully showed that the process can be scaled up to demonstrate large-scale integrated circuits, for instance, an integrated ~497 mm wide CMOS driver and GaN Power FET, "DrGaN." that deploys close to 250,000 transistor gates (Fig. 7.54).

Instead of a two-die solution comprising a CMOS driver chip and separate GaN power chip connected via a parasitic (inductive) package route, the DrGaN combines both in a single chip, thus eliminating the parasitic routing. This realizes the full potential of GaN for higher performance and efficiency. This is the first time that a large-scale integrated circuit of this kind has been successfully fabricated with a 300 mm 3D heterogeneous monolithic process that integrates GaN (group III-V) and silicon (group IV) technologies on the same wafer. With this process, the Intel team further demonstrates a mm-Wave active RF phase-shifter circuit [87], where a seven-bit active phase shifter with an integrated CMOS differential current steering digital-to-analog converter (DAC) is used to control the I/Q Gilbert cells made of GaN transistors. Here the work demonstrates crucially, that it is possible to distribute or "sprinkle," and co-locate where needed, silicon CMOS devices next to a GaN device, at submicron proximity, to optimize performance and area. This would be a

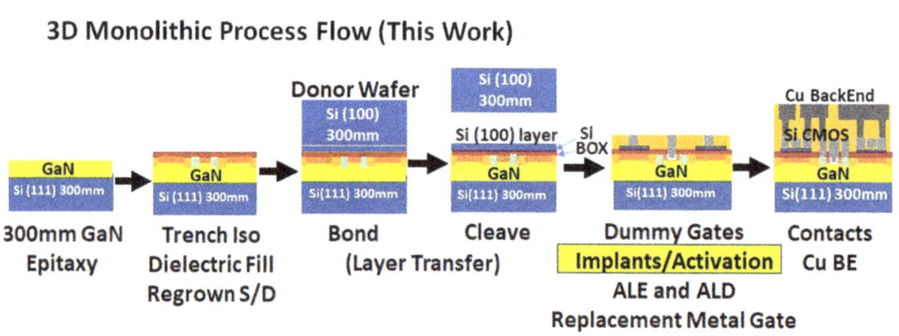

Fig. 7.53 A 3D monolithic integration approach that employs the technique of layer transfer [86]. To preserve the quality of the high-k gate dielectric of the integrated GaN MOSHEMT, the high temperature activation steps for the fabrication of the Si CMOS transistors are completed before the gate dielectric of the GaN MOSHEMT transistors is deposited

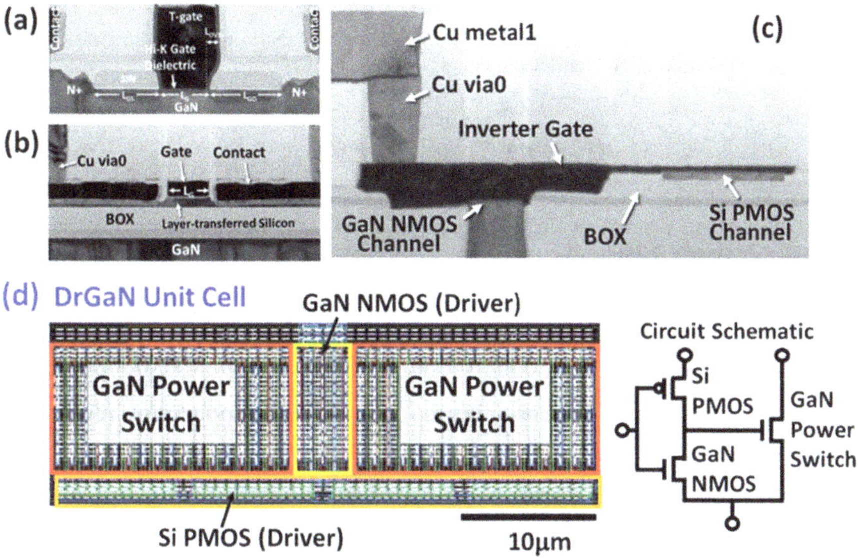

Fig. 7.54 Shows the TEM micrographs of (**a**) GaN MOSHEMT transistor, (**b**) Si PMOS transistor, and (**c**) CMOS inverter, fabricated in this monolithic integrated process on a 300 mm wafer [67]. (**d**) The layout of a 40 × 10 µm^2 DrGaN unit cell consisting of Si PMOS and GaN NMOS driver blocks, and two GaN MOSHEMT power switch transistor blocks. Inset shows the circuit schematic of the electrical connections between the Si PMOS, GaN NMOS and GaN MOSHEMT power switch transistors

unique advantage of a monolithically integrated GaN and silicon CMOS process over wafer-to-wafer bonding or advanced die-stacking approaches. Monolithic integration can achieve integration of two dissimilar best-in-class semiconductor technologies on the same silicon to deliver the best performance at the highest density. Ultimately, industry will likely deploy the entire spectrum of the heterogeneous integration technologies, from advanced packaging (die-to-die, die-to-wafer) to wafer-to-wafer bonding and monolithic integration, to meet the different needs and requirements of a diverse universe of applications.

References

1. Morita, T. et al., 650 V 3.1 mΩcm^2 GaN-based monolithic bidirectional switch using normally-off gate injection transistor, in *IEEE International Electron Devices Meeting*, 2007
2. N. Nain et al., Self-reverse-blocking control of dual-gate monolithic bidirectional GaN switch with Quasi-Ohmic on-state characteristic. IEEE Trans. Power Electron. **37**(9), 10091 (2022)
3. V. Odnoblyudov, Large-diameter, thermal-expansion matched substrates for GaN, in *12th International Conference on Nitride Semiconductors*, July 25th, 2017, Strasbourg
4. T. J. Anderson et al., "Electrothermal evaluation of thick GaN epitaxial layers and AlGaN/GaN high-electron-mobility transistors on large-area engineered substrates," Appl. Phys. Express, vol. 10, no. 12, Nov. 2017, 126501

5. V. Odnoblyudov et al., U.S. Patent, 15,621,335, 15 Feb 2018
6. T. Hikosaka et al., Growth of high-quality >10 μm thick GaN-on-QST with low-dislocation density in the order of 107 /cm^2, Compound Semi Week, 2019
7. M. Alshahed et al., Low-dispersion, high-voltage, low-leakage GaN HEMTs on native GaN substrates. IEEE Trans. Electron. Devices **65**(7), 2939 (2018)
8. M. Tadjer, T. Anderson, *Thermal Management of Gallium Nitride Electronics* (Elsevier, 2022)
9. C.R. Huang et al., Monolithically integrated GaN power and RF ICs on 150mm poly-AlN for envelope tracking power amplifier applications, CS Mantech, 2021
10. A. Vohra et al., Epitaxial buffer structures grown on 200 mm engineering substrates for 1200 V E-mode HEMT application. Appl. Phys. Lett. **120**(26), 261902 (2022). https://doi.org/10.1063/5.0097797
11. D. Kinzer, GaN power IC technology: Past, present, and future," in *Proceedings under International Symposium on Power Semiconductor Devices and Ics*, 2017, pp. 19–24
12. X. Li et al., Demonstration of GaN integrated half-bridge with on-chip drivers on 200 mm engineered substrates. IEEE Electron Device Lett. **40**, 1499–1502 (2019)
13. X. Li et al., Integration of GaN Power ICs on 200 mm Engineered Substrates, in *Procedings of International Conference on Compound Semiconductor Manufacturing Technology (CS-MANTECH)*, May 2020, pp. 241–244
14. J. Hu et al., Materials and processing issues in vertical GaN power electronics. Mater. Sci. Semicond. Process. **78**, 75–84 (2018). https://doi.org/10.1016/j.mssp.2017.09.033
15. K. Mukherjee et al., Challenges and perspectives for vertical GaN-on-Si trench MOS reliability: From leakage current analysis to gate stack optimization. Materials **14**, 2316 (2021). https://doi.org/10.3390/ma14092316
16. Y. Zhang et al., Origin and control of OFF-state leakage current in GaN-on-Si vertical diodes. IEEE Trans. Electron Devices **62**(7), 2155–2161 (2015). https://doi.org/10.1109/TED.2015.2426711
17. A. Tanaka et al., Structural and electrical characterization of thick GaN layers on Si, GaN, and engineered substrates. J. Appl. Phys. **125**(8) (2019)
18. K. Geens et al., Route toward commercially manufacturable vertical GaN devices. IEEE Trans. Electron Devices. https://doi.org/10.1109/TED.2023.3340661
19. R. Kucharski et al., Growth of bulk GaN crystals. J. Appl. Phys. **128**(5), 050902 (2020). https://doi.org/10.1063/5.0009900
20. H.W. Then et al., 3D heterogeneous integration of high performance high-K metal gate GaN NMOS and Si PMOS transistors on 300mm high-resistivity Si substrate for energy-efficient and compact power delivery, RF (5G and beyond) and SoC applications, in *2019 IEEE International Electron Devices Meeting (IEDM)*, (San Francisco, 2019), pp. 17.3.1–17.3.4. https://doi.org/10.1109/IEDM19573.2019.8993583
21. L. Di Cioccio et al., From epitaxy to converters topologies what issues for 200 mm GaN/Si? in *2015 IEEE International Electron Devices Meeting (IEDM)*, (Washington, DC, 2015), pp. 16.5.1–16.5.4. https://doi.org/10.1109/IEDM.2015.7409712
22. B. De Jaeger et al., Au-free CMOS-compatible AlGaN/GaN HEMT processing on 200 mm Si substrates, in *2012 24th International Symposium on Power Semiconductor Devices and ICs*, (Bruges, 2012), pp. 49–52. https://doi.org/10.1109/ISPSD.2012.6229020
23. U. Peralagu et al., CMOS-compatible GaN-based devices on 200mm-Si for RF applications: Integration and performance, in *2019 IEEE International Electron Devices Meeting (IEDM)*, (San Francisco, 2019), pp. 17.2.1–17.2.4. https://doi.org/10.1109/IEDM19573.2019.8993582
24. E. Morvan et al., "6.6W/mm 200mm CMOS compatible AlN/GaN/Si MIS-HEMT with in-situ SiN gate dielectric and low temperature ohmic contacts", 2023 IEEE International Electron Devices Meeting (IEDM), San Francisco, 2023, pp. 38.3.1–38.3.4
25. S. Decoutere, N. Ronchi, Packaging solution for GaN-on-200mm Si power devices, May 30, 2018, www.imec-int.com/en/imec-magazine/imec-magazine-june-2018/packaging-solution-for-gan-on-200mm-si-devices

26. H.W. Then, Stacking GaN and silicon transistors on 300 mm silicon, in *Compound Semiconductor*, 18 May 2020. compoundsemiconductor.net/article/111291/Stacking_GaN_and_silicon_transistors_on_300_mm_silicon/feature
27. K. Shinohara et al., Deeply-scaled self-aligned-gate GaN DH-HEMTs with ultrahigh cut-off frequency, in *2011 International Electron Devices Meeting*, (Washington, 2011), pp. 19.1.1–19.1.4. https://doi.org/10.1109/IEDM.2011.6131582
28. H.W. Then et al., Advanced scaling of enhancement mode high-K gallium nitride-on-300mm-Si (111) transistor and 3D layer transfer GaN-silicon FinFET CMOS integration, in *2021 IEEE International Electron Devices Meeting (IEDM)*, (San Francisco, 2021), pp. 11.1.1–11.1.4. https://doi.org/10.1109/IEDM19574.2021.9720710
29. P. Neininger et al., Advances in GaN devices and circuits at higher mm-wave frequencies. e-Prime **4**, 100177 (2023). https://doi.org/10.1016/j.prime.2023.100177
30. M. Micovic, et al., Ka-band LNA MMIC's realized in Fmax >580 GHz GaN HEMT technology, in *IEEE Compound Semiconductor Integrated Circuits and Technology Symposium*, 2016, pp. 1–4. https://doi.org/10.1109/CSICS.2016.7751051
31. H.W. Then et al., Scaled submicron field-plated enhancement mode high-K gallium nitride transistors on 300mm Si (111) wafer with power FoM (RON xQGG) of 3.1 mohm-nC at 40V and fT/fMAX of 130/680GHz. IEEE Int. Electron Devices Meeting Tech. Dig. **111**, 35.1.1–35.1.4 (2022). https://doi.org/10.1109/iedm45625.2022.10019373
32. R.H. Dennard, F.H. Gaensslen, H.-N. Yu, V.L. Rideout, E. Bassous, A.R. LeBlanc, Design of ion-implanted MOSFET's with very small physical dimensions. IEEE J. Solid State Circuits **9**(5), 256–268 (1974). https://doi.org/10.1109/JSSC.1974.1050511
33. S. Nakamura et al., Jpn. J. Appl. Phys. **31**, L139–L142 (1992). https://doi.org/10.1143/JJAP.31.L139
34. S. Nakamura et al., Jpn. J. Appl. Phys. **31**, 1258–1266 (1992). https://doi.org/10.1143/JJAP.31.1258
35. S.J. Bader et al., Prospects for wide bandgap and Ultrawide bandgap CMOS devices. IEEE Trans. Electron Devices **67**(10), 4010–4020 (2020). https://doi.org/10.1109/TED.2020.3010471
36. N. Chowdhury, Q. Xie, M. Yuan, K. Cheng, H.W. Then, T. Palacios, Regrowth-free GaN-based complementary logic on a Si substrate. IEEE Electron Device Lett. **41**(6), 820–823 (2020). https://doi.org/10.1109/LED.2020.2987003
37. J.P. Kozak et al., Stability, reliability, and robustness of GaN power devices: A review. IEEE Trans. Power Electron. **38**(7), 8442–8471 (2023). https://doi.org/10.1109/TPEL.2023.3266365
38. S.C. Binari, P.B. Klein, T.E. Kazior, Trapping effects in GaN and SiC microwave FETs. Proc. IEEE **90**(6), 1048–1058 (2002). https://doi.org/10.1109/JPROC.2002.1021569
39. M. Faqir, G. Verzellesi, G. Meneghesso, E. Zanoni and F. Fantini, "Investigation of high-electric-field degradation effects in AlGaN/GaN HEMTs," in IEEE Trans. Electron Devices, vol. 55, no. 7, pp. 1592–1602, July 2008, doi: https://doi.org/10.1109/TED.2008.924437
40. G. Meneghesso et al., Reliability of GaN high-electron-mobility transistors: State of the art and perspectives. IEEE Trans. Device Mater. Rel. **8**(2), 332–343 (2008). https://doi.org/10.1109/TDMR.2008.923743
41. J.A. del Alamo, J. Joh, GaN HEMT reliability. Microelectron. Reliab. **49**(9/11), 1200–1206 (2009). https://doi.org/10.1016/j.microrel.2009.07.003
42. M. Meneghini et al., GaN-based power devices: Physics, reliability, and perspectives. J. Appl. Phys. **130**(18), 181101 (2021). https://doi.org/10.1063/5.0061354
43. E. Zanoni et al., Microwave and Millimeter-Wave GaN HEMTs: Impact of Epitaxial Structure on Short-Channel Effects, Electron Trapping, and Reliability. IEEE Trans. Electron Devices. https://doi.org/10.1109/TED.2023.3318564
44. F. Medjdoub et al., "Low on-resistance high-breakdown normally off AlN/GaN/AlGaN DHFET on Si substrate," in IEEE Electron Device Lett., vol. 31, no. 2, pp. 111–113, Feb. 2010, doi: https://doi.org/10.1109/LED.2009.2037719
45. A.J. Green et al., ScAlN/GaN high-electron-mobility transistors with 2.4 a/mm current density and 0.67 S/mm transconductance. IEEE Electron Device Lett. **40**(7), 1056–1059 (2019). https://doi.org/10.1109/LED.2019.2915555

46. P. Döring, S. Krause, C. Friesicke, R. Quay, Theoretical Limits of the Matching Bandwidth and Output Power of AlScN-Based HEMTs. IEEE Trans. Electron Devices. https://doi.org/10.1109/TED.2023.3334224
47. K. Shinohara et al., Scaling of GaN HEMTs and Schottky diodes for submillimeter-wave MMIC applications. IEEE Trans. Electron Devices **60**(10), 2982–2996 (2013). https://doi.org/10.1109/TED.2013.2268160
48. J. Casamento et al., AlScN high electron mobility transistors: Integrating high piezoelectric, high K dielectric, and ferroelectric functionality, in *2023 IEEE BiCMOS and Compound Semiconductor Integrated Circuits and Technology Symposium (BCICTS)*, Monterey, 2023, pp. 132–136. https://doi.org/10.1109/BCICTS54660.2023.10310921
49. S. Rajan et al., Jpn. J. Appl. Phys. **44**, L1478 (2005). https://doi.org/10.1143/JJAP.44.L1478
50. S. Rajan et al., N-polar GaN Electronics, in *2007 International Workshop on Physics of Semiconductor Devices*, Mumbai, 2007, pp. 368–368. https://doi.org/10.1109/IWPSD.2007.4472521
51. Nidhi et al., "fT and fMAX of 47 and 81 GHz, respectively, on N-polar GaN/AlN MIS-HEMT," in IEEE Electron Device Lett., vol. 30, no. 6, pp. 599–601, 2009, doi: https://doi.org/10.1109/LED.2009.2020305
52. M. Sumiya et al., Growth mode and surface morphology of a GaN film deposited along the N-face polar direction on c-plane sapphire substrate. J. Appl. Phys. **88**, 1158–1165 (2000). https://doi.org/10.1063/1.373791
53. E. Akso et al., Schottky Barrier Gate N-Polar GaN-on-Sapphire deep recess HEMT with record 10.5 dB linear gain and 50.2% PAE at 94 GHz, in *IEEE Microwave and Wireless Technology Letters*. https://doi.org/10.1109/LMWT.2023.3345531
54. G. Moore, Cramming more components onto integrated circuits. Electronics **38**(8), 114 (1965)
55. H.W. Then et al., Enhancement-mode 300-mm GaN-on-Si (111) with integrated Si CMOS for future mm-wave RF applications. IEEE Microw. Wirel. Technol. Lett. **33**(6), 835–838 (2023). https://doi.org/10.1109/LMWT.2023.3268184
56. P. Saunier et al., In AlN barrier scaled devices for very high fT and for low-voltage RF applications. IEEE Trans. Electron Devices **60**(10), 3099–3104 (2013). https://doi.org/10.1109/TED.2013.2277772
57. H. W. Then et al., High-performance low-leakage enhancement-mode high-K dielectric GaN MOSHEMTs for energy-efficient, compact voltage regulators and RF power amplifiers for low-power mobile SoCs, in *2015 Symposium on VLSI Technology (VLSI Technology)*, Kyoto, 2015, pp. T202–T203. https://doi.org/10.1109/VLSIT.2015.7223674
58. Q. Yu et al., 5G mmWave power amplifier and low-noise amplifier in 300mm GaN-on-Si technology, in *2022 IEEE Symposium on VLSI Technology and Circuits (VLSI Technology and Circuits)*, Honolulu, 2022, pp. 126–127. https://doi.org/10.1109/VLSITechnologyandCir46769.2022.9830383
59. Q. Yu, H. W. Then, I. Momson, D. Thomson, J. Garrett, S. Rami, A 25.5-31GHz power amplifier using enhancement-mode high-K dielectric GaN MOS-HEMTs in 300mm GaN-on-Si technology, *in 2023 IEEE/MTT-S International Microwave Symposium - IMS 2023*, San Diego, 2023, pp. 931–934. https://doi.org/10.1109/IMS37964.2023.10187966
60. K.-J. Chuang, K.P. Tang, Y.-H. Lin, T.-H. Chen, C.-S. Wu, T.-W. Huang, An efficient and lLinear 24.4dBm Ka-band GaAs power amplifier for 5G communication, in *2021 IEEE International Symposium on Radio-Frequency Integration Technology (RFIT)*, Hualien, 2021, pp. 1–3. https://doi.org/10.1109/RFIT52905.2021.9565269
61. T.-C. Tsai, C. Bohn, J. Hebeler, M. Kaynak, A.Ç. Ulusoy, A linear and efficient power amplifier supporting wideband 64-QAM for 5G applications from 26 to 30 GHz in SiGe:C BiCMOS, in *2021 IEEE Radio Frequency Integrated Circuits Symposium (RFIC)*, Atlanta, 2021, pp. 127–130. https://doi.org/10.1109/RFIC51843.2021.9490461
62. F. Wang, T.-W. Li, S. Hu, H. Wang, A super-resolution mixed-signal Doherty power amplifier for simultaneous linearity and efficiency enhancement. IEEE J. Solid State Circuits **54**(12), 3421–3436 (2019). https://doi.org/10.1109/JSSC.2019.2937435

63. A. Asoodeh, H.M. Lavasani, M. Cai, S. Mirabbasi, A highly linear and efficient 28-GHz PA with a Psat of 23.2 dBm, P1 dB of 22.7 dBm, and PAE of 35.5% in 65-nm bulk CMOS. IEEE J. Solid State Circuits **56**(3), 914–927 (2021). https://doi.org/10.1109/JSSC.2020.3036830
64. B. Lu, E. Matioli, T. Palacios, Tri-gate normally-off GaN power MISFET. IEEE Electron Device Lett. **33**(3), 360–362 (2012). https://doi.org/10.1109/LED.2011.2179971
65. R.S. Howell et al., GaN SLCFET technology for next generation mmW systems, demonstrating pout of 10.87 W/mm with 43% PAE at 94 GHz. IEEE Microw. Wirel. Technol. Lett. **33**(6), 839–842 (2023)
66. R. Kabouche, K. Harrouche, E. Okada, F. Medjdoub, Short-term reliability of high-performance Q-band AlN/GaN HEMTs, in *2020 IEEE International Reliability Physics Symposium (IRPS)*, Dallas, 2020, pp. 1–6. https://doi.org/10.1109/IRPS45951.2020.9129322
67. M.B. Tahhan et al., Passivation schemes for ScAlN-barrier mm-wave high electron mobility transistors. IEEE Trans. Electron Devices **69**(3), 962–967 (2022). https://doi.org/10.1109/TED.2021.3140016
68. E. Akso et al., Record 1 W output power from a single N-Polar GaN MISHEMT at 94 GHz, in *2023 Device Research Conference (DRC)*, Santa Barbara, 2023, pp. 1–2. https://doi.org/10.1109/DRC58590.2023.10187008
69. F. Chiocchetta et al., Role of the AlGaN cap layer on the trapping behaviour of N-polar GaN MISHEMTs, in *2021 IEEE International Reliability Physics Symposium (IRPS)*, Monterey, 2021, pp. 1–2. https://doi.org/10.1109/IRPS46558.2021.9405111
70. E. Acurio et al., On recoverable behavior of PBTI in AlGaN/GaN MOS-HEMT. Solid State Electron. **132**, 49–56 (2017)
71. T. Wu et al., Toward understanding positive bias temperature instability in fully recessed-gate GaN MISFETs. IEEE Trans. Electron Devices **63**(5), 1853–1860 (2016)
72. S. Joy et al., PBTI in scaled oxide submicron enhancement mode high-K gallium nitride transistors, in *2024 IEEE International Reliability Physics Symposium (IRPS)* (Accepted)
73. X. Garros et al., PBTI mechanisms in La containing Hf-based oxides assessed by very Fast IV measurements, in *2010 International Electron Devices Meeting*, (2010), pp. 4–6
74. M. Alam, A critical examination of the mechanics of dynamic NBTI for PMOSFETs, in *IEEE International Electron Devices Meeting 2003*, p. 14, 2003
75. M. Alam, ECE 695A reliability physics of Nanotransistors, in Course Lecture (2013). //nanohub.org/resources/16560
76. C.C. Hu, M.F. Chen, W.C. Chiou, D.C.H. Yu, 3D multi-chip integration with System on Integrated Chips (SoIC™), in *2019 Symposium on VLSI Technology*, (Kyoto, 2019), pp. T20–T21. https://doi.org/10.23919/VLSIT.2019.8776486
77. M.-F. Chen, F.-C. Chen, W.-C. Chiou, D.C.H. Yu, System on Integrated Chips (SoIC™) for 3D heterogeneous integration, in *2019 IEEE 69th Electronic Components and Technology Conference (ECTC)*, (Las Vegas, 2019), pp. 594–599. https://doi.org/10.1109/ECTC.2019.00095
78. D.B. Ingerly et al., Foveros: 3D integration and the use of face-to-face Chip stacking for logic devices, in *2019 IEEE International Electron Devices Meeting (IEDM)*, (San Francisco, 2019), pp. 19.6.1–19.6.4. https://doi.org/10.1109/IEDM19573.2019.8993637
79. A. Elsherbini et al., Enabling next generation 3D heterogeneous integration architectures on Intel process, in *2022 International Electron Devices Meeting (IEDM)*, (San Francisco, 2022), pp. 27.3.1–27.3.4. https://doi.org/10.1109/IEDM45625.2022.10019499
80. Y. -K. Cheng, Embracing the full power of TSMC 3DFabric™ design: Challenges and solutions, in 2023 International VLSI Symposium on Technology, Systems and Applications (VLSI-TSA/VLSI-DAT), HsinChu, (2023), p. 1. https://doi.org/10.1109/VLSI-TSA/VLSI-DAT57221.2023.10134011
81. S.W. Liang, G.C.Y. Wu, K.C. Yee, C.T. Wang, J.J. Cui, D.C.H. Yu, High performance and energy efficient computing with advanced SoIC™ scaling, in *2022 IEEE 72nd Electronic Components and Technology Conference (ECTC)*, (San Diego, 2022), pp. 1090–1094. https://doi.org/10.1109/ECTC51906.2022.00176

82. T.E. Kazior, R. Chelakara, W. Hoke, J. Bettencourt, T. Palacios, H.S. Lee, High performance mixed signal and RF circuits enabled by the direct monolithic heterogeneous integration of GaN HEMTs and Si CMOS on a silicon substrate, in *2011 IEEE Compound Semiconductor Integrated Circuit Symposium (CSICS)*, Waikoloa, 2011, pp. 1–4. https://doi.org/10.1109/CSICS.2011.6062443
83. J. LaRoche, Gallium nitride materials and devices XV. SPIE OPTO **11280**, 112801G (2020). https://doi.org/10.1117/12.2543913
84. H.W. Then et al., Advances in research on 300 mm Gallium Nitride-on-Si (111) NMOS transistor and silicon CMOS integration, in *in 2020 IEEE International Electron Devices Meeting (IEDM)*, (San Francisco, 2020), pp. 27.3.1–27.3.4. https://doi.org/10.1109/IEDM13553.2020.9371977
85. J.W. Chung, L. Bin, T. Palacios, On-wafer integration of nitrides and Si devices: Bringing the power of polarization to Si, in *2009 IEEE MTT-S International Microwave Symposium Digest*, (Boston, 2009), pp. 1117–1120. https://doi.org/10.1109/MWSYM.2009.5165897
86. H.W. Then et al., DrGaN: An integrated CMOS driver-GaN power switch technology on 300 mm GaN-on-Si with E-mode GaN MOSHEMT and 3D monolithic Si PMOS, in *2023 IEEE International Electron Devices Meeting (IEDM)*, (2023), pp. 9–7
87. Q. Yu et al., 300mm GaN-on-Si Technology improvements with industry's first circuit demonstration of monolithically integrated GaN and Si transistors, in *International Microwave Symposium*, 2024 (Accepted)

Patrick Le Fèvre, Chief Marketing and Communications Officer for Powerbox, he is an experienced, senior marketer and degree-qualified engineer with a 40-year track record of success in power electronics. He has pioneered the marketing of new technologies such as digital power and technical initiatives to reduce energy consumption. Le Fèvre has written and presented numerous white papers and articles at the world's leading international power electronics conferences. These have been published over 450 times in media throughout the world. He is also involved in several environmental forums, sharing his expertise and knowledge of clean energy.

Geoff Haynes is a serial entrepreneur. He was a co-founder at SiGe Semiconductor, Mobile Knowledge, GaN Systems, and TTPi. Geoff is now the owner of Inspirit Ventures Ltd, a consultancy dedicated to the encouragement and advancement of Power Electronics Research and Development. In this role, he acts as an adviser to the board at RAM Innovations and QPT.

Kennith Kin Leong is a Lead Principal Engineer at Infineon Technologies Austria AG. He joined Infineon Technologies Austria AG, in Villach as System Concept Engineer with research interests including GaN bidirectional switch, GaN integration, power converter topologies and control. In 2023, he became the Lead Principal, Product Definition, and Concept Engineer for CoolGaN™ BDS. He has over 28 granted patents and numerous technical publications.

Vladimir Odnoblyudov, the co-founder and CTO of QROMIS, Inc. (Silicon Valley, USA), has spent 20 years in optoelectronic and electronic devices, process integration and wide-bandgap semiconductor technologies with executive leadership and technical management positions, from early-stage start-ups to Fortune 500 companies. Prior to co-founding QROMIS, Dr. Odnoblyudov served as Director of R&D and Emerging Technologies of Bridgelux, Inc., and prior to that he served as Director of LED Solutions at Micron Technology, Inc., leading R&D efforts for revolutionary large diameter GaN wafer and device technology in 200 mm CMOS pilot line fab. Before that, he was the Founder and CTO of Quanlight, Inc., an LED start-up focusing on development of novel yellow-red chip technology. Dr. Odnoblyudov has >200 issued and pending US patents and > 30 articles in referred journals and conference presentations. He holds a Ph.D. degree in Applied Physics (University of California San Diego, Department of Electrical and Computer Engineering).

Cem Basceri has spent more than 30 years in materials science and semiconductor technologies with various executive leadership and managerial positions for creating, leading, and scaling businesses in start-up ventures to Fortune 500 public companies, including Motorola, Micron Technology, Intrinsic Semiconductor, CREE/Wolfspeed, Bridgelux and QROMIS. Currently, Cem is the founder and the President & CEO of the privately held company, QROMIS, Inc., headquartered in Santa Clara, California. The Company, the recipient of the EE Times' Global Top 100 List and Frost & Sullivan's 2024 Enabling Technology Leadership Award, and its manufacturing partners/licensees, Shin-Etsu Chemical Co., Ltd. (SEC) and Vanguard International Semiconductor Corp. (VIS), are the commercial QST® substrate, GaN-on-QST® epitaxy wafer, and GaN-on-QST® device foundry service providers in the rapidly growing energy efficient GaN electronics industry by enabling a disruptive, CMOS fab-friendly 200 mm and 300 mm manufacturing, and economies of scale for high performance GaN adoption. Cem holds more than 335 issued US and worldwide patents, a Ph.D. degree in materials science and engineering *(North Carolina State University)* and MS/BS degrees in metallurgical and materials engineering *(Orta Dogu Teknik Universitesi).*

Han Wui Then is a senior principal Engineer at Intel Foundry Technology Research, Intel Corporation, working on advanced transistor and chiplet technologies. He currently leads the process technology research in gallium nitride electronics and 3D CMOS integration where he pioneered the development of 300 mm GaN-on-Si (111) MOSHEMT technology and demonstrated the first 3D monolithic integration of GaN and Si CMOS. He has published over 60 journal and conference papers, and holds >150 US patents on semiconductor technologies. He received his Ph.D. degree in ECE from the University of Illinois at Urbana-Champaign in 2009.

Chapter 8
Positioning and Perspectives of GaN-Based Power Devices

Gerald Deboy and Matthias Kasper

With the adoption of GaN-based power devices, three major technologies, i.e., silicon superjunction, SiC, and GaN, and various device concepts compete specifically in the 600 V/650 V class. This article will introduce key characteristics of the technologies and show, along with reference designs, their value propositions in a number of applications. We will, furthermore, discuss the different device concepts co-existing today in the market on GaN-based power devices.

8.1 Technology

Technology—GaN-based power devices: GaN-based power devices can be classified into vertical and lateral device concepts, as well as normally-on and normally-off devices. Within the normally-on devices, cascode arrangements co-exist with direct-driven approaches. Normally-off devices can be classified by their gate structure into Schottky gate (SG) and gate-injection transistor (GIT) concepts.

While vertical GaN devices—GaN epitaxy built on GaN substrates—can move closer to the Baliga limits of the material, the cost and availability of bulk GaN substrates pose a major hurdle for these devices. Furthermore, vertical JFET structures forming either normally-on or normally-off channels are available in SiC as well. Vertical GaN devices, therefore, do not provide enough value at present to be considered as a path forward.

Lateral device concepts feature all drift zones with very high electron mobility. The combination of layers of GaN with AlGaN creates a discontinuity of the polarization and forms a natural two-dimensional electron gas (2DEG) at the interface.

G. Deboy (✉) · M. Kasper
Infineon Technologies Austria, Villach, Austria
e-mail: Laura.Lesnik@infineon.com

© The Author(s), under exclusive license to Springer Nature Switzerland AG 2024
M. Di Paolo Emilio (ed.), *GaN Technology*,
https://doi.org/10.1007/978-3-031-63238-9_8

These electrons are confined in the vertical direction towards the AlGaN by a heterojunction step in the conduction band and toward the substrate by a triangular potential well. Within this area, electrons show a very high mobility and saturation velocity as they move within a zone that is effectively undoped. This characteristic creates advantages in the figure-of-merit ($R_{DS(on)} \times Q_{oss}$) as fewer electrons are needed for a given load current.

The formation of the two-dimensional electron gas leads naturally to normally-on devices. To make them usable in voltage-source converter architectures, the devices are turned into normally-off circuits by adding a low-voltage MOSFET in series with the normally-on device. The circuit can be either controlled in cascode mode by turning on/off the LV MOSFET with the gate of the normally-on GaN HEMT tied to the LV MOSFET's source, or by controlling the normally-on device directly while keeping the LV MOSFET always on. The latter method allows good control of switching characteristics with respect to dv/dt and di/dt, while the cascode arrangement suffers from a lack of feedback capacitance from the normally-on device to the driving circuit. Another issue in cascode configurations is the addition of Q_{rr} and Q_{oss} charges from the LV MOSFET in reverse or third quadrant operation to the net-zero reverse recovery charge of GaN HEMTs. This dilutes, to some extent, one of the core advantages of GaN-based power devices. On the other hand, the forward voltage drop across the circuit is lower than that in classic normally-off HEMTs while operating in the diode mode. The conduction losses are, therefore, lower before turning on the device in synchronous rectification during the third quadrant mode of operation. Switching frequency and dead time control in the third quadrant operation will determine whether a normally-on or normally-off concept is more suitable for a given application.

In the camp of normally-off devices, Schottky gate and gate-injection (GIT) concepts co-exist in the market. While the first concept requires precise voltage control of the gate, the GIT concept is more forgiving toward imperfect layouts as the p-GaN gate naturally clamps overvoltage to the threshold of its forward-biased diode characteristic. By controlling this forward current into the gate, the effective gate voltage can be well-controlled independent of the voltage drop across the source-to-gate region. The injection of holes from the p-GaN gate furthermore enhances the concentration of electrons in the 2DEG and creates an advantageous conductivity modulation effect. In combination with a hybrid drain structure—a proprietary concept with a p-region adjacent to the drain electrode injecting holes during the switching transient—dynamic $R_{DS(on)}$ effects can be substantially reduced in hard-switching applications. Figure 8.1 shows the Schottky gate and the gate injection transistor concepts, respectively.

For the rest of this chapter, we refer to GaN-on-silicon devices prevailing today in the 600 V class. Device concepts based on GaN-on-sapphire or GaN-on-engineered-substrates, such as QST, are discussed for voltage classes in the 1200 V range.

Technology—SiC MOSFET and silicon-superjunction: In many applications, GaN HEMTs compete with both silicon superjunction devices as well as with SiC MOSFETs. Figure 8.2 shows a comparison of the key figures-of-merit of SiC

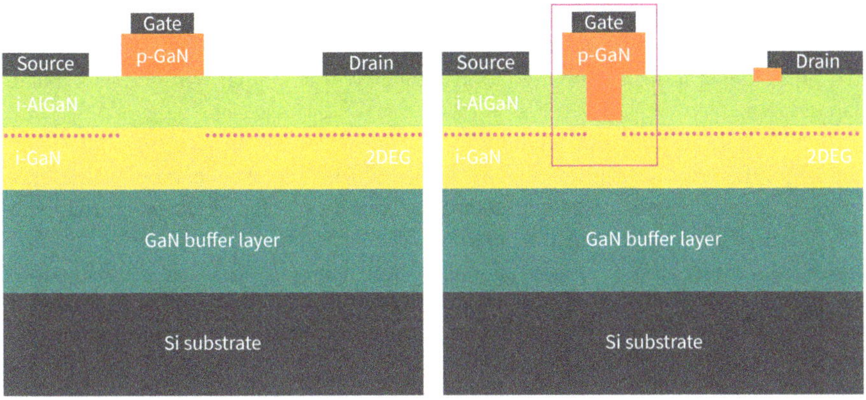

Fig. 8.1 Lateral GaN HEMT concept with Schottky gate (left) and gate injection with hybrid drain structure (right)

Device	$V_{(BR)DSS}$ [V]	$R_{DS(on)}$*E_{oss}	$R_{DS(on)}$*Q_g	$R_{DS(on)}$*Q_{oss}	$R_{DS(on)}$*Q_{rr}
CoolMOS™ C7	600	100%	100%	100%	100%
CoolGaN™	600	84%	6%	13%	0%
CoolSiC™	650	133%	41%	21%	2%

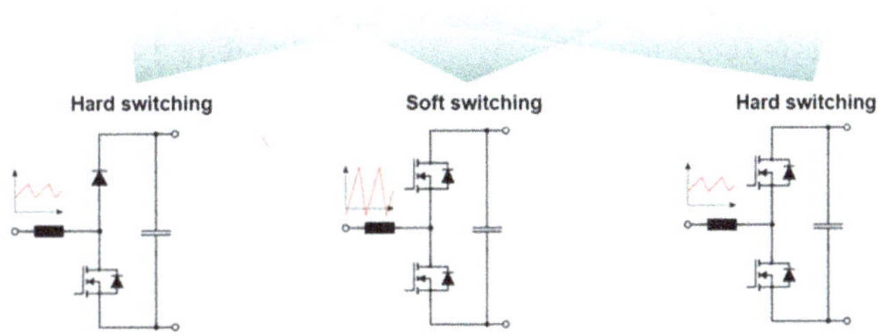

Fig. 8.2 Key parameters of SiC MOSFETs and GaN HEMTs relative to silicon superjunction technology and their relevance for fundamental switching circuits

MOSFET and GaN HEMT vs the established CoolMOS™ C7 Superjunction technology.

The superjunction concept will continue towards smaller dimensions of the underlying p- and n-column structure to further exploit the ever-lower $R_{DS(on)} \times A$ values. The on-state resistance in a standard package will hence reduce further. The CoolMOS™ C8 technology will offer an $R_{DS(on)}$ as low as 16 mOhm in a standard TO220 package, a tremendous improvement from 190 mOhm during the introduction of superjunction devices back in 1998. With the shrink of the active device area, the high-voltage values of the output capacitance decrease and consequently the figure-of-merit $R_{DS(on)} \times E_{oss}$, due to its overproportioned emphasis in the calculation formula. This parameter corresponds to the energy stored in the output capacitance. In single-ended hard-switching applications such as switching with a MOSFET against a SiC Schottky diode, the FoM $R_{DS(on)} \times E_{oss}$ correlates with the minimum switching loss for unipolar devices, the so-called Nakajima limit [1]. Even though both GaN and SiC show better values in this FoM vs the state-of-the-art CoolMOS C7, future generations of superjunction devices will be on par with the wide band-gap devices.

In SiC, normally-on and normally-off devices such as vertical JFET structures and MOSFETs co-exist in the market. The MOSFET concepts can be further classified as planar or vertical-gate structures. Common to MOSFETs is the relatively low increase of $R_{DS(on)}$ with temperature. The increase in resistance of the drift region is compensated by improvements of the channel resulting in a substantially smaller increase in resistance as compared to GaN and silicon-based power devices. This effect is specifically strong in devices with large channel contributions to the overall resistance, such as 600 V-rated devices. Vertical gate structures allow a natural protection of the gate oxide in the blocking state without disadvantages for the on-state resistance and are hence the preferred concept.

The switching losses in half-bridge circuits are dominated by the two FoMs—$R_{DS(on)} \times Q_{oss}$ and $R_{DS(on)} \times Q_{rr}$—as the charge from the device being commutated creates a load-current contribution in the device turning on close to the full DC link voltage [2]. SiC MOSFETs show a significant reduction in the reverse recovery charge $R_{DS(on)} \times Q_{rr}$ due to a very short ambi-polar carrier lifetime, whereas GaN-based HEMTs feature zero reverse recovery charge due to the absence of a p-n junction. Furthermore, both technologies show an output capacitance charge Q_{oss} that is lower by a factor of 5–7. Hence, both SiC and GaN devices enable continuous hard commutations in half-bridge circuits.

GaN-based HEMTs exhibit the lowest gate charge of all device concepts and are hence the first choice in the very high switching-frequency domain, such as power conversion beyond 400 kHz. In this frequency domain, soft-switching with zero-voltage turn-on is the natural choice. In combination with very fast turn-off transitions and the elimination of E_{oss}–associated losses through the resonant circuit, excellent efficiency and power density can be achieved. It is noteworthy in this context that a fast turn-off with ideally zero losses requires to turn-off the load current through the channel at or before the point in time when the device starts to block the voltage. GaN HEMTs with their combination of very low gate charge, steep transconductance, and low driving voltage enable this. Superjunction devices with their strong non-linear output capacitance yielding effectively a time window of 100 ns and more before the voltage rises, are very good in this aspect as well [3].

Positioning of the technologies in major applications: *Power factor correction* is a major application in power electronics and is mandatory for switch-mode power supplies rated above 75 W. Numerous circuits exist, but the two solutions with the highest share in the market are the boost circuit, including a bridge rectifier, and the totem pole circuit without a bridge rectifier. While the classic boost topology is still widely used in cost-sensitive markets, the totem pole circuit avoids the forward voltage drop of two diodes in the bridge rectifier and enables efficiencies higher than 99%.

The boost circuit corresponds to the circuit shown in Fig. 8.2, left. There is only one switch and one diode. In the case of SiC Schottky diodes with zero reverse recovery charge and very low Q_c, turn-on losses can be minimized. Alternatively, ZCS can be applied by switching load currents in boundary mode or discontinuous mode. Switching losses are hence dominated by the energy stored in the output capacitance of the switch, which is transferred into heat during hard turn-ons. In this domain, superjunction devices will prevail in the market. Further progress in the technology will reduce the FoM $R_{DS(on)} \times E_{oss}$ to an extent that it is on par with GaN. With the capability to reduce turn-off losses close to the Nakajima limit, SJ will remain the first choice for hard-switching boost circuits.

In totem pole circuits, two control modes co-exist: continuous current mode at low to moderate switching frequencies (40–60 kHz typically), and triangular current mode at high switching frequencies (300–700 kHz). In the CCM mode, SiC MOSFETs are the first choice. Even though GaN HEMTs show lower Q_{rr} losses, the difference in efficiency is nearly negligible at typical low switching frequencies. The lower increase of $R_{DS(on)}$ with the temperature of SiC MOSFETs and the resulting lower conduction losses at operating temperatures in the vicinity of 100 °C compensate for the slightly higher switching losses and bring both solutions on par with respect to efficiency.

GaN HEMTs dominate in the high-frequency domain, where the triangular current modulation yields zero voltage turn-on at low (negative) currents and turn-off at high (positive) currents. We have demonstrated a PFC circuit using an active bridge and three interleaved half-bridges operating at an effective switching frequency in a fixed/variable switching frequency setup between 1.8 and 2.4 MHz. The power density is at an outstandingly high level of 138 W/in^3. Figure 8.3 shows the circuit (left) and the demonstrator (right).

Isolating DC/DC stages, such as the ones based on resonant topologies like LLC or CLLC, are commonly used in many market segments, such as server/telecom power, EV charging, etc. Similar to power factor correction, we see in this segment a co-existence of both wide-bandgap technologies as well as silicon-based power devices.

Superjunction devices prevail at switching frequencies up to 200 kHz and are used on both 400 V DC link voltage (such as server power supplies) and 800 V DC link voltage (such as EV chargers, in combination with a Vienna rectifier PFC circuit). In the latter case, two converter stages are stacked upon each other and operate between the midpoint and +/− 400 V rails, respectively. 1200 V rated SiC MOSFETs

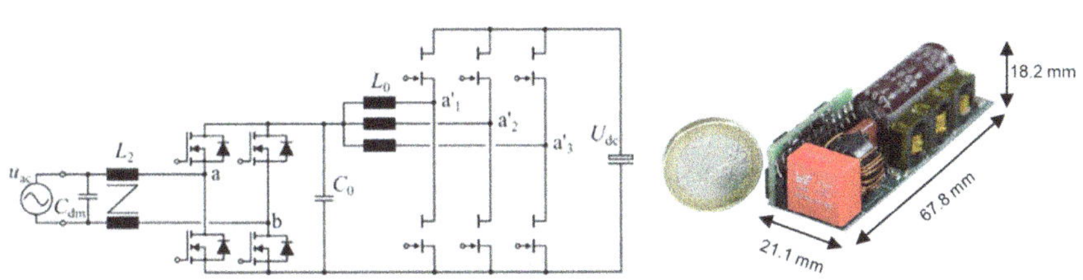

Fig. 8.3 220 W PFC stage operating in triangular current mode between 1.8 and 2.4 MHz

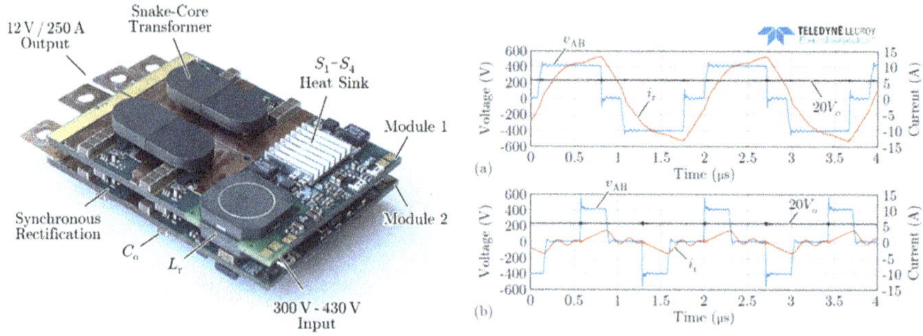

Fig. 8.4 LLC demonstrator operating up to 700 kHz using a stacked matrix transformer construction (left) and a three-level modulation (right)

are employed directly on the 800 V DC link, e.g., in combination with an active frontend such as a B6 or B8 bridge, which is common in onboard chargers.

GaN HEMTs are the preferred choice in the high-frequency domain operating either with a half-bridge or full-bridge configuration on 400 V DC links or in a three-level configuration with two stacked half-bridges on 800 V DC links. Combining zero-voltage turn-on and hard turn-off at moderate current levels creates advantages for GaN HEMTs at high switching frequencies. Furthermore, advanced modulation schemes such as three-level modulation introducing a PWM control to limit the increase in switching frequencies in light loads may lead to hard-commutation events at very low currents. Such an operation would be challenging for superjunction devices even if fast body diode types are used. GaN HEMTs with their zero reverse recovery charge, however, excel in this domain.

We have demonstrated a server power supply unit converting 400 V DC into a regulated 12 V output at a total power level of 3 kW. Due to the large output currents, we employ a matrix transformer configuration with only one magnetic path to cancel circulating currents on the secondary side. By stacking two boards and combining them into one magnetic core, an outstanding power density of 345 W/in^3 has been achieved [4]. Figure 8.4 shows the demonstrator (left) and the three-level modulation (right).

8.2 Outlook

GaN HEMTs, with their lateral device construction offer the unique chance to build monolithic bidirectional switches. These devices are able to block voltages in both polarities and to control currents actively both in the first and third quadrants. Unlike in unipolar switches, where the construction of a bidirectional switch requires two devices in a back-to-back configuration (which doubles the on-state resistance) and another two devices in parallel (to return to the original on-state resistance), a bidirectional GaN switch utilizes the same physical drift layer to block voltages in both

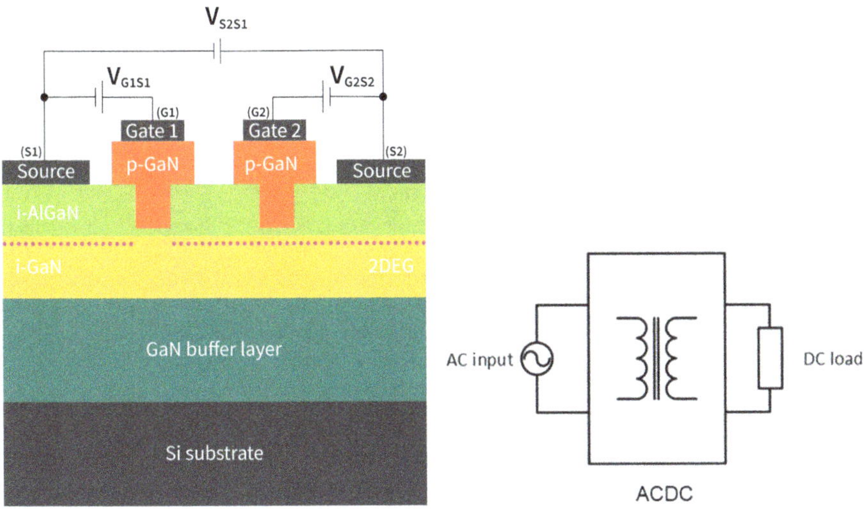

Fig. 8.5 Monolithic bidirectional GaN HEMT (left) and potential use cases in single-stage power conversion (right)

polarities. The device features a second gate module to control currents in both polarities. These gate electrodes are driven with positive voltages referred to the two opposite load terminals, respectively. The substrate needs to follow the lower of the two load terminals to avoid depletion of the two-dimensional electron gas by back-gating effects. Monolithic integrated solutions exist, which keep the substrate at the desired potential.

These new switches open up a realm of opportunities for new circuits specifically for single-stage power conversion from AC to DC or vice versa [5]. These devices will first come to the market in the 600 V/650 V class before expanding into other voltage classes. Figure 8.5 shows the device construction (left) and single-stage power conversion (right).

8.3 Conclusion

We introduced key figures of merit for SiC MOSFETs, GaN HEMTs, and silicon-based superjunction devices. The value proposition of each of these technologies strongly depends on the topology and current control modes chosen for a given application.

Summing up, silicon superjunction devices prevail in single-ended, hard-switching applications, and resonant converters at low to moderate switching frequencies, while SiC creates clear value in hard-switching half-bridge and full-bridge circuits. GaN HEMTs excel in the high-frequency domain both in resonant converters and topologies utilizing triangular current modulation.

References

1. A. Nakajima et al., Power loss limit in unipolar switching devices: Comparison between silicon superjunction devices and wide bandgap devices. IEEE Trans. Electron Devices **56**(11), 2652–2656 (2009)
2. M. Kasper, R.M. Burkart, G. Deboy, J.W. Kolar, ZVS of power MOSFETs revisited. IEEE Trans. Power Electron. **31**(12), 8063–8067 (2016)
3. G. Deboy, O. Haebelen, M. Treu, Perspective of loss mechanisms for silicon and wide bandgap power devices. CPSS Transactions on Power Electronics and Applications **2**(2), 89 (2017)
4. G. Knabben, G. Zulauf, J. Schäfer, J.W. Kolar, M. Kasper, J. Azurza, G. Deboy, Conceptualization and analysis of a next-generation ultra-compact 1.5-kW PCB-integrated wide-input-voltage-range 12V-output industrial DC/DC converter module. Electronics **10**(17), 2158 (2021)
5. J. Huber, J.W. Kolar, Monolithic bidirectional power transistors—Opening new horizons in power electronics. IEEE Power Electronics Magazine **10**(1), 28–38 (2023)

Gerald Deboy received his M.S. and Ph.D. degrees in physics from the Technical University Munich, Germany, in 1991 and 1996, respectively. He joined Siemens Corporate Research and Development in 1992 and the Semiconductor Division of Siemens in 1995, which later became Infineon Technologies. From 2004 to 2009, he was the head of the technical marketing department for power semiconductors and integrated circuits at Infineon Technologies Austria AG. Since 2009, he is heading a business development group specializing in new fields for power electronics. He has authored or coauthored more than 70 papers in national and international journals, including contributions to three textbooks. He holds more than 60 granted international patents and has more applications pending. He is a Senior Member of the IEEE.

Matthias Kasper received the M.Sc. and Dr.Sc. degrees in electrical engineering from the Swiss Federal Institute of Technology (ETH) Zurich, Switzerland, in 2011 and 2016, respectively. During his study, he focused on power electronics, mechatronics, and energy systems. In his Ph.D. at the Power Electronic Systems Laboratory at ETH Zurich, he dealt with multi-cell converter systems for different applications, e.g., PV module integrated DC-DC converters, auxiliary supplies for high step-down ratios, and a highly efficient telecom rectifier module. Since January 2017, he is part of the Systems Innovations Lab of Infineon Technologies Austria AG, where he works on novel topologies, control schemes, and multi-objective optimization routines. He has authored and co-authored more than 20 scientific publications and holds 12 international patents.

Chapter 9
GaN Market

Prabhu Karunakaran and Navdeep Saboo

Designing power electronics with silicon leads to a tradeoff scenario between system performance, efficiency, and form-factor. For instance, switching a semiconductor at higher frequencies requires high power density and reduced device size, but designing with silicon will cause increased power loss and heat dissipation while reducing system efficiency. Conversely, designing a device for high efficiency and reduced power loss is achieved by lowering the switching frequency—yet with larger component size and lower power density. But the power density and efficiency requirements keep growing every 4 years, and silicon's capability is unable to match the needs of today's applications. This led to the search for alternate materials that can exhibit better performance.

GaN, which has a higher bandgap of 3.4 electron volts (eV) (higher compared to SiC [3.2 eV] and Si [1.12 eV]], is intrinsically superior in terms of electron mobility, and possess lower on-resistance, which translates into higher power conversion efficiency in high-frequency and elevated temperature environments—making it an attractive alternative for silicon. Moreover, designing products with GaN benefits designer with a product that has a better thermal performance, which often leads to power systems that have much smaller form-factor as compared to its silicon-based counterpart.

To explore its capabilities, scientists researched and tested GaN material for decades before International Rectifiers (IR) announced the successful development of GaN power transistor prototype in 2008 and launched their first GaN product, iP2010, in early 2010. In the similar time frame, 2010, Efficient Power Conversion (EPC) Corp. announced the launch of enhancement mode GaN-on-Si based power GaN transistor for low voltage products (40–200 V) with a claim of better-switching performance in power conversion compared to the then silicon MOSFET product.

P. Karunakaran (✉) · N. Saboo
Frost & Sullivan, Chennai, Tamil Nadu, India
e-mail: Prabhu.Karunakaran@frost.com

© The Author(s), under exclusive license to Springer Nature Switzerland AG 2024

M. Di Paolo Emilio (ed.), *GaN Technology*,
https://doi.org/10.1007/978-3-031-63238-9_9

EPC's announcement revealed that their GaN transistors were manufactured in a conventional CMOS foundry with minimal modifications to the fabrication processes. This suggests that adding new technology to product portfolio mix is easier, and that encourages competitors in the power semiconductor space to develop GaN products.

Both IR and EPC's GaN power transistors indicated potential applications across several different end-user segments including the mass-markets like laptops, cell phones, Class-D audio amplifiers, and in high-performance requirements like telecoms, and servers as well.

Overall, the power performance superiority of GaN combined with ease of manufacturing, and broader applicability across verticals made it an attractive commercial opportunity to pursue in the power semiconductor space.

Before finding customers in power electronics (which is the key focus of this book), GaN found its first commercial success with Opto-electronics—specifically LED, and then with RF systems.

9.1 Current Status and the Future

Throughout the last decade, the power GaN semiconductor space witnessed the entry of more than 20 companies, including start-ups and established large corporations. These market participants worked towards advancing the technology to a point where power GaN devices won customers' confidence in its performance and reliability. This gradually guided the market toward the commercialization phase, which kicked off in the late 2010s.

Referring to the images below, Fig. 9.1 shows the growth in the GaN competitive landscape over the last two decades, which is a clear indication of the need to meet the growing electrification trend, which also means that end-user OEMs need to focus on higher efficiency power devices and the eventual rise of power GaN design wins. Figures 9.1 and 9.2 also reveal that GaN technology has attracted interest from large established corporations since 2008, which indicates that GaN technology is not just an alternate way for entering the power semiconductor market but is a superior technology to silicon, with promising commercial prospects.

In the initial phases of market development, Fig. 9.3 indicates that GaN market participants primarily targeted applications that required 650 V products at high switching frequencies, where GaN's technological maturity level was convincing customers to move away from Si.

However, referring to Fig. 9.4, in recent times, few companies have announced products (or) are working on developing a product in the over 650 V segment (mainly in the 750 V, 900 V, 1200 V, and 1250 V) based on GaN-on-Si lateral GaN structure. The market for the over 650 V segment is currently at a nascent stage, and the product launch in the category is an indication that market participants are exploring applications beyond consumer and into Industrial applications that require

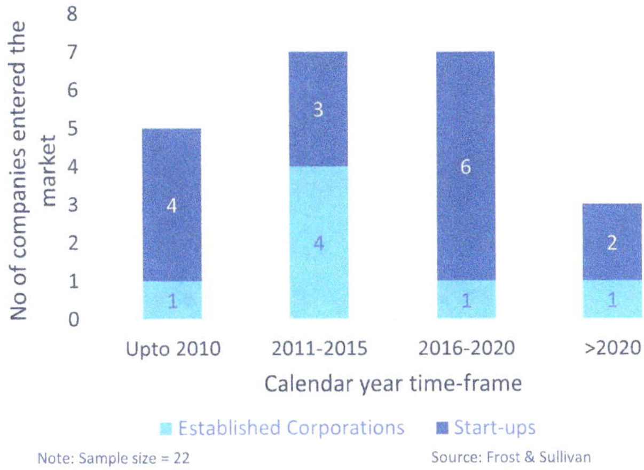

Fig. 9.1 Global power GaN semiconductors market: No. of new market entries by company type and by time-frame, Global, 2008–2023

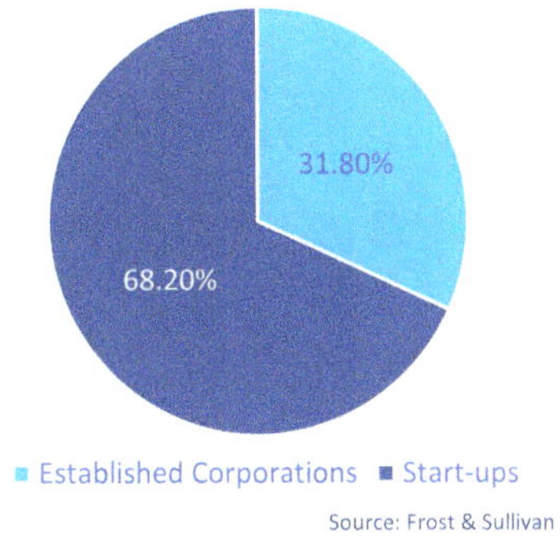

Fig. 9.2 Global power GaN semiconductors market: Percent share of new market entries by company type, Global, 2008–2023. (Note for Figs. 9.1 and 9.2: Established corporations are companies that are founded before 2000 and generated multi-million and/or billion-dollar revenues. Startups: Companies that are founded after 2005, and are in funding stages, and/or generating <50 Mn in revenues)

high-voltage products. However, the over 650 V segment is likely to compete with applications that use silicon carbide (SiC), and hence, design wins for GaN in the higher power segment will require more R&D efforts from the market participants.

Frost & Sullivan also observes that the sub-650 V product segment market will also witness growth in the coming years, mainly to meet the requirements of applications such as audio amplifiers, wireless chargers, electronic warfare, and medical equipment.

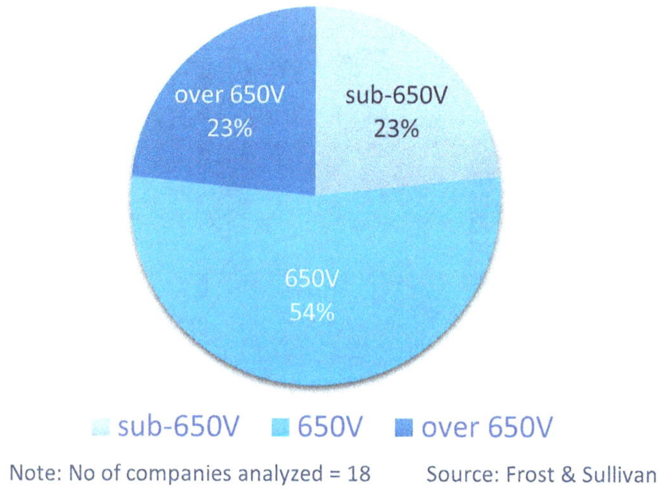

Fig. 9.3 Global power GaN semiconductors market: Product portfolio mix percent ratio by voltage, Global, 2023

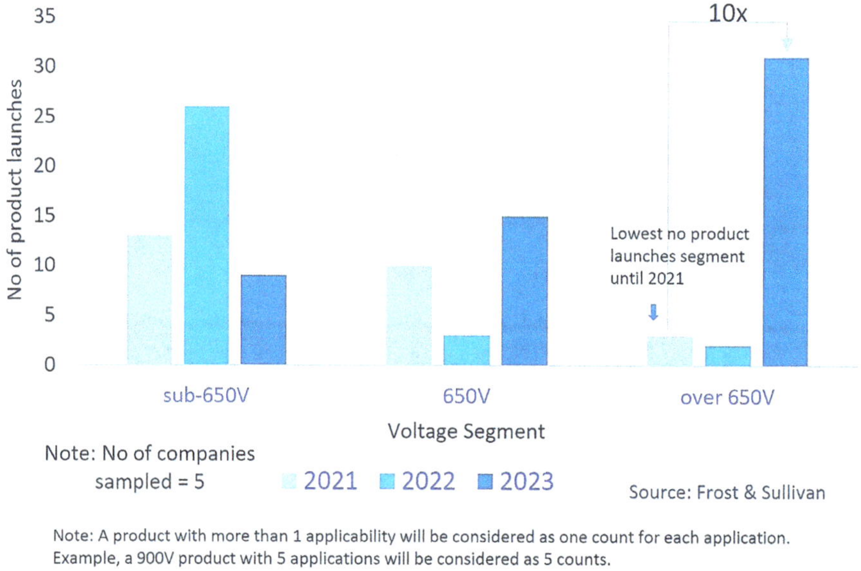

Fig. 9.4 Global power GaN semiconductors market: Product launch announcements by voltage segment, 2021–2023, top five companies, Global

Thanks to the market participants' accelerated technology advancements and customer engagements, power GaN was able to demonstrate its promise of delivering high power efficiency and a small form-factor system with better thermal performance, leading to a significant increase in design wins. These efforts led the market to enter a phase of exponential growth in 2019.

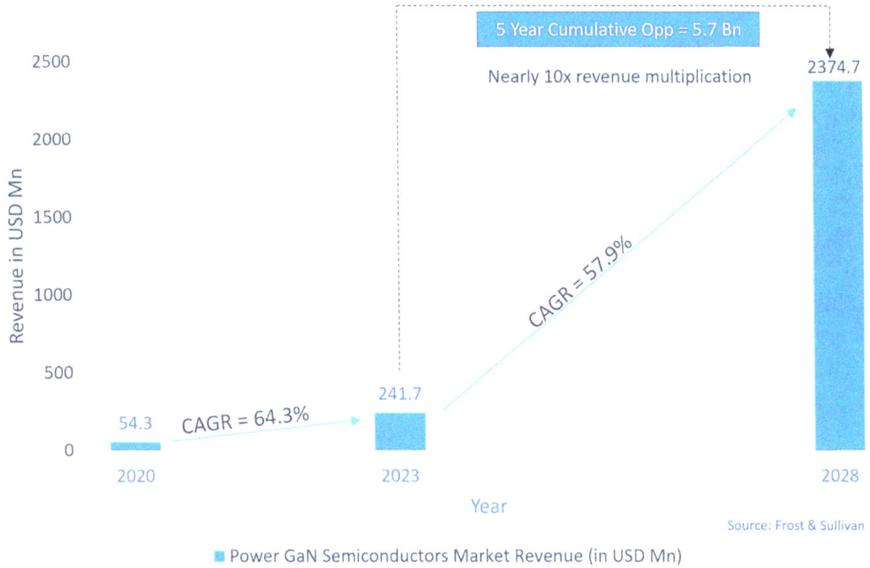

Fig. 9.5 Global power GaN Semiconductors Market: Revenue (in USD Mn), 2020, 2023, 2028 (forecasted revenue in conservative case scenario)

Frost & Sullivan's research and estimates indicate that the power GaN market was USD 54.3 Mn in 2020, multiplied 4.6 times in the last 3 years to reach USD 241.1 Mn by 2023 (refer to Fig. 9.5). The initial market for power GaN was consumer electronics, space, solar, and LiDAR applications. However, the consumer charger market kicked off accelerated growth, specifically the smartphone and laptop chargers at more than 33 W power segment. GaN-embedded advanced chargers enabled smartphone OEMs and after-market charger OEMs to deliver ultra-fast battery charging from 0–100% at a smaller charger form factor—which proved to be a critical factor in winning customers. By 2023, at least nine of the top ten smartphone manufacturers incorporated GaN-based chargers into their product offerings. This signifies the growth potential and attractiveness of the market. Looking into the next 3–5 years, Frost & Sullivan continues to see the growth momentum carry over, translating into a forecasted CAGR of 57.9% (in conservative case scenario) to reach an annual revenue of USD 2.4 Bn by 2028. Over the next 5 years, Frost & Sullivan expects a cumulative revenue opportunity of USD 5.7 Bn from the power GaN semiconductor market. Accelerated penetration rates will lead to growth in the next 5 years in consumer and fast-charging applications, where the power GaN market has proven advantages. We will see an increase in GaN usage in consumer electronics, such as smart speakers, to reduce product size and energy consumption. GaN will also power energy-intensive appliances like washing machines and displays.

Most importantly, 2024 and 2025 will mark the beginning of multi-year high-growth and high-volume sales stemming from electric vehicles and industrial applications where market participants have focused on launching new products over the last 3 years. According to Fig. 9.8, a survey and analysis of product launch activities by the top five companies, new product launches were focused on the industrial, energy, and electric vehicle sectors. In 2023, these segments accounted for 75% of announcements; in 2021, they accounted for only 50%. This increase indicates that GaN market participants are vying for the next business cycle with newer applications that can drive long-term sustainable revenue and growth opportunities. Power GaN devices will make a commercial entry into the EV in 2024 with OBC and DC-DC converter applications, which will create a multi-year high-volume growth, which Frost & Sullivan forecasts to translate into a CAGR of 113.8% from 2023 to 2028. More growth will stem from data center PSUs, solar micro-inverters, and industrial applications. All developments of GaN in these applications are linked to energy efficiency regulations, which force OEMs to reduce their carbon footprint. Data center, e.g., is growing in its requirement to support the ever-growing computing needs today on AI workloads, which leads to increased power consumption. Hence, focusing on components and products that help the data center to turn green and energy efficient becomes mission-critical. GaN products in the server's power supply unit are improving energy efficiency while making the PSUs smaller, saving real estate and, hence, costs. Overall, the power GaN market is poised for healthy and accelerated growth in the next 5 years, with an expected penetration of 5.6% of the overall power semiconductor market by 2028.

9.2 Cost Per $ Watt Inflection Point Nears and Will Be a Key Market Enabler to Drive Volumes and Margins

In its initial stages, the power GaN semiconductor market required a highly performing technology and proven reliability to have discussions with customers and convince them to design trials. Over the last 5–10 years, the market participants overcame this challenge and moved to a point where the high cost of the GaN device became a key market challenge. Initially, the market participants focused on the total cost of ownership, the lesser bill of material, and performance benefits in the long term to overcome this challenge. However, as the technology matured and orders increased, the overall cost of GaN per watt transitioned to a comparable range with silicon. Over the next 1–2 years, the cost per watt will decrease to levels similar to that of silicon (Si). This will be a crucial turning point in the market, as it will be much easier for market participants to persuade customers who were previously hesitant to switch to GaN due to affordability concerns. Figure 9.6 indicates the transition of cost per $ watt in the foresight. This transition will also help market participants improve the current power GaN gross margin levels of 35–40% to increase further to the 50–60% range, comparable to the conventional silicon devices market.

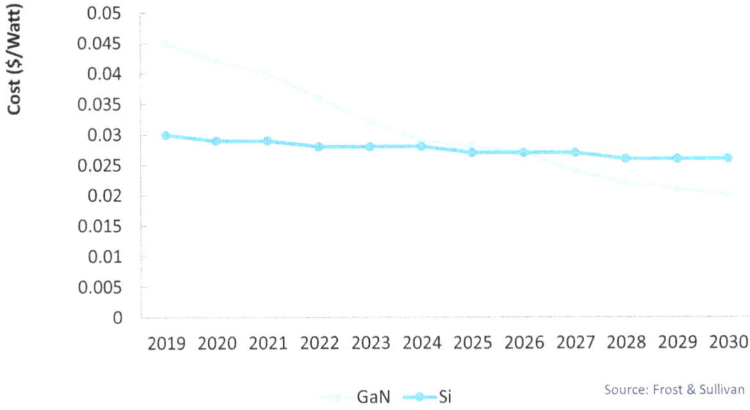

Fig. 9.6 Global power GaN semiconductors market: GaN vs Si Cost ($/Watt), Global, 2019–2030

9.3 Electrification, Sustainability, and Industry Regulations Will Power Up the Growth Trajectory

Rapid electrification across industries as part of Industry 4.0, connected ecosystems, and electric mobility megatrends makes focus on power semiconductors indispensable and essential to enable efficient functionality of critical systems. The automotive industry is rapidly accelerating towards a tipping point in 2026 with the adoption of electric vehicles. To ensure that EVs do not become a bottleneck to the power grid, superior power technologies for use in EVs are becoming necessary. This is prompting automakers to replace traditional power IC technologies, such as Si MOSFET and Si IGBT, with SiC and GaN-based power devices because they require less thermal management.

Second, sustainable mass production of affordable and efficient electronics is critical for reducing carbon emissions. GaN plays a crucial role in enabling sustainability in multiple ways. First, the process of fabricating the GaN device uses much lower CO_2 footprint compared to legacy silicon chips. Implementing GaN devices in power electronics systems results in a significant reduction in overall product size. This reduction in size leads to less material usage and a lower carbon footprint during manufacturing. Moreover, utilizing GaN devices also results in higher energy efficiency, which means less electricity consumption and less coal burning. In summary, using GaN will significantly accelerate the road to sustainability in multiple ways, from manufacturing to end-product usage.

Third, industry regulations that mandate energy-efficient products like ENERGY STAR, 80PLUS Standards, USB PD EPR, and European energy-using Product (EuP) regulations are emphasizing manufacturers to provide more energy-efficient products in the consumer space—which directly boosts the power GaN adoption prospects in the consumer market space, where the technology is proven and set to accelerate in the coming years. Also, from a policy standpoint, not only Governments are focusing on sustainability policies, but also companies. Implementing sustainability policies and practices can provide significant benefits for manufacturing companies. These benefits include reducing costs, improving sustainability-related ratings, enhancing operational efficiency, boosting customer engagement, and gaining a competitive advantage. As a result, companies are more likely to invest in and partner with businesses that prioritize using renewable energy and energy-efficient products. This trend is driving the demand for power GaN products even higher.

9.4 Key Market Challenges

Earlier, we saw that the overall power GaN market is poised to grow exponentially in the next 5 years. However, the market is in its early commercialization stages and faces multiple challenges to overcome. Frost & Sullivan identifies seven critical challenges the industry needs to address to unleash the full growth potential ahead of us.

- *Strong competition from SiC*: The power GaN semiconductor market is currently highly preferred for handling power requirements in low voltage (up to 650 V)/power and high-frequency applications. The technology's performance and reliability at higher voltages are still a subject of discussion, especially with a lateral GaN structure. Though market participants have announced products up to 1250 V to date, the commercial viability of GaN at high-voltage and power levels is yet to be proven. Currently, power requirements over 650 V for high-power applications are met by the emerging SiC-based power devices. Companies with product mix across Si, SiC, and GaN are not likely to push for GaN products in overlapping applications where SiC and GaN will compete. This will be a crucial hindrance for pure-play GaN companies and likely hinder the overall market growth prospects in the mid-to-long term.
- *Risk-averse customer mindset will hinder market growth*: GaN technology has yet to fully prove itself in critical applications such as data centers and automotive, where downtime or failure can have significant consequences. Currently, GaN is used mainly in consumer power adapters, which are less critical and easily replaceable in case of malfunction. Additionally, the operating conditions in consumer applications are less harsh and dynamic than those in critical applications like automotive. The automotive industry, in particular, is known for being risk-averse in switching to newer technologies. Any malfunction in a commercially launched vehicle can lead to costly recalls affecting millions of cars, resulting in billions of dollars in expenses for automotive OEMs. Therefore, automotive manufacturers may only be able to transition from the well-established silicon technology to GaN if they are convinced of its reliability and benefits.

  While GaN has made progress in terms of cost competitiveness with silicon, it may still be slightly higher or at similar cost levels. However, silicon has a proven track record of reliability in automotive applications, which makes automotive OEMs reluctant to switch to GaN without assurance of its reliability and benefits.

Though GaN is making increasing inroads into space applications, these are low-volume (about 2900 satellites were orbited in 2023), and the environment is typically constant throughout their operation, unlike the highly dynamic environment in real-world applications such as automotive. In addition, space applications are generally much more open to transitioning to newer technologies and innovation. Also, they are generally not life-critical.

The challenge for GaN manufacturers lies in convincing the end-user industries of the reliability, performance, and cost-effectiveness of GaN technology for applications such as automotive and data centers. Building trust and providing compelling evidence of the advantages of GaN over silicon will be crucial for broader adoption in critical high-volume applications.

In summary, the power GaN market has come a long way since its early stages of technology advancement, overcame challenges in cost, performance, and reliability, and convinced customers to implement GaN into power electronics design to benefit from the technology's high-power density, efficiency, and small form-factor benefits. This paved the way for the market participants to realize commercial value and guided the market towards exponential market growth in the late 2010s, led by accelerated penetration in the smartphone and laptop charger application. Beyond 2023, the power GaN market will continue to witness significant revenue opportunities from the consumer electronics sector. However, for better return on investment, the power GaN market needs to explore possibilities beyond the consumer sector, which the market participants rightly target. The new product development activities and collaborations with OEMs in the automotive, energy (including solar), and industrial sectors (specifically space, defense, and motors) will lead to a new inflection point of growth in the next 1–2 years, which will add new revenue streams and create a better end-user mix to sustain the market in the long-term.

9.5 Opportunities

The future of the power GaN market depends on the penetration rate within each application. In the current scenario, power GaN has achieved the highest penetration against conventional Si-based power semiconductors in the consumer electronics sector (mainly smartphone and laptop chargers), as seen in Fig. 9.7. Looking at the 2–5-year opportunities ahead, Frost & Sullivan observes increased efforts by the market participants to penetrate the industrial, energy, and automotive end-user verticals as shown in Figs. 9.8 and 9.9. This application diversification efforts will help market participants to accelerate and sustain market growth. Hence, this section aims to understand the prospects of power GaN across various end-user applications.

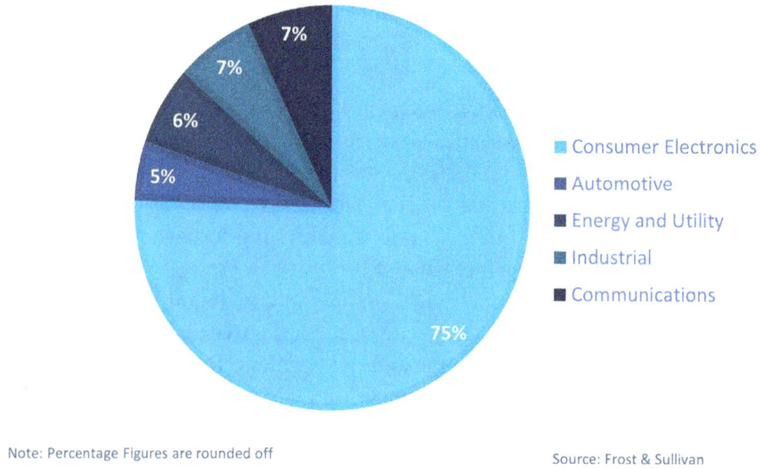

Fig. 9.7 Global power GaN semiconductors market: Percent revenue split by end-users, 2023, Global

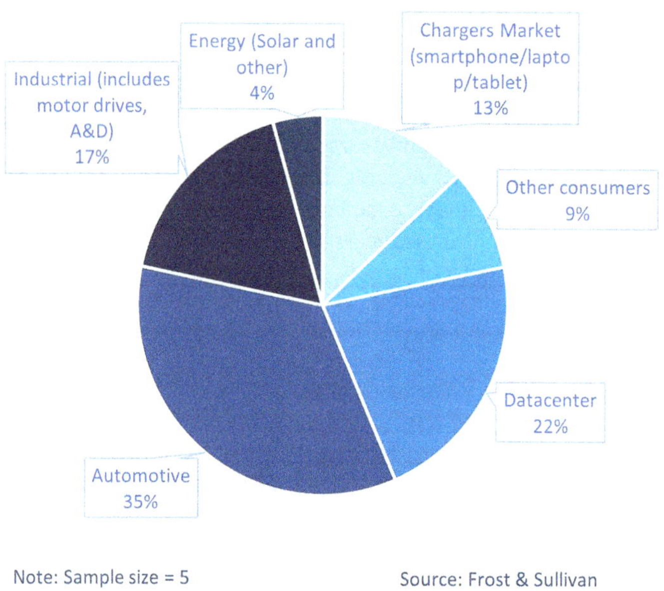

Fig. 9.8 Global power GaN semiconductors market: Percent share of product announcements by end-user applications, 2021, Global

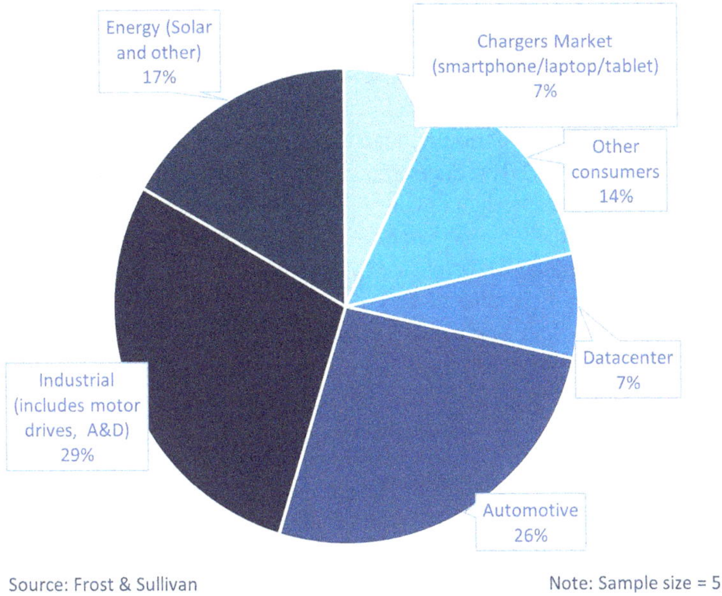

Fig. 9.9 Global power GaN semiconductors market: Percent share of product announcements by end-user applications, 2023–2024 Feb, Global

9.6 Consumer Electronics—Driven by Fast Chargers

GaN-integrated power circuits are being integrated into power adaptors to rapidly charge consumer electronics devices, such as smartphones, laptops, and tablets. This is because GaN power devices significantly increase the power efficiency of chargers, which translates into ultra-fast charging of phones and charges without thermal issues. Furthermore, GaN-based charger designs can result in 20–30% smaller-sized chargers, which are now used by aftermarket charger OEMs to design multi-port chargers. The GaN-enabled chargers are also enabling a shift toward standard USB-type C charging ports across all end products. Smartphone OEMs leverage such capabilities to stop providing chargers in their shipment and encourage customers to procure an OEM-certified multi-port charger. This will be a key trend that will drive the future of the fast-charging market powered by GaN. More importantly, such an initiative will drive down e-waste significantly, which is again a winning factor for power GaN market participants. In the next 5 years, GaN-based fast chargers will account for over 50% of all smartphone chargers shipped globally.

GaN-based power electronics are also deployed in home appliances, such as refrigerators and air conditioners, to save energy. GaN-based inverters can drive compressor motors of refrigerators and air conditioners. Also, GaN FETs will enhance amplifier performance in audio systems due to their low resistance and capacitance, reducing open-loop impedance and improving sound quality.

The fast switching of GaN reduces total harmonic distortion. The product of gate charge and on-resistance product in GaN HEMTs is much lower compared to Si and SiC devices with similar ratings. This combination leads to higher switching frequencies and reduces the size of passive components, which is an obvious benefit from the perspective of circuit size in consumer products. Therefore, power GaN semiconductors will find increased penetration rate in non-mobile consumer products.

9.7 GaN in Space Applications

In the demanding realm of space electronics, where components are subjected to extreme environmental conditions, including radiation, the choice of semiconductor material becomes critical. GaN has emerged as a standout candidate due to its exceptional properties, particularly its robustness against radiation damage. Radiation in space comes in various forms, including gamma rays, neutrons, and heavy ion bombardments. These high-energy particles can wreak havoc on electronic devices, leading to charge trapping in non-conducting layers, crystal damage, and even burnout of components. Traditional semiconductor materials like silicon are susceptible to such radiation-induced damage, making them less than ideal for space applications where reliability is paramount. In contrast, GaN demonstrates superior immunity to radiation-induced damage. This resilience is attributed to the inherent properties of GaN, which allow it to withstand the effects of radiation without significant degradation in performance. Consequently, GaN-based devices exhibit enhanced reliability and longevity when deployed in space environments.

A crucial performance metric used to evaluate power semiconductor devices is the figure of merit (FOM), which is calculated by multiplying the on-resistance (RDS(on)) by the input gate charge (Qg). A lower FOM value indicates that the device is easier to drive at a specific on-resistance level, resulting in improved efficiency and performance. GaN-based devices typically boast significantly lower FOM values than their Si counterparts, making them more efficient and suitable for space applications where power management is critical. GaN devices find extensive use in space for various applications, including DC-DC converters and motor drivers in satellites. Additionally, GaN-based drivers are employed in pumps used for launch and in-space fluid management systems, where reliability and efficiency are paramount.

While SiC is known for its high-temperature tolerance, its efficiency at lower voltages is not as optimal as GaN. Furthermore, SiC devices are not as robust against single-event effects, which are sudden and unexpected changes in device behavior caused by radiation-induced phenomena. As a result, GaN emerges as the preferred choice for efficient and radiation-immune power management solutions in space applications, primarily when operating at voltages below 600 V.

9.8 GaN in Automotive Applications

In the automotive landscape, GaN is gaining popularity in applications such as onboard chargers (OBC) and DC-DC converters. GaN-based power devices are transforming the design of OBCs for electric vehicles. Using GaN technology, OBCs become significantly lighter and smaller, reducing the vehicle weight, increasing driving range, and offering new design possibilities for OBC integration. DC-DC converters are crucial in connecting high-voltage battery packs to low-voltage auxiliary circuits in electric or hybrid vehicles. GaN devices are highly efficient and offer best-in-class power density, reducing power wastage and generating heat. Integrating GaN semiconductors makes the DC-DC converters smaller, contributing to decreased vehicle weight.

The move towards 800 V battery systems in EVs further necessitates the move to WBG technology in EV traction inverters. The move to SiC is being considered inevitable in this shift toward high-voltage battery systems. GaN technology for 800 V systems in EVs is still in the early commercialization phase. Several industry innovators are working on improving the efficiency of GaN devices through advancements, such as vertical GaN structures and superior packaging technologies, for 800 V EV application-feasible GaN FETs.

GaN power devices are also used to power Brushless DC (BLDC) motors in electric vehicles. BLDC motors are used for motion control, positioning, and other applications such as power steering, transmission actuation, and engine cooling systems. Compared to traditional brushed motors, BLDC motors have many advantages, including higher durability, efficiency, and lower proneness to failure. GaN-based power electronics offer even higher efficiency and smaller sizes, making them the perfect fit for BLDC motors in electric vehicles.

Modern infotainment systems demand more power from a vehicle, due to advanced features such as touchscreens, GPS, and gaming. GaN devices are smaller and more efficient than silicon MOSFETs, allowing for smaller, cooler, and less expensive DC-DC solutions. Additionally, GaN FETs improve the audio performance of Class-D audio systems. GaN is vital for developing vertical cavity surface emitting lasers (VCSEL), which are essential for developing LIDAR sensors. LIDAR is important for distance mapping in ADAS, which is the future of the automotive industry. It is also an essential component in many aerospace and defense systems.

Overall, with a multiple applicability in the overall automotive sector, the power GaN semiconductors are poised to create a high-volume and high-growth opportunity in the mid to long-term.

9.9 GaN in Data Centers/Servers

With the rapid proliferation of AI and the growing connected ecosystem, modern-day data centers are becoming more compute-intensive than ever, leading to the need for high efficiency, speed, and power. About 20% of the power consumed by data centers is wasted as heat, while more than 40% is used to keep the server racks cool.

GaN power supplies are expected to bring significant cost benefits with applications in servers, racks, and UPS in data centers. GaN power devices enable smaller and more energy-efficient power supplies that can accommodate more servers in existing racks, reducing expenditure on building new data centers. For example, owing to the smaller size and higher efficiency of GaN transistors, the overall data center server density can be increased by up to 56%, enabling more servers, CPU power, and storage in existing racks.

GaN-based power electronics increase the power supply efficiency in data centers from near 94% in silicon-based power supplies to more than 97%, thereby decreasing energy costs and increasing profitability. Moreover, the global push for green energy and energy efficiency has led to stringent efficiency and energy consumption standards and regulations for data centers. For instance, the European Union's Ecodesign Directive demands data center power supply units to meet specific efficiency and power factor requirements. The use of GaN shows great promise in meeting the prerequisites of government regulations and industry standards due to its superior efficiencies and performance.

In the ever-changing world of computing, keeping data centers updated with the latest technology and innovations is crucial. Failure to do so can put companies at risk of losing their competitive edge. Therefore, transitioning to power electronics based on GaN in data centers would be a wise and sustainable move for the future. Therefore, data center market is poised to be another high-volume and high-growth opportunity for power GaN semiconductors in the mid and long-term.

9.10 GaN in Industrial and Energy Sector

GaN has emerged as an alternative for many power electronic applications within the industrial and energy sectors, such as commercial and industrial switch mode power supply (SMPS), motor drives for conveyors and pumps, DC-DC converters, regulators, and power factor correction circuits for UPS systems, and solar micro-inverters.

Industrial motors and motor drives consume over two-thirds of the industrial power consumption. GaN power devices in industrial motor drives efficiently satisfy the demand for lower voltages, smaller size, and safety. GaN-based motor drives offer increased efficiency and faster response time and require no additional cooling owing to the characteristics of GaN. GaN-based motor drives offer more than 98% energy conversion efficiency compared to about 92% with silicon while enabling a 50% smaller size. It results in about 25% less electricity usage in industrial setups.

In industrial robotics, owing to the high frequency of GaN devices, there is a lesser requirement for passive components such as magnetics, capacitors, and filters. Further, GaN enables lower power losses resulting in smaller motors without the need for heat sinks. Moreover, GaN power also enables efficient wireless charging for industrial robots and drones.

The adoption of GaN devices is on the rise in solar applications for residential and small commercial solar installations due to their reliability and efficiency benefits. The utilization of GaN transistors in inverters for renewable energy leads to enhanced power density, efficient maximum power point tracker (MPPT), reduced heat dissipation, and smaller passive component size. GaN's high switching frequency enables precise power conversion and improved energy harvesting efficiency. Moreover, lower power losses and smaller form factors result in smaller and lighter solar inverters and converters, making installation and maintenance easier. The superior thermal performance of GaN devices make them suitable for stable operation under varying environmental conditions, thereby increasing the overall efficiency and reliability of solar power systems.

Prabhu Karunakaran is a Principal Analyst with Frost & Sullivan and has over 12 years of experience in the semiconductor industry. He has been actively monitoring and analyzing the semiconductor industry and offers valuable insights and thought leadership content across power semiconductors, advanced communication technologies, semiconductor test, and metrology markets. Prior to his current role, he served as a Process Engineer (dry etch) at STMicroelectronics and gained valuable industry experience during his 6-year tenure. Prabhu also holds a master's degree from National University of Singapore.

Navdeep Saboo is a Senior Industry Analyst at Frost & Sullivan, specializing in semiconductors and test & measurement markets within the industrial technologies sector. With nearly 7 years of research and consulting experience, he specializes in power semiconductor technologies, chip manufacturing, OSATs/ATMP landscape, testing equipment, and testing services. Navdeep has sound experience delivering solid and robust insights-driven analysis of trends, market estimation, and segmentation, contributing to go-to-market strategies. With a striking balance of business and technology acumen, Navdeep has solid analytical thinking and an insights-driven approach to problem-solving, ensuring impactful deliverables and successful project management.

Index

© The Editor(s) (if applicable) and The Author(s), under exclusive license to Springer Nature Switzerland AG 2024
M. Di Paolo Emilio (ed.), *GaN Technology*,
https://doi.org/10.1007/978-3-031-63238-9

Printed in the USA
CPSIA information can be obtained
at www.ICGtesting.com
CBHW062203030924
PP15535400002B/3